# Environmental History

Volume 12

**Series Editor**

Mauro Agnoletti, Florence, Italy

More information about this series at http://www.springer.com/series/10168

Bila-Isia Inogwabini

# Reconciling Human Needs and Conserving Biodiversity: Large Landscapes as a New Conservation Paradigm

## The Lake Tumba, Democratic Republic of Congo

Springer

Bila-Isia Inogwabini
Center for Research and Communication
in Sustainable Development (CERED)
The Jesuit Loyola University of Congo
Kinshasa, Congo, Republic

ISSN 2211-9019 ISSN 2211-9027 (electronic)
Environmental History
ISBN 978-3-030-38727-3 ISBN 978-3-030-38728-0 (eBook)
https://doi.org/10.1007/978-3-030-38728-0



This Springer imprint is published by the registered company Springer Nature Switzerland AG
The registered company address is: Gewerbestrasse 11, 6330 Cham, Switzerland

# Acknowledgements

Some important materials included in this book come from field experience I have had while I managed the Lake Tumba Landscape as Program Manager recruited by World Wide Fund (WWF), Central Africa Program. The field work being discussed has been, then, generously funded by the US Agency of International Development (USAID) under its Congo Basin Forest Partnership (CARPE). Ideas, views and statements I make here, however, do reflect neither views of USAID/CARPE nor those of my former employer (WWF). Apart from the funds to support the field work, I have been fortunate enough to get additional financial support from several biodiversity conservation organizations that helped the academic side of the work I did in Lake Tumba. These organizations included African Wildlife Foundation (AWF), British Ecological Society (BES), UN Educational, Scientific and Cultural Organization (UNESCO) and Wildlife Conservation Society (WCS). I thank all of them for the support I received while I was conducting my Ph.D. research whose significant portions are part of this book.

The book is built on published and unpublished elements. All previously published and unpublished materials included in this volume have been fully and appropriately acknowledged to the credits of all who have participated in making them available. This is the case when I compiled both published and grey literatures. As it is indicated above, some parts came from my Ph.D. research work at the University of Kent at Canterbury (UK) but some other materials came from different processes. This is, for example, the case of the historical chronicle of the birth of the idea of Landscape as a conservation paradigm, the ethical questions raised by conserving biodiversity in humanly occupied areas and all the theoretical discussions about threat indexes, etc. For the first case, the work was essentially done through reviewing the literature and reflecting back on instances where I was part of the process. All materials related to the ethics of biodiversity conservation were drawn from my MA Research at the University of Leeds (UK) and discussions on the evolution of Pan, threat index, etc. are my own reflections. I benefited enormously from the support of the field teams I had worked with. Special thanks go to

Prof. Michael Brudford of the Cardiff University and Cheryl Meridth with whom I jointly worked on the genetics of bonobos. I designed the sampling methods and collected the samples that Cheryl Meridth analyzed for her Master Degree supervised by Professor Brudford at Cardiff. However, the interpretation of the results presented here is solely mine though I draw results from Cheryl Meredith's Master thesis, which has been extensively cited throughout that specific chapter. I would also like to thank some of my former colleagues here: Abokome Mbenzo, Bakanza Nzala, Bokika Ngwalo, Dauphin Lomboto, Faustin Tokate, Guy Tshimanga, Matungila Bewa, Mbende Longwango, Philip Colson, Mputu Dianda, Zanga Lingopa who were part of the research teams. Where the field work was previously published with each of them as co-author, the work is entirely cited throughout the text of this volume. This is also the case for unpublished reports and materials. Finally, some materials, particularly those that were part of either my Ph.D. Thesis at Kent or my MA Dissertation at Leeds, were read through and commented on by Prof. Nigel Leader-Williams and Prof. Christopher Megone. Mistakes or errors are solely mine and should not be blamed on people cited here or organizations that provided support in one way or the other.

# Contents

**Part I General Concepts**

**1 From Parks to Landscapes: Reading of a Long Process** . . . . . . . . . 3
1.1 Introduction . . . . . . . . . . . . . . . . . . . . . . . . . . . . . . . 3
1.2 Landscapes: Processes of New Conservation Paradigm . . . . . . 5
1.3 The Ecoregion Program in Central Africa . . . . . . . . . . . . . . . 5
1.4 The Landscape-Species Approach in Central Africa . . . . . . . . 8
1.5 International and Regional Political Processes . . . . . . . . . . . . 11
1.6 Forces Driving the Landscape Approach and Conditions for Its Successes . . . . . . . . . . . . . . . . . . . . . . . . . . . 13
References . . . . . . . . . . . . . . . . . . . . . . . . . . . . . . . . . . . 15

**2 Landscape: Re-assessing the Conservation Paradigm** . . . . . . . . . . 17
2.1 Introduction . . . . . . . . . . . . . . . . . . . . . . . . . . . . . . . 17
2.2 Landscape Defined as a Conservation Vision . . . . . . . . . . . . . 19
2.3 Landscape Is Also a Planning Tool Linking Ecology and Economics . . . . . . . . . . . . . . . . . . . . . . . . . . . . . . 21
2.4 Ecology and Economics: What Is the Place of Conservation? . . . . . . . . . . . . . . . . . . . . . . . . . . . . . . 25
References . . . . . . . . . . . . . . . . . . . . . . . . . . . . . . . . . . . 27

**3 Biodiversity Conservation in Human-Prone Landscapes: Social and Ethical Issues** . . . . . . . . . . . . . . . . . . . . . . . . . . . 29
3.1 Introduction . . . . . . . . . . . . . . . . . . . . . . . . . . . . . . . 29
References . . . . . . . . . . . . . . . . . . . . . . . . . . . . . . . . . . . 43

**4 Landscapes Require New Legal Framework to Conserve Biodiversity** . . . . . . . . . . . . . . . . . . . . . . . . . . . . . . . . . 45
4.1 Introduction . . . . . . . . . . . . . . . . . . . . . . . . . . . . . . . 45
4.2 Dealing with Obsolete Legal Framework in a New Conservation Paradigm . . . . . . . . . . . . . . . . . . . . . . 46

4.3 Using International Agreements and Treaties to Circumvent the Legal Vacuum . . . . . 47
4.4 National Sovereignty Limits the Successes of Regional Planning Processes . . . . . 49
4.5 Could the Biodiversity Diplomacy Help Where the Normal Diplomacy Failed? . . . . . 51
4.6 Concrete Steps Toward New Forms of Legal Frameworks . . . . . 52
References . . . . . 54

**Part II Data to Support the Conservation Action**

**5 Qualitatively Describing Forests of the Landscape** . . . . . 59
5.1 Introduction . . . . . 59
5.2 Methods to Assess Habitats . . . . . 61
5.3 Forest Types in the Lake Tumba Landscape . . . . . 62
5.4 Habitats in the Northern Part of the Lake Tumba Landscape . . . . . 63
5.5 Habitats in the Southern Part of the Lake Tumba Landscape . . . . . 64
5.6 Biological Diversity of the Landscape . . . . . 66
5.7 Plant Species . . . . . 67
5.8 Human Ecology in the Lake Tumba Landscape . . . . . 68
5.9 Habitats of the Lake Tumba Placed in the Perspectives of the Congo Basin . . . . . 74
References . . . . . 76

**6 Bonobos in the Lake Tumba: Describing the Landscape Species** . . . . . 79
6.1 Introduction . . . . . 79
6.2 Bonobo Presence in the Lake Tumba Landscape . . . . . 80
References . . . . . 87

**7 Genetics of Bonobos in the Lake Tumba Landscape** . . . . . 89
7.1 Introduction . . . . . 89
7.2 Materials and Methods . . . . . 91
7.3 Genetic Variations in the Bonobos in the Lake Tumba Forest and Other Populations . . . . . 92
7.4 Genetics of the Bonobos in the Lake Tumba Forests in Perspectives . . . . . 93
7.5 Genetics and the Habitat Types Can Help in Identifying the Species Evolutionary Path . . . . . 94
References . . . . . 98

**8 Forest Refugia Theory, Density Dependence and Stress Syndrome and the Proto-Pan** . . . . . 101
8.1 Introduction . . . . . 101
8.2 Potential Effects of Density Dependence on Proto-Pan During the Glaciation . . . . . 102
8.3 Stress Syndrome Theory . . . . . 104
8.4 All Together: Bonobos, Chimpanzees, Refugia, Density Dependence, and Stress . . . . . 105
References . . . . . 107

**9 Wild Bonobos and Wild Chimpanzees and Human Diseases** . . . . . 109
9.1 Introduction . . . . . 109
9.2 Potential Bi-Directional Human Ape Disease Transmissions . . . . . 110
9.2.1 Anthrax . . . . . 110
9.2.2 Ebola . . . . . 112
9.2.3 Herpes . . . . . 112
9.2.4 Human Immunodeficiency Virus (HIV) and Simian Immunodeficiency Virus (SIV) . . . . . 113
9.2.5 Influenza . . . . . 113
9.2.6 Malaria . . . . . 114
9.3 Bonobos, Zoonoses and Human Epidemics in Perspectives . . . . . 115
References . . . . . 118

**10 Alternative Cheaper Methods to Estimate Bonobos** . . . . . 123
10.1 Introduction . . . . . 123
10.2 Materials and Methods . . . . . 124
10.3 Line-Transects . . . . . 124
10.4 Presence–Absence Data and Estimation of Density . . . . . 125
10.5 Updating Bonobo Estimates in the Southern Lake Tumba . . . . . 127
10.6 Abundance Estimates from Transect and Presence–Absence Survey . . . . . 127
10.7 Comparing Line-Transect and Presence–Absence Estimates . . . . . 128
10.8 Updated Bonobo Population Estimates . . . . . 128
10.9 Estimates of Bonobo Population in Perspectives . . . . . 128
References . . . . . 131

**11 Chimpanzees of the Ngiri Triangle** . . . . . 133
11.1 Introduction . . . . . 133
11.2 The Chimpanzees of Lake Tumba Landscape and the Species Conservation . . . . . 134

11.3 The Habitats of the Ngiri and Bosobele . . . . . . . . . . . . . . . . . 135
11.4 What Interests Were There to Study the Chimpanzees in This Area? . . . . . . . . . . . . . . . . . . . . . . . . . . . . . . . . . . . . 136
11.5 Procedures and Protocols Used to Study Chimpanzees . . . . . . 137
11.6 Chimpanzee Population Estimates in the Lake Tumba Landscape . . . . . . . . . . . . . . . . . . . . . . . . . . . . . . . . . . . 137
11.7 Relative Estimates of Chimpanzee Population in the National and Regional Perspectives . . . . . . . . . . . . . . . 138
References . . . . . . . . . . . . . . . . . . . . . . . . . . . . . . . . . . . . . . . 140

**12 Lions of Malebo: Population and Conflicts with Humans** . . . . . . . . 143
12.1 Introduction . . . . . . . . . . . . . . . . . . . . . . . . . . . . . . . . . . . . 143
12.2 What are the Characteristics of This Area Where Lions Roam in the Middle of the Forests? . . . . . . . . . . . . . . . 144
12.3 What Was Done to Understand the Lions at Malebo? . . . . . . . 145
12.4 Estimating the Lion Population at Malebo . . . . . . . . . . . . . . . 146
12.5 Lions of Malebo in the Perspectives of the Species Conservation Status . . . . . . . . . . . . . . . . . . . . . . . . . . . . . . 147
References . . . . . . . . . . . . . . . . . . . . . . . . . . . . . . . . . . . . . . . 151

**13 Diurnal Primates: Estimates and Conservation Issues** . . . . . . . . . . 153
13.1 Introduction . . . . . . . . . . . . . . . . . . . . . . . . . . . . . . . . . . . . 153
13.2 How Were the Data on Diurnal Primates Collected in Lake Tumba Landscape? . . . . . . . . . . . . . . . . . . . . . . . . . 154
13.3 Diurnal Primates: Species and Estimated Populations . . . . . . . 156
13.4 Diurnal Primates of Lake Tumba in the Perspectives of Primate Conservation . . . . . . . . . . . . . . . . . . . . . . . . . . . 157
References . . . . . . . . . . . . . . . . . . . . . . . . . . . . . . . . . . . . . . . 160

**14 Elephants in Lake Tumba Landscape: Malebo, Ngiri, and Bolombo-Losombo** . . . . . . . . . . . . . . . . . . . . . . . . . . . . . . . . 163
14.1 Introduction . . . . . . . . . . . . . . . . . . . . . . . . . . . . . . . . . . . . 163
14.2 Elephant Survey in the Forests of Lake Tumba Landscape . . . . . . . . . . . . . . . . . . . . . . . . . . . . . . . . . . . 164
14.3 Methods to Document Elephants in the Lake Tumba Landscape . . . . . . . . . . . . . . . . . . . . . . . . . . . . . . . . . . . 165
14.4 Dung Deposition and Decay Rate . . . . . . . . . . . . . . . . . . . . 165
14.5 Survey Area and Sample Effort . . . . . . . . . . . . . . . . . . . . . . 166
14.6 The Distribution and Abundance of the Elephant in the Lake Tumba Landscape . . . . . . . . . . . . . . . . . . . . . . . . 169
14.7 The Distribution and Abundance of the Elephant in the Lake Tumba Landscape in Perspectives . . . . . . . . . . . . 169
References . . . . . . . . . . . . . . . . . . . . . . . . . . . . . . . . . . . . . . . 171

**15 Developing a Threat Index for Documented Large Mammal Species** . . . 173
15.1 Introduction . . . 173
15.2 Data on Large Mammals . . . 174
15.3 Assumption and Methods on Threat Index . . . 175
15.4 Parameterization and Definitions of Threat Levels . . . 176
15.5 Negatively Contributing Factors . . . 176
15.6 Positively Contributing Factors . . . 177
15.7 Factoring the Unknown in Threat Index: Penalty for the Unmeasured Effects . . . 178
15.8 Putting All of It Together . . . 182
15.9 Mammalian Diversity . . . 182
15.10 Species Abundance . . . 183
15.11 Threat Index by Species . . . 183
15.12 Species Threat Index in the Biodiversity Conservation Perspectives . . . 184
15.13 Effects of Human Activities in the Southern Lake Tumba on Wildlife Species . . . 185
References . . . 189

**16 Synopsis of Freshwaters, Species Diversity, and Conservation Issues** . . . 193
16.1 Introduction . . . 193
16.2 Rivers and Lakes of the Lake Tumba Landscape . . . 194
16.3 Fish Species Diversity in the Rivers and Lakes of the Lake Tumba Landscape . . . 197
16.4 Wildlife Species Depending on the Rivers and Lakes of the Lake Tumba Landscape . . . 198
16.5 Freshwaters of the Lake Tumba Landscape: Threats and Current Status . . . 199
16.6 Freshwaters of the Lake Tumba Landscape: Prospects for the Future . . . 202
References . . . 204

**17 An Abridgement of the Birds Throughout the Diversity of Habitats** . . . 207
17.1 Introduction . . . 207
17.2 How Insights on Birds of the Region Gathered? . . . 208
17.3 Bird Species in the Region . . . 209
17.4 Bird Species Described in the Region in the Biodiversity Conservation's Perspectives . . . 210
References . . . 211

**Part III Not Only Biodiversity but Also Human Communities**

**18 The Political Economy of the Landscape** . . . 215
18.1 Introduction . . . 215
18.2 Defining the Concept of Stakeholder for the Context of Biodiversity Conservation . . . 216
18.3 Methodological Approach for the Political Economy Analysis . . . 218
18.4 Who Is Where and Who Has What Weight in the Lake Tumba Landscape? . . . 220
18.5 Who Wins and Loses What in the Lake Tumba Landscape? . . . 229
18.6 Political Economy, Landscape, and Total Economy for Landscape Communities . . . 239
18.7 Uncovering the Obvious . . . 245
18.8 Overall Who Wins in the Political Economy of the Landscape? . . . 247
18.9 Limitations of the Political Economy Analysis in the Context the Landscape . . . 248
References . . . 250

**19 Assessing the Needs in Lands in the Lake Tumba Landscape** . . . 253
19.1 Introduction . . . 253
19.2 Defining Land and Land Needs . . . 254
19.3 Approaches to Land Need Assessment . . . 255
19.4 Theoretical Land Needs Assessment . . . 255
19.5 Theoretical Land Needs Assessment and the Reality of Claims Over Land Rights . . . 257
19.6 Combining the Theory and the Reality of Claims of Rights to Make Sense . . . 259
References . . . 262

**Part IV Using the Data to Strategize and Manage the Landscape**

**20 Planning for the Management of the Landscape** . . . 265
20.1 Introduction . . . 265
20.2 The Situational Appraisal for the Implementation of the Landscape . . . 266
20.3 The Geography and the Biological Diversity of the Lake Tumba Landscape . . . 266
20.4 Rapid Human Population Growth . . . 267
20.5 Rampant Poaching . . . 267
20.6 Logging, Oil Extraction and Other Issues of Importance to Biodiversity Conservation . . . 268

20.7 Setting Global Priorities for the Lake Tumba Landscape . . . . . 268
20.7.1 Forests Protection . . . . . . . . . . . . . . . . . . . . . . . . . . . 269
20.8 Freshwater Should Not Be Ignored in the Planning Process . . . . . . . . . . . . . . . . . . . . . . . . . . . . . . . . . . . . 274
20.9 Species of Conservation Concern and Others Should Be Conserved . . . . . . . . . . . . . . . . . . . . . . . . . . . . . . . 275
20.10 Clearly Laid Logics of the Intervention . . . . . . . . . . . . . . . . . 278
20.11 Actions Planned to Achieve the Global Biodiversity Conservation Priorities . . . . . . . . . . . . . . . . . . . . . . . . . . . . 285
20.12 Management of Natural Resources by Local Communities . . . . . . . . . . . . . . . . . . . . . . . . . . . . . . . . . . 288
20.13 Active Conservation of the Landscape—Species in Lake Tumba Landscape . . . . . . . . . . . . . . . . . . . . . . . . . 292
20.14 Communicating Conservation Activities and Lessons Learnt to the Global Society . . . . . . . . . . . . . . . . . . . . . . . . . 295
20.15 Reconciling Divergent Needs: Actions for the Pilot of Sustainable Uses . . . . . . . . . . . . . . . . . . . . . . . . . . . . . . . 297
20.16 Resources Needed to Achieve Project's Mission . . . . . . . . . . . 298
20.17 Human Resources Needed . . . . . . . . . . . . . . . . . . . . . . . . . . 298
20.18 Training Needs for the Project to Operate . . . . . . . . . . . . . . . . 300
20.19 Financial Needs for the Project to Operate . . . . . . . . . . . . . . . 302
20.20 Exit Scenarios . . . . . . . . . . . . . . . . . . . . . . . . . . . . . . . . . . 303
References . . . . . . . . . . . . . . . . . . . . . . . . . . . . . . . . . . . . . . . 303

**21 Setting Habitats Aside for Biodiversity Conservation** . . . . . . . . . . . 307
21.1 Introduction . . . . . . . . . . . . . . . . . . . . . . . . . . . . . . . . . . . 307
21.2 Definitions, Principles, Processes, and Alternative Ways of Saving Habitats . . . . . . . . . . . . . . . . . . . . . . . . . . . . . . 308
References . . . . . . . . . . . . . . . . . . . . . . . . . . . . . . . . . . . . . . . 313

**22 Protected Areas: Defining the Optimum Law Enforcement Resources** . . . . . . . . . . . . . . . . . . . . . . . . . . . . . . . . . . . . . . . . . 315
22.1 Introduction . . . . . . . . . . . . . . . . . . . . . . . . . . . . . . . . . . . 315
22.2 Scenarios, Materials, and How to Use the Approaches to Allocating Patrol Efforts . . . . . . . . . . . . . . . . . . . . . . . . . 317
22.3 Staffing for Effectiveness . . . . . . . . . . . . . . . . . . . . . . . . . . . 317
22.3.1 Guard Density Model . . . . . . . . . . . . . . . . . . . . . . . . 317
22.3.2 Strategic Deployment Model . . . . . . . . . . . . . . . . . . . 318
22.4 Budgeting for Effectiveness . . . . . . . . . . . . . . . . . . . . . . . . . 318
22.5 Effective Law Enforcement in the Newly Created Protected Areas . . . . . . . . . . . . . . . . . . . . . . . . . . . . . . . . . . . 320
22.6 The Protected Areas of the Lake Tumba and the Broader Scenes of Protected Areas . . . . . . . . . . . . . . . 321
References . . . . . . . . . . . . . . . . . . . . . . . . . . . . . . . . . . . . . . . 324

**23 Planning the Mobilization of Resources via Sustainable Tourism** . . . . 327
23.1 Introduction . . . . 327
23.2 Natural Assets: Bonobos, Birds, Landscape Sceneries, and Bush Trekking . . . . 328
23.3 Cultural Assets: Dancing Fair and Net Hunting . . . . 329
23.4 Navigating Kasai-Congo River and Kayaking Along the Rapids . . . . 330
23.5 Necessary Investments: Infrastructure . . . . 330
23.6 Joint-Venture: Government, Conservation, Private Sector, and Communities . . . . 332
23.7 Revenue Sharing . . . . 333
23.8 Investment Plans: Invested Funds and Available Assets and Further Needs . . . . 333
23.9 Timeframe of the Investment . . . . 334
23.10 Tour Operators and Immediate Tourism Activities . . . . 334

**24 Decent Knowledge for Future Directions in the Landscape Management** . . . . 335
24.1 Introduction . . . . 335
24.2 Needs in Knowledge for the Lake Tumba Landscape . . . . 336
24.3 Biological Diversity Knowledge . . . . 337
24.3.1 Systematic Herpetological Surveys in the Lake Tumba Landscape . . . . 337
24.3.2 Systematic Bird Survey in the Lake Tumba Landscape . . . . 339
24.3.3 Fish Species Diversity and Biomass Survey in the Lake Tumba Landscape . . . . 340
24.3.4 Insect Systematic Survey in the Lake Tumba Landscape . . . . 341
24.4 Mammalian Studies in the Lake Tumba Landscape . . . . 341
24.4.1 Mammalian Diversity Assessment, Abundance and Distribution . . . . 342
24.4.2 Comparative Ecology of Bonobos and Chimpanzees . . . . 342
24.4.3 Forest Elephants and Key Food Species . . . . 343
24.5 Botanical Systematics in the Lake Tumba Landscape . . . . 344
24.6 Assessment of the Biophysical Environment in the Lake Tumba Landscape . . . . 345
24.6.1 Assessing Freshwater Habitats and Conditions in the Lake Tumba Landscape . . . . 345
24.7 Aims of This Part of the Multi-resolution Study of Lake Tumba Landscape . . . . 347

24.8 Climate Vulnerability of Freshwater Resources in Lake Tumba Landscape . . . . . . . . . . . . . . . . . . . . . . . . . . . 349
24.9 Hydrological Characterization . . . . . . . . . . . . . . . . . . . . . . . . . 350
24.10 Human Perceptions and Biodiversity in the Lake Tumba Landscape . . . . . . . . . . . . . . . . . . . . . . . . . . . . . . . . . . . . . . . . . 351
References . . . . . . . . . . . . . . . . . . . . . . . . . . . . . . . . . . . . . . . . . . 352

**Part V Ethics of Biodiversity Conservation and the Needs of Local Communities**

**25 Are There Ethical Reasons to Preserve Biodiversity Against Local Communities?** . . . . . . . . . . . . . . . . . . . . . . . . . . . . . . . . 355
25.1 Introduction . . . . . . . . . . . . . . . . . . . . . . . . . . . . . . . . . . . . . 355
25.2 What Is Biodiversity and What Kind of Biodiversity Should Matter? . . . . . . . . . . . . . . . . . . . . . . . . . . . . . . . . . . . . . 357
25.3 What Kind of Biodiversity Matters? Are All the Species and Organisms of Equal Value? . . . . . . . . . . . . . . . . . . . . . . . 359
25.4 Why Does Conserving Biodiversity Matter? . . . . . . . . . . . . . . 361
25.5 Instrumental and Intrinsic Values of Biodiversity . . . . . . . . . . 361
25.6 Is It Ethically Admissible to Exploit Natural Resources in Detriment of Biodiversity? . . . . . . . . . . . . . . . . . . . . . . . . . 365
25.7 Needs of Local Communities and Causing the Least Possible Harm to Biodiversity . . . . . . . . . . . . . . . . . . . . . . . . 370
25.8 Non-harmful Human Activities to Biodiversity . . . . . . . . . . . . 371
25.9 Biologically Harmful Activities to Biodiversity but Essential to Human Basic Needs . . . . . . . . . . . . . . . . . . . 372
25.10 Conclusions . . . . . . . . . . . . . . . . . . . . . . . . . . . . . . . . . . . . . 378
References . . . . . . . . . . . . . . . . . . . . . . . . . . . . . . . . . . . . . . . . . . 379

# Part I
# General Concepts

This part of the book is about concepts used in implementing the concept of Landscape as biodiversity conservation paradigm across Central Africa. It has four chapters, each of them focusing on a specific general area of conservation practice. Chapter 1 describes the historical process that has led to the adoption of the concept and agreements that were reached across Central Africa to get the concept to become an important way of thinking about biodiversity conservation. Chapter 2 of the book comparatively explores different conservation fundamental concepts such as ecoregions and protected areas and what they are in likenesses and dissimilarities with the idea of conservation landscapes as defined and tried in Central Africa. The core of this chapter is an effort to demonstrate that though landscapes are, in many ways, similar to other conservation entities—particularly the notion of ecoregion—they are also, in their essence, a different spirit of conservation practice: they are a fresh vision on how to do conservation. The argument here is that the history of conservation across Central Africa, which is briefly chronicled throughout Chap. 1, has been characterized by tensions not only between local communities and conservationists but also between opposed conceptual frameworks within the community of biodiversity conservationists. Chapter 3 deals with the ethical issues, which are mainly about how to reconcile human needs and the imperatives to conserve the biological diversity within a landscape. Indeed the main argument of this chapter is that landscape is primarily an attempt to reconcile, at least partially, the genuine demands that communities have for their sustainable livelihoods and the moral obligations to conserve biological diversity. The chapter argues that the landscape paradigm is not just a conservation entity, as it is the case for a protected area of any internationally recognized category but it is more of a new conservation vision. As such, its focus is not on wildlife species and habitats versus human communities and their livelihoods. It is rather both: wildlife species and habitats with human communities and their livelihoods. Viewed from this angle, the chapter argues, landscapes are a channel through which the long opposition between protected areas and communities could be leveled down. As a practical essay, Chap. 3 differs significantly with Chap. 25, which a much more theoretical appraisal of ethical reasons to pursue one road or the other. Chapter 25 poses a rather radical question on whether to sacrifice biodiversity for humans or humans for biodiversity, should humanity be brought to that point if current trends are not dealt with properly. Chapter 4 takes on issues of the legality of the landscape concept and how the practical legal issues could be solved. Truly, one of the major obstacles that the landscape, as a new conservation vision, has to clear before becoming a true practical concept is the legal status of these massive areas of which some are transboundary entities. As it will be seen, this is not just a question of theoretical interest but it is one vexing question with many implications, including diplomatic ones. Finally, the chapter ends up with a positive note, which is that biodiversity diplomacy will have to prevail in those transboundary landscapes and can work, as some historical precedents have shown in the region. The only condition for this to happen is to have similar rules and regulations, particularly in the area of law enforcement.

# Chapter 1
# From Parks to Landscapes: Reading of a Long Process

**Abstract** This chapter deals with the history of the biodiversity conservation landscape in Central Africa. Principally, it covers the background information on both political and biological practicalities of the landscape as a vision, while comparing them with traditional biological methods that defined such concepts as hotspots of biodiversity. Processes-wise, the chapter also presents the rationale that led to the current concept of landscape conservation in Central Africa, how their initial geographic configurations came to be drawn and, politically, the landscape approach had attained its current predominance in the conservation of the Congo Basin. As an historical documentation, the chapter goes in a step-wise procedure, commencing from the conceptualization of the 'biodiversity conservation landscape' to funding mechanisms and the implementation of activities via the mapping exercises that set out to define these priority conservation areas in Central Africa.

**Keywords** Landscape · Ecoregion · Hotspots · Species richness · Ecological integrity · Ecological eigenvalue

## 1.1 Introduction

Biodiversity conservation in the Congo Basin, following conservation paradigms, has begun from a species-oriented mindset. It has, therefore, been mostly locked within the confines of protected areas of diverse management categories, albeit all aligned with the militaristic approach (Inogwabini and Leader-Williams 2013). However, facing the truth that species that were targeted by the *fence-and-fine* approach all declined over decades and despite significant effort and investment in major protected areas, studies of most African countries holding the most important biodiversity assemblages have led to the clear conclusion that a complete shift of conservation paradigms is needed.

First, studies indicate that most of the wildlife species were located outside of protected area networks, even when excluding large-sized home range species such as elephants, cheetahs, wild dogs, etc. Second, studies from the social sciences have drawn the attention of the proponents of conservation biology to the fact that human populations were a key determinant in both the distribution of biodiversity and its



B.-I. Inogwabini, *Reconciling Human Needs and Conserving Biodiversity: Large Landscapes as a New Conservation Paradigm*, Environmental History 12,
https://doi.org/10.1007/978-3-030-38728-0_1

long-term sustainable survival. A third element that has pleaded for a shift in the conservation paradigm was the moral concern that it was not justifiable for conservation organizations to be spending financial resources in trying to preserve the wildlife and wild habitats while human populations were suffering from hunger and were being denied access to any natural resources located within a protected area just adjacent to their settlements. This point was supported by a purely human rights perspective that was linked to ways and means in which most protected areas were created (Chap. 2 in this volume), which implied prohibiting local people from accessing the forests, lands, resources, and waters they inherited from their ancestors.

This chapter deals with the processes that led to the current concept of landscape conservation in Central Africa, their configurations, and their predominance in the conservation of the Congo Basin. As such, the chapter provides the background information on both political and biological practicalities of the landscape as a vision, while comparing them with traditional biological methods that defined such concepts as hotspots of biodiversity. We present some critical results of the various steps that helped conceptualize, define, and draw maps of what are now known as priority conservation areas in Central Africa, also known as landscapes (Fig. 1.1).

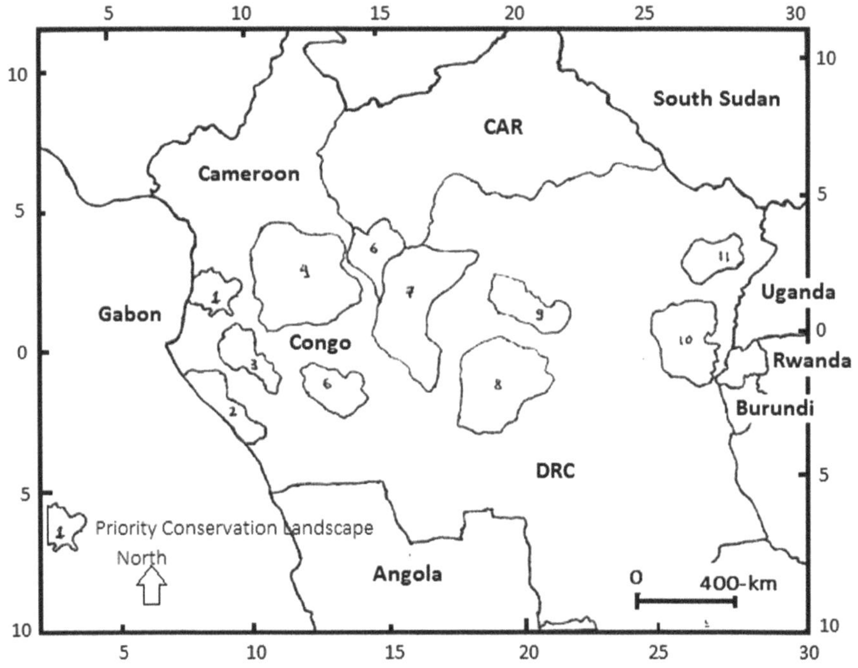


**Fig. 1.1** Congo Basin and its 12 Priority Conservation Landscapes. Lake Tumba is the DRC side of Landscape 7, which is officially known as Lac Tele—Lac Tumba Landscape

## 1.2 Landscapes: Processes of New Conservation Paradigm

Landscapes are defined as cultural, biological, and physical entities with spatial functions that assume that the global needs of human populations are fully and sustainably met while ecological processes and the biodiversity found therein are preserved (Chaps. 2, 3 and 25 in this volume). They have not come to be from a paradigm vacuum. They are a result of a long-term process, which can be easily retraced to three different dynamics that have been occurring in the Congo Basin since 1990: (1) the ecoregion program of the World Wide Fund for Nature (WWF), (2) the Landscape-species approach of the Wildlife Conservation Society (WCS), and (3) the international political willingness to help preserve the forests in Central Africa. These three elements have been also linked with the definition of biodiversity hotspots, a concept principally promoted by Conservation International. The two concepts (ecoregions and landscape-species) can be said to have evolved in parallel and both have had a sense of territorially extending the protected area's concept to include regions that would mediate the long-term conservation of species or habitats where they occur, whereas the notion of biodiversity hotspots is essentially concentrated on regions of the world that show high numbers of species.

## 1.3 The Ecoregion Program in Central Africa

The concept of ecoregion has been a topic of intense discussion among conservation biologists since the birth of conservation biodiversity as an academic discipline. WWF's ecoregion program in Central Africa was closely linked with the global 200 ecoregions of the world, a program that was launched to select the biologically integral and integrated regions on earth in order to launch a dramatic campaign to save the tremendous variety of life on earth (Dinerstein and Olson 1999) by preserving its most significant biological components. In conservation biology history, this concept was seen to replace and broaden the concept of ecosystems. Ecosystems have a more functional definition but lack an operational definition in a sense that they are difficult to concretely delineate. Hence ecoregions would have rough limits that would easily include ecosystems that help protect both long ranging species and other mutually exclusive ecological functions. In this sense, one ecoregion may consist of more than just one ecosystem, coupled with ecological processes that they share.

As a process, the concept of ecoregion in Central Africa has been one of the most ambitious conservation endeavors both in terms of efforts and knowledge gathering. In terms of efforts, the process leading to the geographical delineation of ecoregions of Central Africa was a collective adventure that saw the participation of 167 experts in the region and from around the world (Fig. 1.2). The meeting was convened after a long preparatory process that saw available data from different conservation and research organizations put together for the first time.

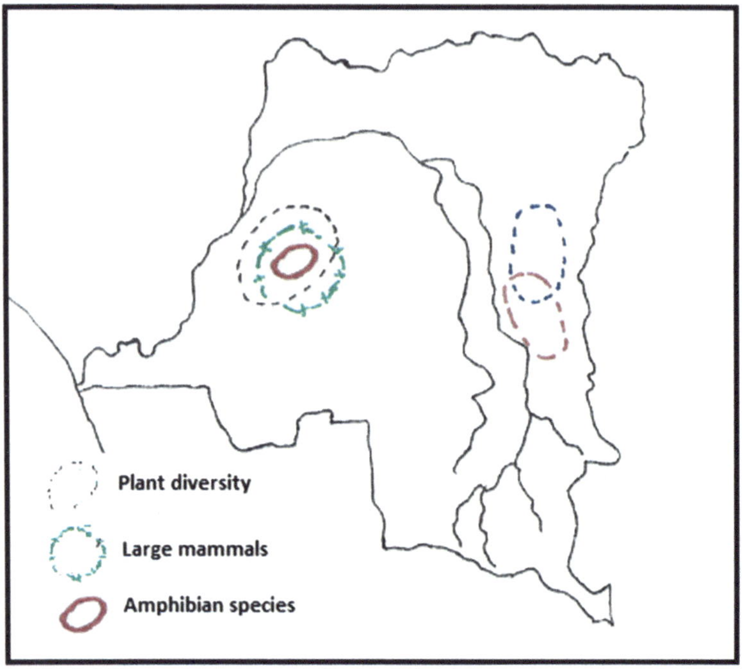


**Fig. 1.2** Example of taxonomic data being deployed on a map. This is an illustration of how taxonomic specialist groups assembled their knowledge on the diversity across the Congo Basin. The image is about DRC only and it does not take account of the other important biodiversity conservation priority areas. It does illustrate the process taking two geographic areas of DRC, namely the Lake Tumba swamp forests and the mountain zones

The data was synthesized and bio-geographic regions and sub-regions were mapped (Toham et al. 2006). It is important to note the inclusion of social anthropology and socioeconomic data sets in the process of defining ecological regions. This is, in terms of conservation biology, a practical example to demonstrate that this new and evolving science is at the juncture of both natural sciences (biology, geography, mathematics, physics, etc.) and social sciences (social anthropology, history, demography, etc.). These maps were developed based on taxonomic, socioeconomic, and anthropological data (Fig. 1.3). At the Libreville meeting, experts from numerous fields (taxonomic groups, social anthropologists, governmental agents, and donor) were asked to identify areas they knew and felt, based on the best knowledge available, to be the most important to support viable biodiversity assemblages (Toham et al. 2006).

Using spatial modeling, particularly the nearest-neighborhood method, distances were measured between different highly important taxonomic areas and other taxonomic groups. Average distances between neighboring taxonomic groups were then compared with complete spatial randomness (CSR; Fortin and Dale 2005). As suggested by Dale (1999), Fortin and Dale (2005), and Haining (2003), neighboring

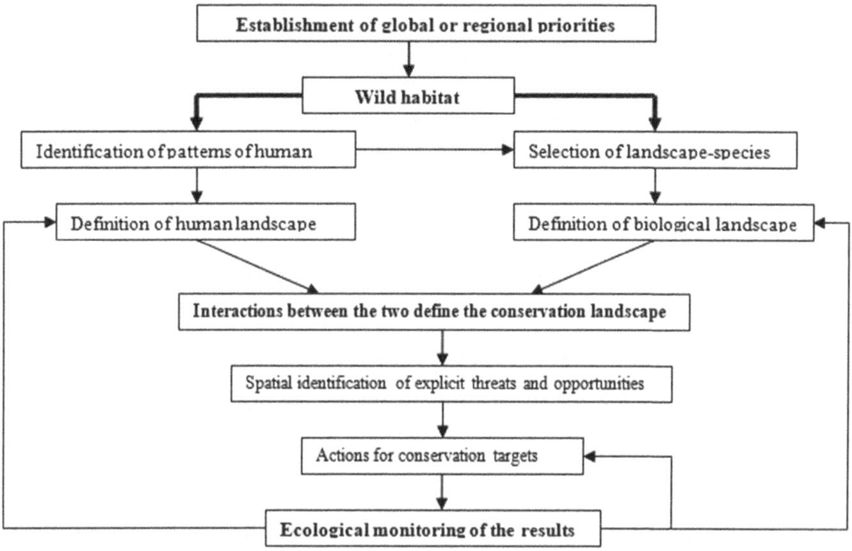


**Fig. 1.3** Conceptual Model used for landscape-species conservation approach. An important aspect of this model is that patterns of how human activities affect landscape are *pari passu* with the landscape species. This means that the work will have to take both into account in order to bear positive results on the landscape species

taxonomic groups with average distance lower than the CSR were lumped together and defined what was called 'ecoregion'. This process was among the first cases in Central Africa where the powerful tool of the then emerging spatial analysis (Haines-Young et al. 1993) was used to define ecologically sensitive areas (Inogwabini 1997) and, as such, contributed to demonstrating the practical uses of new theories and thoughts in conservation biology. From this process of making maps by lumping nearest neighbors, 77 important biodiversity areas emerged (Toham et al. 2006). Then a ranking process had to follow. Ranking was established from several approaches.

Because the species richness had already been captured in the process of choosing the 77 areas, the first approach for spatially delineating ranked areas was to define scales at which each of the 77 defined important areas would contribute more in protecting global biological diversity. Scale that plays a critical role in priority setting (Nally 2005; Osborne et al. 2001; Wilson and Keeling 2000) was crucial in any process intending to firmly establish the conservation action for coming generations.

The second approach was to define the uniqueness of each area and the magnitude of loss at the global scale if biodiversity in such a unique area became extinct. Of course, this was a specific touch to include the long cherished ideas of charismatic species and beautiful scenery, which are very often among the aesthetic reasons for biodiversity conservation (Callicott 2006; Primack 2000). Furthermore, including the uniqueness of any given area was also valuable in that it accounted for endemism and cultural values attached to certain zones. Both biodiversity indexes and species endemism were also used in such processes as the one that led to the delineation of global biodiversity hotspots, therefore, this inclusion was not a ground-breaking idea as it may have appeared. However, defining cultural values was new in the conservation practice in Central Africa where all focus has been on wildlife and wild habitats throughout the history of conservation biology. This addition of cultural values has come principally from the participation of social and cultural anthropologists involved in the process.

The third most important criterion was the ecological integrity of each area. It is difficult to find a clear definition of the concept of 'ecological integrity' on which all proponents would agree. Very often an ecosystem that is still integrated is defined in terms of maintenance of ecological functions such as water cycles, nutrient cycles, and ecological interactions between different entities. In practical terms, and given the lack of knowledge the experts possessed at that time on the different ecosystems of Central Africa, they have decided to use a surrogate concept, which was defined as habitat's intactness.

Implementation-wise, the first consideration brought the experts to concentrate on two scales: the global and regional. Hence, areas that were only locally important had to be in the vicinity of those that were either regionally or globally important in order to be included. However, values obtained from that process were weighted by the introduction of the value of high uniqueness. Therefore, areas included in the ecoregions capture both regional importance and local uniqueness. Combining scale, ecological integrity, and uniqueness defined the eigenvalue for each area (see Chap. 21), which can be included with the notion of intrinsic value, another conservation biology concept that can be measured only through surrogates. These eigenvalues were translated into the final map of 29 areas that were considered to be the most representative of the biodiversity of Guinean-Congolian forests (Toham et al. 2006). The final product was zones extending outside conventional protected areas, even though most of the 29 priority conservation areas had protected areas within their limits. Of course, experts acknowledged that data that led to these 29 areas had some limitations, among which the most important were absent data on some taxonomic groups and some regions, poor quality of data available on some areas, and incomparability of existing data from different sites.

## 1.4 The Landscape-Species Approach in Central Africa

The landscape-species approach was principally drawn from the mainstream conservation biology and was promoted by WCS through its Living Landscape Program. Its idea is simple and it is based on the use of ecosystem surrogates, particularly species whose habitat and ecological requirements encompass large areas and diverse micro-ecosystems (WCS 2001). Landscape-species were defined as those whose needs in habitats are vulnerable because of various human activities in a space-time continuum. They are species that need large and heterogeneous home ranges to meet their ecological requirements. As such, these species were those whose home ranges were not necessarily confined within political boundaries. Most, if not simply all, identified landscape-species were not locked within the limits of given protected areas. Therefore, identifying the ecological needs of those landscape-species would guide the management of the extended conservation priority areas (WCS 2002). They were also defined as species threatened by human activities within their home ranges, and could be easily used as flagship species. A species-focused approach, the landscape-species approach is based on the species conservation paradigm, which had always been at the heart of conservation biology even before the emergence of the subject as an academic realm.

The landscape-species approach followed five steps to define priority conservation areas: (1) defining the area of interest based on preliminary field experiments; (2) identifying a set of landscape-species residing within the pre-selected zone; (3) defining and mapping species' home ranges with special emphasis on micro zones harboring sufficient resources for the long-term sustainability of viable populations of landscape-species; (4) defining and mapping human activities and their intensities; and (5) examining intersections between biological landscapes defined by step (3) and human landscape defined by step (4).

Identifying areas of interest based on field knowledge means either having a collection of field data on which to base the analysis or to work using areas already defined in other prioritization processes. Basing the analysis on field data is a laudable approach in that it does bring the priority setting exercise to collate with field realities: action needs to be based on sound scientific knowledge. However, it does also bias results toward those sites where action has been already ongoing. Using priority areas already defined by other processes has the merit of marrying priority zones defined through landscape-species approach with national, regional, or international priorities, even when these may be based on slim scientific evidence. In Central Africa, the WCS Living Landscape Program identified Nouabalé-Ndoki as its area of interest based on a set of long-term data.

As a set of landscape-species candidates, forest elephant (*Loxodonta africana cyclotis*) was the species whose ecological requirements needed an extended home range. It was also both a world-widely acknowledged flagship species due to both its imposing mass and threatened status. Other large mammals and charismatic species of the Nouabalé-Ndoki included the bongo (*Tragelaphus eurycerus*), the chimpanzee (*Pan troglodytes*), the forest buffalo (*Syncerus caffer nanus*), and the lowland gorilla (*Gorilla gorilla gorilla*). To these mammals was added the dwarf crocodile (*Osteolaemus tretraspis*). It was felt that juxtaposing ecological niches of those six species covered a variety of habitats that would maintain functional ecosystem processes and protect a cohort of other wildlife species. It was also thought that preserving viable populations of those species would be a tangible means to gauge the evolution of ecosystem's health.

Data were collected on elephant home range and its habitat uses over a long period in the Nouabalé-Ndoki (Blake 2002). These included not only field survey data to define abundance and species distribution but also telemetry data to refine the available knowledge on elephant home sizes (Blake et al. 2001, 2008). Taking advantage of technological advances, the telemetry data set was a key element in defining ecological boundaries of the Nouabalé-Ndoki landscape. Coordinates from GPS telemetry were used to define individual home range for elephant (Blake et al. 2008; Frair et al. 2004; Mauritzen et al. 2002). The home range was defined as 95% minimum convex polygons using the ArcView extension for animal movements (Blake et al. 2001, 2008; Frair et al. 2004). The association of fixes with months that typically constitute rainy or dry season provided seasonal ranges. Seasonal mean positions and seasonal home ranges were compared using regression equations (Frair et al. 2004). Habitat selection by elephant and by season followed a two-step process (Mauitzen et al. 2002). First, the entire zone was classified into its different broad vegetation classes by the use of image processing packages. Second, animal locations by seasons were compared with available habitats (random locations) within the Nouabalé-Ndoki area (Blake et al. 2001). Furthermore, to define minimum ecological requirement, mean elephant seasonal home ranges were plotted on the map against seasonal fruit availability indexes and showed that elephant ranging patterns correlated with fruiting seasons (Blake 2002). Interestingly, but not surprisingly, the elephant home range comprised individual home ranges of all other species chosen as landscape-species, as it was larger than what most of ecologists would have expected and included different types of microhabitats that were specific niches for all species of wildlife and plants (Blake et al. 2008; Blake 2002).

Accounting for human landscape in the process of designing conservation priority areas by use on landscape-species has been done in several steps. The first of these included collecting demographic data in the selected region of Nouabalé-Ndoki followed by a thorough assessment of various land-use practices, land rights, and existing political structures. The process also included mapping access roads, major human settlements, major markets, etc. Assessing land-use practices has put a greater emphasis on the identification of numerous stakeholders, including a variety of actors: from commercial logging companies, to hunters, fishermen, and small scale farmers. The impacts of those different actors have resulted in delineating micro zones with their functional attributes, including logging concessions, communal areas reserved for agriculture, etc.

The examination of intersection areas was accomplished by calculating the conventional overlap coefficient. This implied that only areas with a weak overlap coefficient were given higher priority while those that have higher overlap coefficient were given lower priority. Saying so means only that priority conservation zones were defined as those zones located outside of irreversible human encroachment in a given habitat. This equaled the same as choosing by principles of a habitat's intactness, which is similar to the ecoregion process described above. Biological and human maps developed through the processes described above were overlaid and produced a complex mosaic of micro zones with different functional attributes.

Implementation-wise, the process has placed much emphasis on developing conceptual models Fig. 1.3), which were thought to be the best way to analyze both threats and opportunities for species conservation (Painter and Wallace 2006; Sanderson et al. 2002). These models were essentially centered on measurable results meaning abundance of landscape-species over time. The idea behind this was that conservation investment would be worthwhile only when species are maintained as stable or increase in their numbers. This focus on numbers was a sound attempt to provide indications on whether conservation efforts were matching with successes on the ground, which has always been a difficult question for conservation organizations operating in Central Africa.

The most interesting output of this process was the effect of the landscape-species concept on the management requirements of logging concessions where bongos, chimpanzees, forest buffaloes, and gorillas occurred. Stakeholders agreed that there should be clear conservation zones within the logging concessions adjacent to the core Nouabalé-Ndoki conservation area, which were established with the help of conservation organizations. This result is very important in that it has now become part of the timber certification process in Central Africa. The other important result of the landscape-species approach has been the fact that the majority of actors agreed on the need to establish viable ecological corridors for forest elephants, a species that was circulating between three different countries (Central African Republic, Republic of Cameroon, and Republic of Congo).

## 1.5 International and Regional Political Processes

Prior to the Libreville meeting organized by the WWF Central African Regional Program and launching of the WCS Living Landscape Program, political discussions had started to mobilize the political will in Central Africa to manage wild habitats, wildlife, and other natural resources. International political discussions on the Central African forests emerged from such international processes as the 1992 Rio de Janeiro Convention on the Biological Diversity (CBD) and the Kyoto Protocol. Regionally, political discussions resulted from different regional and world summits of which the most important were the Yaoundé Summit and the Johannesburg World Summit.

Contrary to what many people are led to believe, the Yaoundé Forest Summit preceded the Libreville meeting. In fact, some participants in the Libreville meeting were mandated by the Yaoundé Summit, as part of a follow up process. The Yaoundé Forest Summit was held on 17 March 1999 and was attended by Heads of Central African States (Chad, Cameroon, the Central African Republic, the Republic of Congo, Equatorial Guinea, and Gabon) and chaired by Prince Phillip, Emeritus President of WWF. The main aim of the summit was a call for far-reaching commitments on forest conservation, including establishment of numerous new protected areas, control of illegal logging and the bushmeat trade (De Capua 2005). The Yaoundé summit also was about establishing an institutional structure to implement the global biodiversity conservation policies included in the final Yaoundé declaration (De Capua 2005).

Generally, the Yaoundé summit made of the harmonization of forest policies is one of the common denominators for the cooperation between Central African countries (Koyo and Foteu, undated). The summit also led to shifts in ways people perceived conservation in the region by insisting that rural populations and other stakeholders should be involved in the planning and management of forest resources and their use and linking logging and other activities that extract forest resources with local, provincial, and national sustainable development programs. The Yaoundé declaration also promoted the industrialization of forest products, the creation of regional networks of technical and scientific exchange in order to update the scientific data on the biodiversity and forest collection and encouraged the identification and the development of new yet sustainable financial mechanisms for biodiversity conservation, with forestry development as one of the backbones.

The most important outcome of the Yaoundé Summit, which since 1999 has been known as the Yaoundé Process, was the creation of the Commission of Central African Forests (COMIFAC—Commission des Forêts d'Afrique Centrale), a Regional Central African Body to coordinate regional initiatives among governments, and the conservation and sustainable management of forests. The importance of issues related to forest natural resources was becoming more and more global and inter-country coherence was critical to the way each country of the region managed its resources. The COMIFAC treaty was signed by heads of state in 2005 in the second meeting of the Yaoundé Process, which made conservation and sustainable management of natural resources a priority of national policies (article 1 of the treaty) and declared the inclusiveness of all potential stakeholders as the way toward the implementation of a successful conservation program across the region.

The world summit on sustainable development (Johannesburg, South Africa in 2002) was also important in the decision making process that led to the concretization of the priority conservation landscape in Central Africa. Those who met at the World Summit on Sustainable Development had one global objective: to review progress since the Rio conference in 1992 and to agree globally on sustainable development (Von Schirnding 2005). Rather than seeking new treaties and targets, the primary concern of the summit was the implementation of programs that were agreed upon in the Rio de Janeiro Conference (Von Schirnding 2005). Clearly, most experts felt that little has been achieved on the ground since 1992, particularly in Africa. In the end, and despite so much of the dispute between representatives of the global civil society and political leaders, environmental sustainability was agreed upon as an area that needed global improvements in implementation (UNMP 2005). Of particular emphasis was the clear recognition of Africa's environmental and development needs. Biodiversity was one of the eight priorities and was included in the Implementation Plan (UNMP 2005). Strong financial commitments were needed from the international community because one of the major causes of the failures to implement the objectives set out for Africa at the Rio de Janeiro meeting included the lack of financial resources.

## 1.6 Forces Driving the Landscape Approach and Conditions for Its Successes

From a conservation biology point of view, conservation biological diversity in large landscapes can only be achieved if multiple driving forces come together. First of all, there needs to be a strong political commitment. That political dynamism, in the case of Central Africa, has come through a conjunction of events of which the most important was the internationally perceived problem of the environmental effects of climate changes. The second force was the acknowledgment by all that biological diversity resources were resources on which many people depend for their survival and, hence, people needed access to these resources but with the idea of sustainability as the common divider. For centuries, African cultures coexisted with biodiversity but these ages are now long gone; growing human populations require more and more resources (Inogwabini 2007). In many ways, the landscape as 'biodiversity conservation paradigm' in Central Africa can be defined as a way of trying to reconcile the need to fulfill basic human needs with the conservation of biodiversity. Reconciling the two is one of the major challenges for Africa, if its biological diversity is to survive through the ages. The landscape concept stems from historical perspectives on the central African protected areas, which all were created at a time when African mammals appeared to face a bleak future (Hall 2000). In several cases, creating protected areas involved the removal of people from their homelands (Chap. 2). However, in the late 1980s, protected areas alone were felt insufficient to preserve biodiversity (Petersen and Huntley 2005). Programs intended to preserve biodiversity while serving economic needs were therefore introduced, and conservation programs became ipso facto development programs (Oates 1999).

Unfortunately, however, in the early 1990s, conservation and development programs started to fail (Oates 1999), largely because development concepts, models, and practices cannot simply be imposed on people without free consent on their part. New methods were needed, requiring changes in conservation paradigms and including approaches from a stance of partnership (McNelly 1995). All new conservation approaches noticed that in order to be successful, they had to rely on social mobilization to include previously excluded peoples in conservation activities (Cowling and Wilhelm-Rechmann 2007; Mutamba 2004; Adams and McShane 1996). Social mobilization expects local communities to participate in activities related to their natural resources and, in reality, this is not very different from the traditional conservation approach. Even with social mobilization, people are invited to participate only in pre-defined projects; they are not necessarily encouraged to initiate projects and actions that affect their lives (Mutamba 2004).

As it will be lengthily discussed in this volume, the conservation landscape approach is, theoretically, meant to empower local communities by including both natural and social capitals in the multiple unknowns of the conservation equation (Sayer et al. 2006), though it is fully understood that empowering people is a long process best nurtured in a collective, cohesive, and coherent social venture (Inogwabini 2007). However, such empowerment can lead to unexpected results because in Africa, it often occurs without the full involvement of national governments (Mutamba 2004). The other side of this is that African governments are often perceived as undemocratic, preventing local communities from exercising their rights, which is partly understandable, given the poor performance of many African governments on human rights and good governance. However, it is a fact that social mobilization will not work without functional states. Africans will only adopt sustainable management schemes through an empowerment process that makes them the true decision makers, and this will only be effective if African institutions become strong enough to implement the rule of law. In this sense, conservation success depends on both good governance and the engagement of local communities. That is what the conservation landscape approach is meant to implement, in addition to the classic protected area management style: the long processes that involve local communities from the inception in any conservation action through the sharing of the potential incomes. Therefore, the rights acknowledged by the convention on biological diversity also impose the duty to participate in the design and the implementation of conservation action. The bottom line is in acknowledging that sovereign, democratically elected governments must decide the course to take. Countries in Central Africa need to prove their ability to apply sufficient authority over these landscapes to implement agreed-upon rules that define local community's duties and rights by using agreed-upon tools for good governance. Reconciliation of the fulfillment of basic human needs and the conservation of biodiversity depends on good governance, and this depends on the overall health of democracy. Of course, good governance will make sense only if landscapes will incorporate socioeconomic dimensions into the natural resources management.

Incorporating socioeconomic dimensions means also to bring sufficient and significant resources (Chap. 23), which is the third force that can enable a landscape conservation paradigm to succeed. That financial assets are needed to implement conservation programs is not a new idea; all approaches have faced this reality since the early years of conservation biology as a scientific branch (Primack 2000). As it is discussed in Chap. 23 of this volume, landscapes need more resources than can be imagined at first glance. For the ongoing efforts in Central Africa, this has been possible only after the commitment by the US Government in the World Summit on Sustainable Development in Johannesburg. Like it or not, the commitment announced by General Colin Powell, in his capacity as the American Secretary of State, gave a strong impulse that set in motion the implementation of ongoing scientific and political conservation processes in Central Africa. By committing 53 million US dollars to biodiversity conservation (WWF undated), the US provided the basis from which all the plans leading to large scale conservation could be attempted throughout the 12 priority conservation areas of the Congo Basin. Of course, the funding granted by this commitment was far less than the scope of the field work required by such a mega initiative but it constituted the greatest chunk of contribution ever received to fund biodiversity conservation in Central Africa. That commitment by the US Government to the 7th of the eight millennium goals was of critical importance to make the long process, started by the Yaoundé Summit, become a true field conservation action (WWF undated). Of course, as will be argued in Chap. 15, without significant financial resources, landscape as a conservation paradigm will be difficult to implement and this calls for international solidarity and fairness. Wild habitats, wildlife, and human cultures in areas selected as priority conservation areas in Central Africa provide significant environmental services to the global humanity; they deserve more attention in order to preserve the natural equilibrium. Fortunately, that American initiative has been emulated by other countries that contributed, at varied scales, resources toward making the whole effort continue.

## References


Adams JS, McShane TO (1996) The myth of Wild Africa: conservation without illusion. University of California Press

Blake S (2002) The Ecology of forest elephant distribution and its implications for Conservation. Ph D Thesis submitted to the University of Edinburgh. Edinburgh, United Kingdom

Blake S, Deem SL, Strindberg S, Maisels F, Momont L, Inogwabini BI, Douglas-Hamilton I, Karesh WB, Kock MD (2008) Roadless wilderness area determines forest elephant movements in the Congo Basin. PLOS ONE (10): e3546, 1–9. www.plosone.org

Blake S, Douglas-Hamilton I, Karesh WB (2001) GPS telemetry of forest elephants in Central Africa: results of a preliminary study. African J Ecol 39:178–186

Callicott JB (2006) Conservation values and ethics. In: Groom MJ, Meffe GK, Carroll RC (eds) Principles of conservation biology, Third Edn. Sinauer Associates, Inc. Publishers, Sunderland, pp 111–135

Cowling RM, Wilhelm-Rechmann A (2007) Social assessment as a key to conservation success. Oryx 41:135–136

Dale MRT (1999) Spatial pattern analysis in plant ecology. Cambridge University Press

De Capua J (2005) Saving the Congo Basin rainforest. Voice of America, 1 February 2005. http://www.voanews.com/english/news/a-13-2005-02-01-voa6-67379422.html

Dinerstein E, Olson D (1999) The Global 200: preserving a rich tapestry of the earth's most outstanding wildlife habitats. In: Lanting F, Rowel G, Doubilet D (eds) World wide fund living planet: preserving Eden of the earth. Crown Publishers, p 252

Frair JL, Nielsen SE, Merrill EH, Lele SR, Boyce MS, Munro RHM, Stenhouse GB, Beyer HL (2004) Removing GPS collar bias in habitat selection studies. J Appl Ecol 41:201–212

Fortin MJ, Dale MRT (2005) Spatial analysis: a guide for ecologists. Cambridge University Press

Haines-Young R, Green DR, Cousins SH (eds) (1993) Landscape ecology and GIS. Taylor and Francis

Hall JS (2000) A review of myth and reality in the rain forest: how conservation Strategies are failing in West Africa, by John F. Oates (1999). J Polit Ecol 7:1–6

Haining R (2003) Spatial analysis: theory and practice. Cambridge University Press

Inogwabini BI (1997) Using GIS to determine habitat use by large mammals and to define sensitive areas of Kahuzi-Biega National Park, Eastern Congo. MSc. Dissertation submitted to the University of Kent. Canterbury, United Kingdom

Inogwabini BI (2007) Can biodiversity conservation be reconciled with development? Oryx 41:2–3

Inogwabini BI, Leader-Williams N (2013) Conservation paradigms seen through the lenses of bonobos in the Democratic Republic of Congo. In: Navjot S, Raven P (eds) Essays in conservation biology—perspectives from practitioners in tropical environments. Blackwell-Wiley

Mauritzen M, Derocher AE, Wiig Ø, Belikov SE, Boltunov A, Hansen E, Garner GW (2002) Using satellite telemetry to define spatial population structure in polar bears in Norwegian and Western Russian Artic. J Appl Ecol 39:79–90

McNelly JA (1995) Partnerships for conservation: an introduction. In McNelly JA (ed) Expanding partnerships in conservation. Island Press, pp 1–10
Mutamba E (2004) Community participation in natural resources management: reality of rhetoric? Environ Monit Manag 99:105–113
Nally MR (2005) Scale and an organism-centric focus for studying interspecific interactions in landscapes. In Wiens JA, Moss MR (eds) Issues and perspectives in landscape ecology. Cambridge University Press
Oates JF (1999) Myth and reality in the rain forest: how conservation strategies are failing in West Africa. University of California Press
Osborne PE, Alonso JC, Bryant RG (2001) Modeling landscape-scale habitat use using GIS and remote sensing: a case study with great bustards. J Appl Ecol 38:458–471
Painter L, Wallace R (2006) Landscape conservation in the greater Madidi Landscape, Bolivia: planning wildlife conservation across different scales and jurisdictions. In: Groom MJ, Meffe GK, Carroll RC (eds) Principles of conservation biology, Third edn. Sinauer Associates, Inc. Publishers, pp 453–458
Petersen C, Huntley BJ (eds) (2005) Mainstreaming biodiversity in production landscapes. Global Environment Facility, Washington, DC, USA
Primack RB (2000) A primer of conservation biology, 2nd edn. Sinauer Associates, Inc., Publishers
Sanderson EW, Redford KH, Vedder A, Coppolillo PB, Ward SE (2002) A conceptual model for conservation planning based on landscape species requirements. Landsc Urban Plann 58:41–56
Sayer J, Campbell B, Pethera L, Aldric M, Pere MR, Endamana D, Dongmo ZLN, Defo L, Mariki S, Doggart N, Burgess N (2006) Assessing environment and development outcomes in conservation landscapes. Biodivers Conserv. https://doi.org/10.1007/s10531-006-9079-9
Toham AK, D'Amico J, Olson D, Blom A, Trowbridge L, Burgess N, Thieme M, Abell R, Carroll RW, Gartlan S, Langrand O, Mussavu RM, O'Hara D, Strand H (2006) A vision for biodiversity conservation in Central Africa: biological priorities for conservation in the Guinean: Congolian forest and freshwater region. World Wide Fund
United Nations Millennium Project (UNMP) (2005) Investing in development: a practical plan to achieve the millennium development goals, New York
Von Schirnding Y (2005) The world summit on sustainable development: reaffirming the centrality of health. globalization and health 2005, 1: 8. https://doi.org/10.1186/1744-8603-1-8
Wilson HB, Keelin MJ (2000) Spatial scales and low-dimensional deterministic dynamics. In: Dieckmann U, Law R, Met JAJ (eds) The geometry of ecological interactions: simplifying spatial complexity. Cambridge University Press
Wildlife Conservation Society (WCS) (2001) L'approche 'Espèce-Paysage': Un outil pour la conservation in situ. Programme Paysage Vivants—Bulletin №. 2
Wildlife Conservation Society (WCS) (2002) Les roles des 'espèces-paysages' dans la conservation sur terrain. Programme Paysage Vivants—Bulletin № 3

# Chapter 2
# Landscape: Re-assessing the Conservation Paradigm

**Abstract** The chapter builds a case for the novelty and the appropriateness of biodiversity conservation landscape in Central Africa. It departs from the foundational definition of conservation in Central Africa that has been, for years, essentially concentrated on protected areas, which were legally spatial entities locked away from human activities and delineated in a gradient ranging from total exclusion to allowed human activities. These isolated spaces were supposed to preserve wildlife species and their habitats intact by means of strict law enforcement. It argues that one reason, among many others, for minimal successes of conservation had been in this strict fence-and-fine approach. As support to that massive claim, it brings in crucial factors that prevent strict law enforcement to function adequately, including such facts as the existence of long-distance ranging species such as forest elephants that need spaces beyond borders of protected areas, increasing human populations, lack of resources, and advanced human poverty and the superimposition, in some areas, of important resources with high biological diversity that conservation proponents want to protect. The factors lead to clashes between basic human needs, governmental obligation to use those resources to develop regions in which those resources are located and biodiversity conservation. Looking at other tried approaches to bridge the above clash and why they reached only mute results, the chapter argues that this was so because these other approaches failed because either they were neither sufficiently inclusive of major stakeholders nor apt enough to functionally and geographically delineate functions that are, by their very nature, opposed to each other.

**Keywords** Conservation paradigm · Planning process · Data needs · Identification of stakeholders · Stakeholders assessment · Planning tool

## 2.1 Introduction

The concept of landscape is new in African conservation in general and particularly in the case of Central Africa. Over years, conservation has been defined as essentially concentrated on protected areas. In Central Africa, protected areas can be easily defined as a set composed of spatial entities locked to human and delineated in a gradient of allowed activities. They are, in decreasing prevention from human activities,



B.-I. Inogwabini, *Reconciling Human Needs and Conserving Biodiversity: Large Landscapes as a New Conservation Paradigm*, Environmental History 12,
https://doi.org/10.1007/978-3-030-38728-0_2

composed of parks, natural reserves, and game reserves. Within those isolated spaces, it was supposed that wildlife species and their habitats would be specially protected from purely law enforcement perspectives. As indicated in chapter one of this book, however, conservation in Central Africa deals with long-distance ranging species such as elephants, which need spaces beyond borders of protected areas. It also deals with increasing human populations, lack of resources, and advanced human poverty (Putz et al. 2000). In some areas, important resources are superimposed with high biological diversity that conservation proponents want to protect, clashing with both basic human needs and governmental obligation to use those resources to develop regions in which those resources are located.

These ecological and human realities pose to conservation practitioners types of questions that are different from those that are raised in other ecological contexts (Putz et al. 2000), namely those in the developed western countries. This statement is seldom new; many conservation thinkers have acknowledged it a long time ago and attempts to reconcile human and conservation needs have been tried in several African environments. Before 1960 already, wildlife biologists called attention to the fate of Africa's mammals to raise awareness and prevent these mammals from extirpation (Hall 2000). These efforts led to the coalition of the first formal conservation efforts on the African continent and, to some degree, wildlife conservation organizations (Hall 2000). First trials included the creation by newly independent states of new protected areas (Terborgh 2004) followed by conservation-development programs (Oates 1999). First, the scramble for still more and more protected areas meant the erasure of human populations from the depiction of ecosystems wherein they had evolved for millennia (Adams and McShane 1996). Human communities saw the lands of their ancestors confiscated (Adams and McShane 1996), which led to crumbled cultural roots. The *fence-and-fine paradigm* adopted to make conservation work showed its limits and needed another paradigm (Inogwabini and Leader-Williams 2013a). Second, acknowledging that conservation would not work as a stand-alone activity in the middle of a sea of human population, conservation efforts were transformed to include development aspects. Unfortunately, *development-as-conservation-tool paradigm* also led conservation objectives astray and, in many cases, ended up as major forces working against conservation (Hall 2000; Oates 1999).

It is out of these felt or real failures in implementing different old conservation paradigms that new ideas and approaches are being generated and tested. There are such ideas as pro-people, pro-poor approaches to sustainable development, which are thought to be part of biodiversity conservation paradigms. In reality, concept such as sustainable development and pro-poor emanate from development conceptualization of the world and are, therefore, used in conservation as taken-for-granted with very limited serious thinking of what they bear as historical and philosophical backgrounds. That is why, however, strongly these new concepts are grounded in the actual conservation practices and thinking, there is a need to re-evaluate them. There is also a need to use the accumulated main stream conservation knowledge to think of other paradigms and approaches that can insert conservation in the new context of the world. Landscapes, as conceptualized in Central Africa, are such a drive to think beyond conventional conservation niche.

## 2.2 Landscape Defined as a Conservation Vision

The landscape concept has been used even before the Christian era by philosopher such as Plato, when he wrote that removing trees from Attica had led to ongoing loss of soil from high ground and left the landscape looking like a human skeleton, a body wasted by disease (Inogwabini and Leader-Williams 2013a; Thirgood 1981; Plato 360 B.C.E). However, in the recent, the classical conservation biology theory, the concept of landscape is intimately related with the emergence of the industrial era in England (Arvill 1967) when it was defined in relation to proper land-use planning thought to be both as applied human ecology and as an enlightened conception of how to consciously control the environment (Fairbrother 1972). Practically, *landscape as an ecological paradigm* was already seen as a comprehensive vision to account for the totality environment conditions in one given region of England undergoing the industrially induced urbanization (Fairbrother 1972). In that sense, landscape had to equal the total of habitat types and different human population classes. As such, the concept meant to symbolize the totality of ecological relationships but was to be realized in different types of scenery (Fairbrother 1972). Landscape as an ecological paradigm has, therefore, occupied a distinct niche in ecological thinking since the early times of the industrial era when humans began to escape from the natural laws that governed their development and started losing their grip on the natural responses to their different actions.

To demonstrate its importance in the development of the main stream ecological sciences, the concept of landscape has given birth to an entire area of ecological research, known as landscape ecology. Landscape ecology has been defined in different ways in recent literature without a consensus (Fahrig 2005). As part of ecological research, landscape ecology is focused on looking at the effects of landscape layouts on abundance and distribution of biodiversity (Fahrig 2005), which implies a combination of both the inquiry in geomorphology and ecology (Allaby 2004). Despite the important insights, however, *landscape as ecological paradigm*, it is important to note that this conceptualization of ecological links and ways to address different questions has not been put in practice in tropical conservation biology. Furthermore, it is clear from ways in which landscape ecology and landscape as an ecological paradigm are conceptually dealt with in the main stream conservation is a narrower perspective compared to the experiments in Central Africa. Even more narrower is the fact that in some circles of thinkers, landscape ecology is even thought of as an essentially architectural concept, hence confined in spatially demarcated space, i.e. where human infrastructures are to be built.

Landscapes, as defined under the circumstances of Central Africa, are zones in which conservation action has seen a major shift in its classical paradigm of natural reserves. These are zones in which cultural, economic, and biological values have to be managed sustainably. As such, landscapes in Central Africa add three different main elements in the traditional *landscape as conservation paradigm* definition provided above. First, geographical extends are far larger than what can be envisaged in the western world. Lake Tumba, for example, is 80,000 $km^2$, which is more than the size of Rwanda or Burundi. Second, there is a cultural dimension (Taylor 2005; Sayer et al. 2006) that is added to the environmental conceptualization that for long been sole in designing conservation programs. This is a tremendous addition to the main stream conservation thinking in that landscapes are supposed to protect more than just biodiversity but also cultural diversity. The cultural dimension addition has come from the recognition that homogenizing concept such as development, happiness, basic needs, human well-being, and relationships with human immediate environment will lead to loss of not only the biodiversity but also human cultures and, therefore different types of knowledge possessed by different human cultures. Third, and finally, the landscape concept as used in Central Africa thrives to integrate economic values into the conservation equation (Dunning et al. 2006). It proceeds from a utilitarian perspective of conservation and the premises that conservation will only succeed if human well-being is achieved through preserving the biodiversity (Inogwabini 2007). This does not, however, mean that for Africans conservation is only valuable when people find their daily concerns about their livelihood addressed by conservation organization but rather that realistically it is only when people have means to live on that they care about their environment: *landscape as a conservation paradigm* puts specific interest of each group at the center of conservation thinking.

Surely, the *landscape as a conservation paradigm* goes beyond the extension of the geographic dimension of conservation units, which is what the *eco-region paradigm* introduced in the last decade of the twentieth century. Eco-region is a concept that was introduced based on the fact that parks would not suffice to protect species of conservation. It was thought that by preserving habitats and systems wherein species occurred, conservationists would preserve the biodiversity therein found, inclusive of species of conservation concern. This was a leap forward for conservation thinking but the eco-region paradigm was essentially biodiversity-focused, with elements of biogeography, and tried to divide natural environment into discontinuous units (Groombridge and Jenkins 2002). This way of thinking ignored thus the reality that current patterns in the distribution and abundance of species, persistence of ecological functions and habitats very often reflect the historical and current human ecology. The re-emergence of the concept landscape in the context of Central Africa is, therefore, a trial to bring human ecology together with other components of conservation biology in order to attempt solving problems inherent to decreasing biodiversity consequent of habitat loss. The question that remains is only how to transform the *landscape as a conservation paradigm* in a concrete articulation of a vision that ensures that the human ecology side of the paradigm does not impede the conservation biodiversity.

Bringing together human ecology and other conservation articulations to address loss of habitats and biodiversity is a shift in paradigm and requires many revolutionary changes in our collective thinking about conservation. First, conservation has to be looked at as the mean to ensure equilibrium between different natural conditions by its opponents. Second, utilization of natural resources has to become what it really is:

the right of communities that reside within those ecosystems that harbor intrinsically valuable biodiversity segments. Ecology and economics are inter-related. Third, at the methodological and pedagogical levels, the emergence of new actors becomes an obligation. Actors of applied conservation biology are not only those professional conservation biologists but also different social groupings. This implies bringing conservation biology back to its fundamental character: being at the junction of different areas of scholarship.

Concretely, the overall vision for the Lake Tumba Landscape was defined accounting for all these as a space wherein sustainable natural resource management measures should be implemented in order to conserve the Lake Tumba Swamp Forest through a concerted land-use planning that attributes appropriate functions to each land and water unit included in the landscape segment. A specifically important component of this vision is that all charismatic species residing in the landscape segment such as bonobos, forest elephants, lions, and their habitats are protected in perpetuity within the Lake Tumba Landscape. Combining these lines has been a difficult exercise since in the Lake Tumba landscape, little was known before the inclusion of the zone within the network of priority conservation zones of Central Africa. Of course, that could be done only through a thorough analysis of problems.

## 2.3 Landscape Is Also a Planning Tool Linking Ecology and Economics

Inserting the landscape's economic dynamics into the conservation equation was the most intriguing endeavor though the collected socioeconomic data had helped simplify the problem. In summary, these socioeconomic data indicated that the collapse of the national economy and paid employments that had pushed all human social categories had fallen back to extracting natural resources as means to obtain financial incomes (Colom et al. 2006). That has led to some sort of classical commerce, with only on average 25% of resources being used for local subsistence, as illustrated in the Fig. 2.1 (Colom et al. 2006).

Clearly, grasping the nature of economic dynamics in the landscape is not sufficient. There was a need to orchestrate a scenario that could be both accepted by the majority of people and work for both local communities' livelihood or human well-being and biodiversity conservation. That leads to creating a democratic process through which different views would be presented at the negotiation table. Nonetheless, this vision had its own drawbacks. The most fundamental of these problems was that of planning process at different hierarchical levels; from the local level to the national level via the decentralized province. The second most important issue was that of a fair representation in an environment where democratic processes are at their very best still embryonic. The third question that both the understanding of the landscape as a conservation paradigm and the vision of Lake Tumba as a landscape posed was that of non-organized local communities. How would disparate communities,

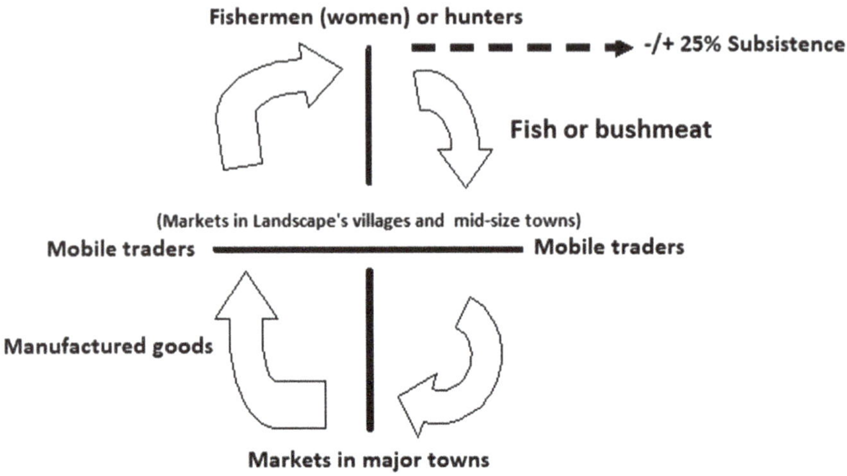


**Fig. 2.1** Economic circuits for the landscape planning process. A special attention has to be paid for the critical place occupied by mobile traders who link the landscape and markets in large towns. Mobile traders take fish, bushmeat and other natural goods from the landscape and trade those with manufactured goods in markets and large towns and vice versa

with multiple varied views of their varied interests, make their voices be heard by other stakeholders, including state organizations? The strategy that was elaborated is illustrated in a simple conceptual model presented below (Fig. 2.2), which was meant to allow the integration of these questions in the planning processes at the local level. Beyond the integration at the local level, there was also the question of integrating

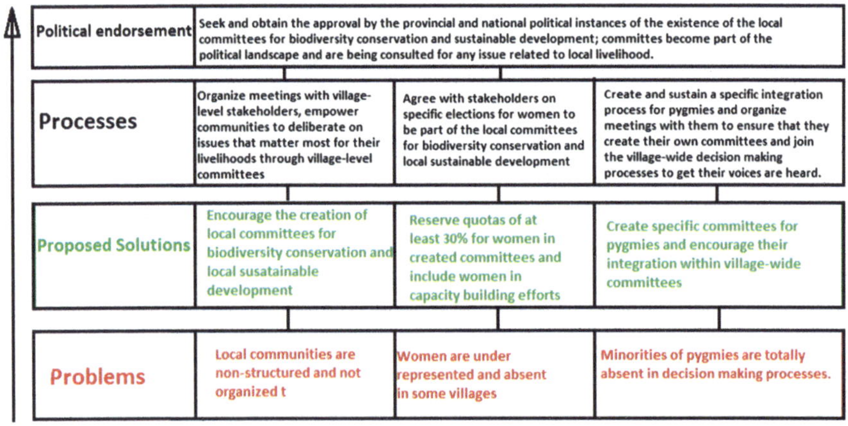


**Fig. 2.2** Simple conceptual Model on the integration of communities. This model is to be read as a theory of change for integration of women and autochthonous people. The framework can be used for other issues identified on the ground

processes at the landscape level to the provincial and national macroeconomic and political processes.

As such, landscapes need for global search with cultural, biological, and physical attributes defining spatial functions that assume that the global needs of the entire landscape are met. This implies that the relative location of patches of functional area are weighted in such a way that properties that ensure the quality of delineated micro-ecosystems do not undermine the existence of other functions and, in all cases, maintain ecological processes' functions. The methodological approach, negotiations, and different interactions are undertaken as schematically described below (Fig. 2.3). This schematic process has been put forward bearing in mind the global political processes happening in the country (Democratic Republic of Congo), particularly those concerning the devolution of some power to provinces and the creation of new political and administrative entities.

In this process, interactions happen through (1) lobbying and/or information exchange sessions, (2) legal consultative means, and (3) legal decisional mechanisms. In the planning process, conservation non-governmental organization had led the formatting and analysis of information. They also share that information with other stakeholders, principally, development non-governmental organization, private sector, and local communities represented by local committees of conservation and development. Communication is the most important tool used by socioeconomic teams working with conservation organization during the process of organizing local communities in different committees. There was also a need to dig out the existing legal instruments in order to help people not only to be aware of the legal framework in which their activities should be envisaged but also to form their own opinion about the potential and problems of that their forests hold. Of particular importance were the availability of the forest code promulgated in 2002 and the conservation law of 1982. These were supplemented by all international instruments; signed and endorsed by the Democratic Republic of Congo.

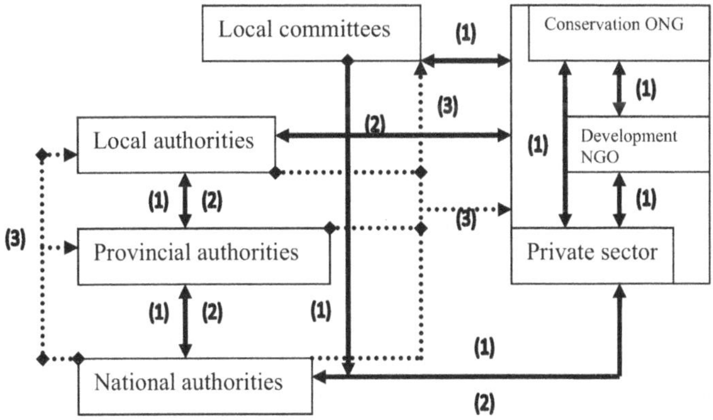


**Fig. 2.3** The scheme of different interactions between stakeholders

In order for the above chart to function properly, there was a need to identify stakeholders operating in the region. Stakeholders, as defined by Meffe et al. (2006), included those groups that had a real or perceived interest in the resources in the region of Lake Tumba, are dependent on these resources, believe that management decisions will affect them, are located in or near the landscape, pay for or influence the decision to be taken, and have the position of authority to review the decision.

Thirteen of more or less internally homogenous stakeholder categories were identified in the landscape on which should be added Central African states as regional activities have an impact on the global conservation on the landscape. The Table 2.1 that follows indicates the level of competency of each group and the domain in which that expertise can be brought into play during the planning process. One of the major problems that emerged from pulling together a list and carrying out the analysis of different stakeholders was the low local capacity to bear the burden of the landscape as a conservation paradigm.

**Table 2.1** Stakeholder categories and their respective levels interventions

| Actors | Relation to resources | Level of intervention | Representation |
|---|---|---|---|
| Traditional chiefs | *Managers, users* | Village, district | *Direct* |
| Men | Users, traders | Village, district | Committee |
| Women | *Users, traders* | Village, district | *Committees* |
| Pigmy communities | User and dependent | Village, district | Committees |
| Traders | *Lucrative commerce* | Village, national | *Development NGO* |
| Logging companies | Lucrative commerce | Village, province, national | Direct |
| Agricultural companies | *Lucrative commerce* | Village, province, national | *Direct* |
| Transport agent | Facilitation of commerce | Village, national | Development NGO |
| Development NGO | *Actors in improved livelihood* | Village, National | *Direct* |
| Conservation NGO | Plan for sustainability | Village, National | Direct |
| Territorial agents | *Managers, users* | Village, territoire | *Direct* |
| Provincial governments | Managers, users | Village, district, province | Direct |
| Central government | *Managers, users* | National, Central Africa | *Via province* |
| Central African States | International mobilization | Central Africa | COMIFAC |

## 2.4 Ecology and Economics: What Is the Place of Conservation?

As defined above, landscapes need for global search with cultural, biological, and physical attributes defining spatial functions that assume that the global needs of the entire landscape are met. This implies that the relative location of patches of functional area are weighted in such a way that properties that ensure the quality of delineated micro-ecosystems do not undermine the existence of other functions and, in all cases, maintain ecological processes' functions. It is here, in describing spatial functions that conservation has its entire place; not only because within the landscapes there are protected areas but also because each functional space should have a mode of usage that will integrate the principle of durability. Land use becomes, therefore, a means through which to integrate conservation and sustainable livelihood. However, one needs to acknowledge that we are in a human-dominated landscape. That means conservation stakeholders had to evaluate not only the viability of proposed zoning and their effects on biodiversity across this large spatial scale, but also to project ecological, social, and economic influences that would alter the equilibrium of interactions between human and biological diversity across the landscape in a long-term perspective (Dunning et al. 2006). Therefore, to ensure proper planning, there was a need to collect a minimum data set necessary to inform the zoning process. These included socioeconomic and biological data, physical habitat description, etc. The chapter evaluates processes that led to the collection of data, proposes new paths for the future, and it provides an overview of what has been collected.

Because of the emphasis on the planning process, the field teams had to collect socioeconomic data first. The first set of data to collect was a number of available maps. The most striking feature of gathering maps was how to obtain difficult maps on this part of the country. Those that were available were simply outdated and most of locations and features were off. However, it was a gratifying event to find very old and very detailed maps of ethnic groups produced by catholic missionaries and stored by the *Annales Aequatoria*, an internationally respected journal produced in the Democratic Republic of Congo and focused on social and cultural anthropology. The extensive use of *Annales Aequatoria* emphasizes the importance of documentation and literature review in the process of developing a comprehensive data collecting process and a vision for the landscape that accounts for historical, cultural, and social conditions in the geographic extent of the landscape. It does also highlight the fact that while the concept of landscape as a conservation paradigm is new in Central Africa, historical data records are key in conceptualizing different processes that need to be in place to help a coherent implementation of different plans. Historical records, however, are not sufficient; they need to be updated by new data from cultural anthropology, economics, and sociology of the areas. Therefore, structured interviews were designed to be conducted in randomly selected villages identified from those maps. Villages were selected from various maps with existing data showing which different ethnic groups inhabited the Lake Tumba Swampy

Forests (Inogwabini 2010), supported by detailed text descriptions on land occupation since colonial times (Inogwabini 2010; SIL 2008; Hulstaer 1993a, b; Heenen 1959). These data were combined with the actual tribal distribution map, based on the administrative map of DRC (IGZ 1986). Formal interviews were complemented by informal focus group discussions, wherein people were left to say what they thought and knew without any formal constraints on the length and content of their interventions (Kreuger 1988). Three researchers conducted the interviews, took relevant notes, and filled in the interview's forms. Both informal and formal focus groups were organized throughout many different villages. With opportunistic questions, focus groups served to triangulate responses given by individuals over the course of different interview sessions (Strauss and Corbin 2007; Russell and Harshbarger 2004; Schensul et al. 1999). What missed in this first round of data collection was a thorough assessment of cultural patterns in the region; something that could have been conducted using methods other than conventional socioeconomic investigation. These methods would have included historical, linguistic, and philological approaches such as the use of keywords (Vansina 1990; Sumner 1988; Langley 1976; Turnbull 1973; Soja 1970) and collection of archaeological samples. Assessing cultural patterns would have helped in the definition of the vision and functions of each macro zone. That work was, however, undertaken in a different context (Inogwabini and Leader-Williams 2012b). Using these methods, a clear understanding of how human history and cultural patterns influence the distribution of some key wildlife species such as the bonobos, elephants, and even abundance of fishes in stream and other water bodies of the landscape. As an illustration of what types of socio-cultural data sets are necessary for a sound landscape planning process in the context of Central Africa, the following exercise provided an interesting pattern of biodiversity distribution in relationship with the bonobo distribution in the Lake Tumba Landscape.

The second important data set collected was that of biodiversity. This included data on the distribution and abundance of large mammals, data on major habitat types, and data on freshwater fish species. Arriving in Lake Tumba as landscape, there was very limited knowledge on the biodiversity of the region. Lists of large mammals that were dug out of the Mabali Research Center were either obsolete, geographical restricted, or completely lacking any sense. Therefore, we had to start with gathering every single piece of knowledge nearly from scratch, using a variety of techniques described in Chap. 10 of this volume.

A third set of data that was needed was long-term ecological understanding of forest dynamics in the region. This type of data is necessary in order to help understanding human interactions with the environments of the landscape over the historical period. This type of data is necessary not only to understand the current outlook of the landscape at the broader and longer time perspectives but also to project different scenarios of what would happen if the current trends in human populations and uses of resources would continue to remain constant. They would also help envision what would be the responses of the different biota (forests and waters) of the region would climate change dramatically change as it is being currently projected.

These multiple types of data were gathered to help understand and plan for the landscape. Indeed, as it will be seen from part three of this volume, most of the data gathered had served in the planning process and helped identify future directions of research and identify the needs for the landscape as a whole.

## References

Adams JS, McShane TO (1996) The myth of Wild Africa: Conservation without illusion. University of California Press

Allaby M (2004) The Oxford dictionary of ecology. Oxford University Press, Oxford, United Kingdom

Arvill R (1967) Man and environment: crisis and strategy of choice. Penguins Books Ltd

Colom A, Bakanza A, Mundeka J, Hamza T, Ntumbandzondo B (2006) The socio-economic dimensions of the management of biological resources, in the Lac Télé—Lac Tumba Landscape, DRC Segment: a segment-wide baseline Socio-Economic study's report. Submitted to the World Wide Fund for Nature

Dunning JB, Groom MJ, Pullian RH (2006) Species conservation and landscape approaches to conservation. In: Groom MJ, Meffe GK, Carroll RC (eds) Principles of conservation biology, Third edn. Sinauer Associates, Inc. Publishers, Sunderland, pp 419–465

Fairbrother N (1972) New lives, new landscapes. Penguins Books Ltd., Harmondsworth

Fahrig L (2005) When is a landscape perspective important? In: Wiens JA, Moss MR (eds) Issues and perspectives in landscape ecology. Cambridge University Press

Groombridge B, Jenkins MD (2002) World atlas of biodiversity: earth's living resources in the 21st Century. University of California Press

Hall JS (2000) A review of myth and reality in the rain forest: how conservation Strategies are failing in West Africa, by John F. Oates (1999). J Polit Ecol 7:1–6

Heenen G (1959) The Belgian Congo—Volume I and II. Office de l'Information et des Relations Publiques pour le Congo Belge et le Ruanda-Urundi. Inforcongo. Bruxelles, Belgique

Hulstaert G (1993a) Etudes dialectologiques Mongo. Etudes Aequat 12: 15–4006

Hulstaert G (1993b) Liste et carte des dialectes Mongo. Ann Aequat 14: 401–406

Inogwabini BI (2010) Conserving great apes living outside protected areas: the distribution of bonobos in the Lake Tumba landscape, Democratic Republic of Congo. PhD Thesis, University of Kent at Canterbury, United Kingdom

Inogwabini BI (2007) Can biodiversity conservation be reconciled with development? Oryx 41:2–3

Inogwabini BI, Leader-Williams N (2012) Using the keywords to explain the bonobo distribution as an effect of human perception of the species. Am J Hum Ecol 1(4): 102–110

Inogwabini BI, Leader-Williams N (2013a) Conservation paradigms seen through the lenses of bonobos in the Democratic Republic of Congo. In: Navjot S, Raven P (eds) Essays in conservation biology—perspectives from practitioners in tropical environments. Blackwell-Wiley

Inogwabini BI, Thompson JAM (2013b) The golden-bellied Mangabey (Cercocebus chrysogaster): distribution and conservation status. J Threat Taxa 5(7):4069–4075

Institut Géographique du Zaïre (IGZ) (1986) République du Zaïre: carte politique et administrative. Edition Saint Paul Afrique

Kreuger RA (1988) Focus groups: a practical guide for applied research. Sage Publications, Newbury, United Kingdom

Langley P (1976) Approche ethnolinguistique de l'environnement rural et son utilité pour l'amenagement. In: Richards P, Harris N (eds) Environnement Africain: problèmes et perspectives. Dossier spécial 1. Institut International Africain (Londres), pp 95–108

Meffe GK, Groom MJ, Carroll RC (2006) Ecosystem approaches to conservation—response to a complex world. In: Groom MJ, Meffe GK, Carroll RC (eds) Principles of conservation biology, Third edn. Sinauer Associates, Inc., pp 467–507

Oates JF (1999) Myth and reality in the rain forest: how conservation strategies are failing in West Africa. University of California Press
Putz FE, Redford KH, Robinson JG, Fimbel R, Blate GM (2000) Biodiversity conservation in the context of tropical forest management. Biodiversity Series—Impact Studies: PAPER N$\underline{\text{O}}$. 75. The World Bank Environmental Department
Russell D, Harshbarger C (2004) Ground work for community based conservation. AltaMira Press, Walnut Creek
Sayer J, Campbell B, Pethera L, Aldric M, Pere MR, Endamana D, Dongmo ZLN, Defo L, Mariki S, Doggart N, Burgess N (2006) Assessing environment and development outcomes in conservation landscapes. DOI, Biodivers Conserv. https://doi.org/10.1007/s10531-006-9079-9
Schensul S, Schensul J, Lecompte M (1999) Essential ethnographic methods. Ethnographer's Toolkit No. 2. Altamira Press
Société Internationale de Linguistique (SIL) (2008) The languages of Democratic Republic of the Congo. http://www.ethnologue.com/showcountry.asp?name=cd. Accessed 12 mai 2008
Soja E (1970) Prologue. In: Paden J, Soja E (eds) African experience—essays. Northerwestern University Press
Strauss A, Corbin J (2007) Basics of qualitative research: techniques and procedures for developing a grounded theory. Sage Publications
Sumner C (1988) Aux sources éthiopiennes de la philosophie Africaine, philosophie de l'homme. Editions des Facultés de Théologie Catholique de Kinshasa. République du Zaire
Taylor P (2005) Beyond conservation: a wildland strategy. Earthscan, United Kingdom
Terborgh J (2004) Reflections of a scientist on the World Parks congress. Conserv Biol 18(3):619–620
Thirgood JV (1981) Man's impact on the forests of Europe. J World Forest Resour Manag 4:127–167
Turnbull CH (1973) Africa and change. Alfred A. Knopf, New York, United States of America
Vansina J (1990) Paths in the rainforests—toward a history of political tradition in equatorial Africa. University of Wisconsin Press

# Chapter 3
# Biodiversity Conservation in Human-Prone Landscapes: Social and Ethical Issues

**Abstract** Preserving the World's biodiversity is part of that ecological disquietedness, and worries to preserve the world's biodiversity are one the most stringent narratives of the modern global society. However, the general narrative to preserve biodiversity is often hotly debated; it opposes preservationists against conservationists in a publicly obscured debate caused by its technicality, metaphors, and the scientific biodiversity conservation jargon. Ethical questions are spoken about in many instances but remaining so often left behind during planning exercises aimed at implementing biodiversity conservation activities. The chapter notes that debates on the ethics of biodiversity conservation often lead to strangled considerations with fierce opposition between conservationists and other stakeholder groups such as extractive industries, wildlife poachers, and biodiversity-dependent communities. The chapter argues that the landscape approach, through its stakeholder's inclusive planning process brings not only a new perspective on ethical questions posed by conservation activities but also clarified, owned (shared), and responsibly agreed upon ways to account for human ethical considerations into the global ethics of biodiversity conservation, which will ultimately reduce tensions.

**Keywords** Biodiversity conservation ethics · Rights of local communities · Fears over conservation maps · Wildlife-human conflicts · And participative process

## 3.1 Introduction

Ecology has emerged as one of the most important concerns at the end of the twentieth century and will continue to be an area where the global world community will have serious disquietedness over a large part of the twenty-first century. Preserving the World's biodiversity is part of that ecological disquietedness, and worries to preserve the world's biodiversity have become one the most stringent narratives of the modern global society. However, the general narrative on preserving biodiversity has been often an area of heated debate opposing strands of people varying from those who are technically known as preservationists against conservationists. In the public arena, however, the debate is often obscured because of its technicality and confusions brought by the metaphors and the jargons used by the conservation community.



B.-I. Inogwabini, *Reconciling Human Needs and Conserving Biodiversity: Large Landscapes as a New Conservation Paradigm*, Environmental History 12,
https://doi.org/10.1007/978-3-030-38728-0_3

Ethical questions, though spoken about in many instances, are often left behind when planning for and implementing biodiversity conservation. When they are taken on board and discussed, debates on the ethics of biodiversity conservations often leads to strangled considerations with fierce opposition between conservationists (inclusive of preservationists) and other stakeholder groups such as extractive industries, wildlife poachers, biodiversity-dependent communities.

Increasing human populations and subsequent increases of needs for development led to habitat losses, species extinction, and overall biodiversity erosion. This gloomy picture is particularly true in biologically important areas that are also the most densely human-occupied niches. Landscape, as an approach whereby land is apportioned to different functional units would pose even more problems in view of this gloomy picture if it would promote solely the exclusion of several communities from accessing lands that they possess by virtue of their very presence in those areas for immemorial eras. Indeed, these communities are, per the principles of the first occupants, the right holders for lands on which they live and from which they have been literally drawing the substance of their very existence. Of course, in the context such as the one prevailing in the Democratic Republic of Congo where Lake Tumba is located, the ethical affirmation that local communities are right holders for lands and forests they occupy comes across a legal opposition as for this country, as it is the case for many countries across Africa, land is owned by the state. The land laws in most of the African countries are impregnated with the remnant of the Lockean Theories of properties, particularly the principle that land should be owned by the one who labors on it. This principle has been applied in the African countries, where the context was historical different and land was previously owned by communities. In that context, the idea of empty lands to be seized by the governments was inexistent (Inogwabini 2014). Because of the precedents of difficulties and social hardships in the creation of certain protected areas, communities were right in fearing to see large areas such as landscapes (Fig. 1.1) become areas of biodiversity conservation. The communities were even more right in this feeling particularly as the idea of the landscape approach, despite being a better way to factor human communities in the biodiversity conservation equation, was not clarified to the common understanding of different communities before being adopted by governments of central Africa. Many communities feared that the maps of these large areas would present a further insecurity threat to communities' livelihoods, particularly for those forest-dependent and land-dependent of them, if appropriate measures are not adopted. Indeed, a look at the spread of the geographic extent of landscapes across Central Africa (Fig. 1.1) would scare anyone who does not know what they stand for. In fact, the landscapes of Central Africa look like countries within countries; some of them are even larger than the physical limits of some countries: the Maiko–Tayna–Kahuzi-Biega Landscape, for example, is larger than both Rwanda and Burundi combined; three landscapes (Lac Tele–Lac Tumba, the Sangha Tri-national (TNS), and TRIDOM (Fig. 1.1) are linked to the point of making one single unit; they span across four countries (CAR, DRC, Congo, and Gabon) and are larger than Gabon or Congo. And, if they are to be exclusively left for biodiversity conservation alone, they would create serious problems. Fortunately, as we will see below and as per the definition of landscape that has been adopted, this is not the case. True: landscape approach will mean some privations for communities; the existence of these landscapes has brought about the creation of new protected areas. This is particularly true of the case of Lake Tumba where, the Tumba-Lediima Natural Reserve and Ngiri Triangle Natural Reserves, two new protected areas have been erected where they were not previously in existence. But it should also be recognized that, indeed, landscape by the very fact that they are supposed, if properly understood and properly implemented, helps to alleviate the problems of land rights and forced occupancy of lands that do not factor the demands of other stakeholders. In an essay on the good reasons for people to use natural resources within and outside of protected areas when the use of these resources is detrimental to biodiversity conservation, Inogwabini (2016), argued that in most cases, reconciliation between local human needs and needs to conservation biodiversity is possible. Even in cases where the two were truly irreconcilable, partial reconciliation is plainly possible by hierarchizing harms, accounting for scales and identifying trade-offs by agreeing with local communities, in a participative process, on what can be extracted by local communities. While basic rights principle, the rule of minimum wrong, the Jus soli and the rights of the first occupants can support communities having an upper hand on the uses of natural resources, partial reconciliation is possible through the scales. Indeed, landscapes provide such scales to ensure that multiple usages for different combinations of activities and prohibitions happen without necessarily infringing on communities' rights to survival.

Other instances raising ethical questions are those when wildlife species cause damage to activities that render vital services to human communities or to human physical lives. Instances of wildlife damaging vital activities to human communities include crop-raiding by wildlife species such as elephants and other ruminants. The stories of crop-raiding by wildlife are abundantly widespread across Africa; they have been reported from countries of different ecological profiles and different biodiversity management schemes and bear an interesting name of Human–elephant conflicts (HEC), when the crop-raiders are elephants. Indeed, instances of HEC have been thoroughly documented throughout Eastern and Southern Africa (e.g. Wasilwa 2003; Hoare 2000; Hoare 1998, 1999; Bhima 1998; Osborn 1996; Ville 1996; Osborn and Rasmussen 1995; Kangwana 1995) and in Western Africa (e.g. Vanleenwe and Lambrechts 1999; Waithaka 1999; Sam and Barnes 1998; Sam et al. 1998). For the Central African forest elephants, similar cases have been reported from Cameroon, Gabon, Central African Republic, Congo Brazzaville, Democratic Republic of Congo (e.g. Tchamba 1996; Lahm 1996, N'sosso 1996; Tchamba 1995; Kinzonzi 2004; Inogwabini et al. 2013).

Barnes (1996) warned that HEC is a serious conservation problem and other authors such as Kangwana (1993, 1995) and Dublin (1996) agree that HEC has become a topic of major concern in elephant conservation not only because it has immediate negative effects on elephants but also, and primarily because it affects humans. Both Kangwana (1993, 1995) and Dublin (1996) argue that HEC is frequently a precursor to further decline in the African elephant range. Despite that wider acknowledgment of effects of wildlife on truly ethical rights of human communities, HEC and other Human–elephant interactions in Africa have been often analyzed within the paradigm of expanding human populations and their needs for development. Indeed, human expansion and modern development change patterns of land use, with the consequent fragmentation of wild habitats (Wasilwa 2003; De Boer and Ntumi 2001; Waithaka 1996; Tchamba and Nshombo 1996; Inogwabini et al. 2013). In fact, these studies ignore difficult situations through which humans have to go through when their crops are raided by elephants or any other species of wildlife. In the case of Lake Tumba, studies have documented instances of crop-raiding by elephants and other ungulates in the southern part of Lake Tumba. Inogwabini et al. (2013) documented higher damage Indexes for crops that are vital human communities in that area such as manioc (damage index $I = 60.1\%$) and bananas ($I = 11.4\%$). The same study has documented that the mean annual financial loss caused by crop-raiding in individual fields was USD 400 (USD 97–1,005). For a country whose per capita income was estimated by the World Bank to be 380 USD in 2014, the loss in economic terms is almost unbearable for community's members. Obviously, discussions held with communities were fierce on the ethical issues of why foods for elephants and other ungulates would be given moral precedence over the needs for communities to cultivate staples they can live on and make money from.

The Landscape approach, as described above (Chaps. 1 and 2), would offer some solutions to this type of ethical question. Indeed, the study by Inogwabini et al. (2013) suggested as a solution to this question, the creation of a broader community conservation scheme. This scheme would provide a space where different stakeholders can come together, discuss, and agree on what can be done in this type of case. Indeed, the multi-stakeholders discussions agreements, which are promoted by the landscape approach, would be the best conduit to channel such hot proposals as the transformational proposition made by Inogwabini et al. (2013) to redraw the agricultural map of the entire region. One should recall here that among the major pillars on which the landscape approach rests on, there are participative processes, land planning (attributing functions to each piece of land), and democratic management of natural resources. All the three would provide the answer to the question raised by communities about the prevalence felt being given to the survival of elephants, which they see impede the subsistence and the livelihoods of human communities. A truly participative process would help communities formulate their own solutions and discuss these with solutions brought on the negotiation table by other stakeholders (i.e. conservationists). This would lead both parties to agree on what to do in cases elephants and other wildlife species would raid on vital agricultural staples such as manioc or bananas. The same engagement would also allow communities to be part of designing schemes to avoid further conflicts with wildlife. Indeed, Inogwabini et al. (2013), identified that human communities wanted to be included in a community conservation whereby alternative livelihoods activities could be funded by conservation initiatives in exchange of completely redrawing the agricultural maps of the region, which would move fields from areas where there were permanent elephants trails leading to other critical resources for elephants such as water. This would not end the conflict but it would be a good example of what Inogwabini (2016) calls partial ethical reconciliation.

A participative process, on itself, would nullify or at least smooth the entrenched opposition between ‘preservationists’ and ‘conservationists’. ‘Preservationist approaches’ are the view that all biodiversity components should be maintained intact on earth. Consequently, they adopt the fence-and-fine methods that propose erecting protected areas wherever some species occur and lock these areas from human access and anyone caught within the locked area is prosecuted by the law and pays penalties (Inogwabini and Leader-Williams 2013). Biodiversity preservationism was advocated first by John Muir who used a form of transcendental philosophy of nature to support his claims that biodiversity should be totally preserved. Preservationism has roots that are entrenched in an old historical and philosophical tradition and can be better summarized in the pantheism that had been heralded by some major philosophers such as Plato, Proclus, Eriugena, Eckhart, Nicholas of Cusa, Boehme, Schelling, Bruno and Spinoza, and various other theologians and philosophers. Pantheism ontologically identifies God and the world. In its understanding by Spinoza, everything that exists is a different manifestation of God; God is present in the entirety of his creation. With this view, every single created material object should be preserved as it does represent God in its own way and in its own right.

It is mild as a statement to qualify this view as an extreme one in the current conditions of greater demographic increases, greater democratic demands, and when corporate social responsibilities are requested by the global human community as a solution to numerous issues that are brought about by the management of natural resource. A common historical consequence of the implementation of the preservationist views is that throughout the tropical regions, communities were dragged out of forests and ecosystems where they previously lived (Jepson and Whittekar 2002); human–wildlife conflicts emerged in which local communities oppose biodiversity conservation and increased poverty ensued. Indeed, the background of the creation of parks and other protected areas in the recent history has been that of observed declines in biodiversity, combined with fears of subsequent consequences on human welfare. This has meant that the very science of conservation biology, as an area of scientific inquiry, was born out of fear and out of the rush to find ways in which catastrophic loss in biodiversity can be stopped from happening. Its epistemological foundations were laid on the background of fear and efforts to try to avoid species extinctions. In this sense, conservation biology was born as a science of scarcity and doom. Very little questioning has been happening about how the fear of doom and the idea of scarcity of resources would affect the creation of the scientific knowledge in this new domain of scientific research. Of course, one would argue, partially rightly, that conservation biology is not a new science as it stands in an inter-disciplinary stall. Indeed, conservation biology sits at the juncture of biological and physical sciences (ecology, environmental sciences, applied statistics, geology, forestry, etc.), social and human sciences (anthropology, cultural studies, sociology, philosophy, etc.). However, the body of knowledge it has produced has its own aims and goals of which the most important is to provide applied elements of how species and their habitats can be preserved throughout the world and sustained for as long as possible.

This body of knowledge is separable from the knowledge of its mother sciences (biological and physical sciences and social and human sciences) in its aims, production processes and the use being made of it.

The place of each individual species in the large web of relationships is the core business of ecological research, which is one of the major themes in conservation biology. As such the debate is reduced whether humanity is part of nature or not. Proponents of preservationism as well as those of conservationism agree that humans are part of the ecosystems in which they are but with a difference that preservationists would limit themselves in this statement and conclude that humans are yet another species among these relationships and have no right to destroy the rest of the system. Conservation, looked from the perspective of this debate, would support the view that humans are part of the nature but they are at the same time more than being part of that nature.

The social environment created by the process of creating protected areas out of fear left very slim margins to think through carefully about different aspects of biodiversity conservation and human basic needs. Conservationists, on the other hand, think that biodiversity can be used wisely to accommodate human needs. Landscape, as a paradigm of the prudent use of nature's bounty (as opposed to unrestricted extraction of natural resources, is therefore essentially a part of the conservation philosophical line of argument, which was the argument that Gifford Pinchot defended in the debate that opposed him against John Muir and (Callicott 1990). According to Naylor (2005), Gifford Pinchot sought to apply scientific principles to the maintenance and use of forests and rivers for the good of humankind and was more concerned about 'the greatest good for the greatest number for the longest time' and battled against monopolistic corporate abusers of the land, which led people to call him a crusader for the public good. Obviously, the insistence on 'the greatest good for the greatest number for the longest time' is a utilitarian view; which to many light thinkers might lack a vision that would transcendent purely hedonistic needs. That is far from being the reality in this case and asserting so equates to having a short sight over other motives that drove Gifford Pinchot's motivation toward conservation. Indeed, Gifford Pinchot was immersed in the widespread evangelical Protestantism of his time (Naylor 2005) and most likely his views on conservation were also influenced by a different reading of the Bible. All this is to say that both views (preservationism as opposed to conservationism) of 'how to deal with biodiversity' are in many ways related to views humanity holds of the place of nature in the web of relationships with the spiritual and the place of the humanity in that large web of relationships. From this brief presentation, it can be argued that current landscape idea, at least as it understood in the case of Central Africa, is driven by needs to have human needs met while conserving biodiversity. In this sense, landscapes were biodiversity preservation entities with the elements of conservation. This is true as one looks at the current geographic deployment of the priority conservation landscapes across Central Africa (Fig. 1.1). All of them, with the exception of Lake Tumba, were constructed around existing protected area of IUCN categories ranging from 1 to 3. It is fair to assert that the addition of conservation elements in previously strictly protected areas has come as a response to several pressures, chief among them the needs for local communities to access natural resources within their areas. Other pressures that landscape has come to address are those coming from the extractive industry to extract either timber or other natural resources such as minerals. Looking from the perspectives of Lake Tumba, however, landscape was a conservation entity with an element of preservation: here the move was from barely no protected areas to finding means to create new protected areas in areas that were found to hold either significant populations of landscape important species (such as bonobos, lions, elephants) or with habitats that were unique to the whole geographic context of Central Africa. Hence were created two natural reserves and a delineation of a Ramsar Site. The fact that there was no new park or any protected area of IUCN strict conservation category is a testimony to several factors, including the wariness of the entire conservation community to create still more exclusion entities. Indeed, the history of protected areas in the Democratic Republic of Congo (Inogwabini 2014) was lurking in the minds of all stakeholders in the process of creating both the Tumba-Lediima and Ngiri Triangle Natural Reserves. But the very same fact also testify a genuine implementation of a different conservation paradigms; and, indeed, populations located within these reserves were not asked to move and preserved the rights to access natural resources conditional to certain principles such as respect of seasonality, use of traditional means, avoidance of accessing the core conservation area (an element of preservationist view). All put together, landscape as conservation paradigm has a potential to bridge the morally incommensurable genuine human needs for sustainable livelihoods and legitimate demands for conserving biodiversity.

Instances of wildlife damaging physical human lives consist mostly of cases where wildlife species kill either accidentally or willing humans. These cases are rather rare but are of the utmost impact on ways people perceive wildlife in the environments around their villages. Chimpanzees have been reported to kill people in several occasions (e.g., Abraham and O'Callaghan 2012) for several reasons, including limited geographic home range, habitat fragmentations, competition over resources, etc. Elephants too are known to kill people in some very extreme cases. Reasons for these incidents when elephants kill people are either self-defense for elephants being poached or threatened by poachers or simply groups of humans. Obstruction over their permanent paths linking vital resources such as permanent water points has been also suggested as an explanation.

In the southern part of the Lake Tumba Landscape, incidents of humans being killed by lions were documented between 2006 and 2010 (Chap. 12 in this volume). Incidents of lions killing cows and roving around in the region were reported during the dry seasons. I have suggested that lions in the southern Lake Tumba lacked prey species in dry seasons because water points dried out in savannahs and all duikers and other species retracted to forests where water was still available. Because lions could not follow their prey species in forested zones, they shifted hunting cows, goats, sheep, and other species that were raised by humans. In extreme cases of hunger, they would then attack people. This explanation and all other reasons evoked above for wildlife attacks on humans sound very neat in ecological theories; all these reasons sound sensible and defendable and matter to conservationists. However, when such lethal incidents happen, the first question raised by communities is whether the life of humans is less than that of wildlife species. Without going as far as endorsing the naïve 'myth of wild Africa', it is reasonable to assert that human communities in these high conservation value areas are aware of the value of the biodiversity that happens where they are; they cherish it and even know that they sustain their own livelihoods using biodiversity. In the case of Southern Lake Tumba, it has been proven that communities had become conservation-prone even before the landscape itself was created and their conservation activities around the bonobos preceded any formal conservation laws being enforced in the region. Inogwabini et al. (2013) found that this will to preserve the bonobos was in great parts due to the traditional taboos rather than anything else. So, the question here is not whether people know and want to protect wildlife species but that of valuation of their own lives versus that of wildlife species. To put it bluntly, as it is often the case when discussing with local communities, the issue is that when lions kill people, why people should not retaliate and kill lions. The responses to this type of question are not straightforward and would need more than just a yes or no type of answer. Often, people understand that death can come in any form; it can be an accident as it can come from a long lasting lethal disease. People also know that there must some sort of natural hierarchy of importance; they know without reading Dworkin (1993) and even when acknowledging the intrinsic value of wildlife species such as it is the case for bonobos in the Southern Lake Tumba, that there are distinctions of categories of intrinsically valuable things: those that are intrinsically valuable because of their instrumental value and those that are intrinsically inviolable (sacred). This distinction, combined with the idea of degrees of sacred (Dworkin 1993), is a better way to grade species according to their importance and provides a solid ground for selecting which species should be totally inviolable and which one can be partially violable. Indeed, there are degrees of sacredness […] and our convictions on inviolability are selective (Dworkin 1993). So, even when claiming that biodiversity is intrinsically valuable, only some species are inviolable. That is why few people would care about extinction of mosquitoes, for example, while they will be in great sorrows to hear that great apes have gone extinct.

In this sense, people come to view that their very lives as being more valuable than the life of elephants, lions, and other species. Their claim is that they should be valued more than wildlife species. But when death comes, communities know perfectly well that despite everything that they can do, no one would bring back the deceased. That is why often the discussion is around compensating the families, friends, and the communities. The moral justifications of such a claim are similar to the justifications provided for ensuring that enterprises and communities are bound by the corporate social responsibility principles. This clearly means that landscapes, as other forms of applied biodiversity conservation, should factor these social responsibilities in the planning processes. Indeed, the appeal of landscape as an applied conservation paradigm is its focus on open planning and decision making process, which should include not only the needs for biodiversity conservation but also those of communities to live decently and with the maximum security possible.

In order to understand what the notion of the corporate social responsibility would have to do with biodiversity conservation within landscapes and how that would be undertaken I discuss major objections laid against the idea of the corporate social responsibility through the work of Milton Freedman.

Profit-making businesses in the Landscape would repeat exactly what Milton Friedman wrote 'The Social Responsibility of Business is to Increase its Profit' (Fiedman 1970). Equally, in the context of this process, biodiversity conservation should be looked at as any type of business but whose objective is, contrary to making profit, to maintain biodiversity on earth and particularly across Central Africa. Of course, looked at in this perspective, one would be tempted to paraphrase Milton Freidman and say that the 'Social Responsibility of Conservation is to increase biodiversity'. The 1970s were a time when the free-market capitalism and socialism as economic ideologies were at their heights. This period was also the one in which biodiversity conservation was primarily seen as opposed to the human well-being. Within that background, the social corporate responsibility for enterprises was viewed through the lenses of the prevailing ideologies. Since then ideological lines have moved significantly and discussions around the social corporate responsibility are now at the core of the idea of doing business around the world. However, describing what is it that we call social corporate business responsibility is still a very hot debate, particularly when it comes to multinationals that function as supra-territorial entities and have codes are that are often loose to be implemented. Also, applying the social corporate responsibility in an era when to preserve the planet from adverse effects of climate change, biodiversity conservation has become an international venture, with high pressures on nations with higher biodiversity to do whatever it would take to conserve it for the benefits of the humanity. I argue here that that despite the fact that there is nothing wrong with capitalistic businesses to make profits business **also** bears the social responsibility to the global human well-being. The same argument applies for biodiversity conservation; although it is laudable for biodiversity to be conserved, doing conservation also bears the social responsibility to ensure that human communities continue to be well and in their diversity.

As for profit-making businesses present in the landscape, the first important point in discussing this statement of Freidman is to distinguish businesses by their objectives. There are businesses whose prime objective is to deliver social services such as drinking water, electricity, health care. This type of business is designed to care about the social well-being. The second type of business is the profit-driven businesses that seek the maximization of profits at all costs. The first type needs financial margins to allow businesses to maintain the provision of essential social services and goods. They do not need to indefinitely increase their profits and their social responsibility is embedded in the motives that sustain their very existence. This type of enterprise was clearly not in Freidman's thinking and missing this separation has led Freidman to ignore other aspects of social responsibility that preside over the creations of some businesses. Furthermore, ignoring these businesses also impacted the interpretations some people had on Freidman's idea about the social responsibility of businesses. In biodiversity conservation too, there are conservation businesses: the idea that wildlife conservation areas (protected areas) are there to serve as breeding sites for wildlife that would ultimately come out these niches and be used by communities has been supported for a long time and was a strong argument for people to accept protected areas near where they lived. But there were also wildlife conservation territorial entities that were private. Indeed, when Virunga, the first National Park of the Democratic Republic of Congo, was created it was the reserve of the Belgian crown. That has led to laws such as the 50-km pursuit provision that was that people caught hunting within the 50-km distance from the border of the parks were poachers as the 50-km width was supposed to buffer the effects of human communities on the protected areas. Ironically, in not only in the case of Virunga but in many cases across the country, large human communities remained within the immediate vicinities of protected areas.

In many parts of the world, profit-driven enterprises, to maximize profits, construct residential areas where their employees live. They argue that these residential camps are important for financial efficiency and other reasons. This is the case for the mining industry in the African countries; logging companies, in a slighter extent, do the same while conservation organizations are known to build somewhat adequate headquarters for wardens and eco-guards. In some of these residential camps, particularly in the history of mining industry, staff members have no right to be with their families and other dependents because of the lack of basic facilities such as health and education facilities. These facilities are not provided because they would increase expenses and, because these enterprises think it is not their obligation to care about responsibility toward families of their employees, no institutional framework asks businesses to act on this front. However, from a human right point of view, employees of these enterprises have the right to create families, provide the necessary care for bringing up their children, and just live up their expectations as humans. It is the responsibility of the employers who have moved employees to these locations to provide the much needed social services in these areas to ensure that employees and dependents to live and access to the normal social lives. Apart from this human right perspective, the mining and logging enterprises have this responsibility because they are the ones that create these news suburbs and camps; they have the legal obligation to ensure that these areas are equipped with the minimum social services.

When Friedman wrote that 'businessmen who talk this way are unwitting puppets of the intellectual forces that have been undermining the basis of a free society these past decades' he stated his ideology. This was in a period when capitalistic economy was opposed to the planned economy. Friedman was not just a common world citizen. With a Nobel Prize in Economics in 1976, he was an influential economist of the twentieth century and a member of the Chicago School of Economics. As such, Freidman ideologically and logically rejected Keynesianism as an economic system. He thought that the laissez-faire economic policy was more desirable than government intervention in the economy. Laissez-faire is in stark opposition with Keynesianism, which argues that inefficiencies of private enterprises and markets require public governments to stabilize the economies. For Freidman and other members of the Chicago School of Economics, free-market capitalism and the invisible hand should bring social benefits for hardworking people and striving communities. When Friedman talks about 'the intellectual forces that have been undermining the basis of a free society', he postulates that his idea of the Social Responsibility of Business is heavily burdened by his ideology. Fighting against the notion of 'Social Responsibility of

Business', in this sense, is truly a fight against world view that would bring more interventions from the public governments in the economy. Despite the fact, this is what has been being implemented over decades in countries like the Democratic Republic of Congo where extraction of natural resources is the sole significant economic activity, voices are raised that some sort of Keynesianism is needed to support the initial creation of social capital that would support the private sector. And, indeed, in many cases, capital investments by national organizations, international institutions, and private foundations are the only way to support conservation throughout Central Africa these days; talks about protected areas being able to self-sustain their own preservation through either tourism or any other mechanisms such as payment of ecological services they render to the global human community are, at best, still very negligible and prospects for these mechanisms to become major sources of financial means to achieve biodiversity conservation goals, including providing alternative resources for local communities, are not that promising. Indeed, biodiversity conservation in Central Africa, and particularly the Democratic Republic of Congo, will remain funded by some public investments, the international aid and private supports for decades from now. But the best way for biodiversity conservation to be paid is through the participation of the profit-making enterprises that operate in some of the areas where biodiversity is highly valuable; the same mechanisms should apply for human communities that reside in those areas.

Friedman (1970) believes that planning is not good for the economy and that any public intervention in the market is disruptive to the whole economic system. Global economic planning systems (socialism) currently showed their limits and are vanishing in many areas of the world. However, there is nowhere where there is no minimal planning for economic activities. Even in countries like the US and UK, there are areas of economy where planning is not necessary but desirable. These areas include such important economic domains as the infrastructures, building of the social capitals and providing a normative legal framework within which the private free-enterprise is to thrive (Sachs 2011). The view that laissez-faire government policy is more desirable than government intervention in the economy was too extreme. Governments have, at the minimum, the obligation to provide secure environments for private businesses to develop; they have to provide education to build the general capacities for each individual to compete with equal chance in a free world, etc. An example of how governmental interventions are vital to save the businesses was the 2008 financial crisis. According to The Economist (2013), the September 2008 collapse of Lehman Brothers almost brought down the world's financial system. To save the world economy from a total collapse, western governments had to bail-out the financial institutions to shore up the industry. This was done by taking huge amounts of taxpayer money to rescue the businesses. If taxpayer money can rescue the business, then the businesses need to care about more than just increasing their profits. That something more includes the responsibility toward taxpayers. Rejecting the whole idea of Social Responsibility of Business for ideological backgrounds is less sensible because minimal planning is necessary. The intervention of governments in areas where private enterprises are not necessarily the best to deliver global services to all is indispensable for the good of businesses and businesses too have to bear the burden for these goods.

Freidman posed the question of who is responsible and according to him 'a corporation is an artificial person and in this sense may have artificial responsibilities but business as a whole cannot be said to have responsibilities, even in this vague sense.' The important point raised is the notion of corporation itself. At first look, corporations are more or less anonymous and look somewhat artificial. This is palpable today that multinationals operate in different contexts, cultures, and countries; they look faceless and operate in geographical locations where their founders never dreamt to be operating and will very likely never arrive in their life's time. But that artificiality is not as artificial as one would think. As Werhane (1985) says corporations are associations given legal status [...] to operate as a single unit with limited liability over a [...] a period of time. [Accordingly], corporations are [...] created by a group of individuals (shareholders) for a specific purpose or purposes. They all are moral agents that agree to form an enterprise for their collective well-being. When shareholders obtain their legal status from the governments, they also acquire a legal identity, which comes with operating principles and social norms. In this sense, shareholders are collectively morally responsible for whatever their joint-venture happens to do. However, given the complexity of doing business in the modern world, shareholders employ managers (Executives) to lead their corporations for them, conferring to managers the responsibilities that are theirs. Viewed from this perspective, the answer to the question 'what does it mean to say that "business" has responsibilities' is that this responsibility is held legally by the shareholders. They become legally responsible for whatever comes out of their joint-venture through the incorporation process. Because the legal statuses of corporations are impregnated with the social norms, social responsibility defined in the business framework becomes also part of the agreement for businesses. Yes, indeed, 'only people can have responsibilities' and in this case, the responsibility lays with the Executives recruited not only to care about increasing the profits but also many other things that concur for the production of profits and expected and unexpected externalities. These other duties of the Executives include, for example, the solvency of the business, innovation, competitiveness, and well-being of employees. Friedman is right in saying that 'businessmen, meaning individual proprietors or Executives, are responsible'. Strangely, Freidman left these sound answers aside to ideologically reject the idea that corporations as social entities are bound the respond to the social norms of the societies where they operate.

True: an Executive is employed by the owners of the business and has direct responsibilities to conduct the business according to the desires of his employers and to make more money. Businesses are not philanthropic institutions and, to justify their presence, they have to create wealth. But this vision of corporations is limited and does not convey less than the reality of free-enterprises. It bases all its assumptions on the financial capital assets and neglects other forms of capitals. Corporations also need human and natural capitals without which financial capital will hardly multiply the wealth. Human capitals are key competencies such as knowledge, practical skills and talents, experiences, habits, personal values, etc., that enable employees to perform their duties. Human capital is essential for corporations to thrive. Good corporate executives will not make the dearly needed profits for their businesses if the human capital is in social conditions that would not allow employees to fully exercise their talents. This is the minimum social responsibility of business, which is to acknowledge the importance of human capital.

In this view, enterprises are not a property of the sole shareholders but equally that of a variety of stakeholders, among whom the employees. Friedman was partly right when he asserted that 'a corporate executive is an employee of the owners of the business' and that 'he has direct responsibility to his employers'. This is a mechanistic view of business, opposed to the idea that corporations are also the place where employees should fulfill their own life's goals, a place where employees grow in their own beings and live up their values. Values are necessarily and explicitly a part of doing business (Freeman et al. 2004). Employees constitute the immediate social environment that Executives deal with; as participants in the creation of wealth for the corporations, good Executives should attract good people and maintain them within the organization. Hence, it is sensible to add to Freidman's (1970) assertion that an Executive also has direct responsibility to his employees if he has to increase the profit of the enterprises. This means that the social responsibility of business, at least as far as employees are concerned, is not only to increase its profits but also to keep employees flourishing in order for them to contribute to the creation of the wealth. I believe Freidman would have agreed with this. At a larger scale, businesses use the natural capital (resources such as minerals, oils, and timbers.) that belongs to the communities. The depletion of the natural capital for private profit decreases the total capital for each individual in the community. Because private businesses deplete the natural capital that would benefit the community, businesses should bear social responsibility to provide the minimum that would compensate for the loss incurred by the society.

Surprisingly, Friedman (1970) neglected employees and the natural capital that are also part of the business. This is noticeable if one remembers the social events of 1967 (publication of Populorum Progressio by Pope Paul VI) and 1968 (students demonstrations in Paris). These events, respectively, demonstrated the finite nature of the natural capital and vividly argued for social questions, demanded more humane capitalism, and strengthened trade unions. The chief characteristics of that new form of humane capitalism were more social responsibility and more participation (Meschkat 2008). Friedman (1970) omitted these important historical developments once more, may be because of ideology. But it seems fair to attribute this omission to a real limitation because sustainable development explicitly requiring social fairness and inclusiveness appears only several decades after 1970. Indeed, sustainable development and sustainability rose to the prominence of mantra [...] following the 1987 publication of the UN report (Our Common Future) but the social inclusiveness took its shape after the 1992 Rio de Janeiro conference (Mebratu 1998). Against Freidman's view, sustainability requires Executives 'to make expenditures on reducing pollution' and to clean up all the mess generated through the process of making profit.

Freidman wrote: 'Insofar as his actions in accord with his "social responsibility" reduce returns to stockholders, he is spending their money. Insofar as his actions raise the price to customers, he is spending the customers' money. Insofar as his actions lower the wages of some employees, he is spending their money.' These three statements are a 'no-winner scenario' whereby every actor loses if enterprises are to bear their social responsibility. There is a problem with equating benefit with money. Taking money as the only measure of any good in economic terms is hardly surprising for someone from the Chicago School but that vision is flawed. People do not measure their well-being only in terms of money. People also value things such as safe neighborhoods, impeccable health facilities, good roads, clean water, clean air, beautiful mountains, and handsome rocky sceneries. Some of these goods can be convertible in money but others cannot. People have will to pay a higher monetary price in return for such non material goods such as clean water, clean air, beautiful mountains, and handsome rocky sceneries. Examples of such cases include tourism as social good whereby people pay a high price just to get to see beautiful areas, living an experience with traditional societies in very remote areas of the world. Those people who pay for such items would not see activities related to the social responsibility as 'imposing taxes […] and deciding how the tax proceeds shall be spent […].' Rather they would take doing so as doing the right thing for the global good.

Friedman claims that the role of free-enterprises in a free society is to create wealth for those who work hard. He sees social responsibility as an impediment to the political principle and governmental functions. However, free-enterprises, as biodiversity conservation organizations, operate in environments that are both natural and social. Enterprises need natural resources that they draw from the physical environment and make goods that they sell to make money. Actions of enterprises on the physical environment have wider impacts on their neighborhoods and, in some cases, on distant locations. An example of such cases is that of the acid rain, whose main source is the coal-produced electricity generated by large industrial complexes and travels long distances away from their origins and cause not only widespread ecological mess but also true economic damages. These impacts are directly linked to the production activities and constitute the direct responsibility of these enterprises; they should pay for damages caused. That is the essence of the social responsibility of any enterprises: paying for the negative externality that the global society incurs from the activities of the business (Jafy 2014); people also incur negative externalities when some of biodiversity conservation activities are implemented. Therefore, the role of the participative processes in decision making, which is one of the major features of the landscape approach, is to ensure that some of these negative externalities are taken care of by either enterprises making profits from natural resources of the landscape or by conservation organizations that are working to create a better social environment to ensure that biodiversity is protected. Socially, enterprises operate in societies that have norms and expect them to operate in accordance with these norms. The global consensus that emerged from Rio de Janeiro since 1992 is that enterprises should strive to be sustainable; they should be not only environmentally sensitive but they should also factor social concerns. The narrative that emerged retained the main touchstones of traditional free-enterprise theory, such as the importance of diversity and freedom, while neatly integrating social justice concerns, outsourcing unease, corporate abuse allegations, and well-documented observations about the dangers of excess concentration [...] (Goerner et al. 2009). This is a leap forward grounded on the recent social evolution, which was not the case when Friedman wrote.

## References

Abraham C, O'Callaghan T (2012) Chimps attack people after habitat loss. New Scientist. 11 October 2012

Barnes RFW (1996) The conflict between humans and elephants in the Central African forests. Mamm Rev 26:67–80

Bhima R (1998) Elephant status and conflict with humans on the western bank of Liwonde National Park, Malawi. Pachyderm 25:4–8

Callicott JB (1990) Whiter conservation ethics? Conserv Biol 4(1):15–20

De Boer F, Ntumi C (2001) Elephant crop damage and electric fence construction in the Maputo Elephant Reserve, Mozambique. Pachyderm 30:57–64

Dublin HT (1996) Elephants of the Masai Mara, Kenya: seasonal habitat selection and group size patterns. Pachyderm 22:25–35

Dworkin R (1993) Life's dominion: an argument about abortion and euthanasia. Harper Collins Publishers

Freeman RE, Wicks AC, Parmar B (2004) Stakeholder theory and the corporate objective revisited. Organ Sci 15(3):364–369

Friedman M (1970) Social responsibility of business is to increase its profits. The New York Times Magazine

Goerner SJ, Lietaer B, Ulanowicz RE (2009) Quantifying economic sustainability: Implications for free-enterprise theory, policy and practice. Ecol Econ 69:76–81

Hoare RE (2000) Project of the Human-Elephant Conflict Task Force (HETF): results and recommendations. Pachyderm 28:73–77

Hoare RE (1999) Determinants of human–elephant conflict in a land-use mosaic. J Appl Ecol 36:689–700

Hoare RE (1998) Human–elephant interactions at the ecosystem level. Pachyderm 25:41–42

Inogwabini BI (2014) Conserving biodiversity in the Democratic Republic of Congo: a brief history, current trends and insights for the future. PARKS 20(2):101–110

Inogwabini BI (2016) Congo Basin's shrinking watersheds: potential consequences on local communities. In: Rao P, Patil Y (eds) Reconsidering the impact of climate change on global water supply use and management. IGI Global, Pennsylvania, USA

Inogwabini BI, Leader-Williams N (2013) Conservation paradigms seen through the lenses of bonobos in the Democratic Republic of Congo. In: Navjot S, Raven P (eds) Essays in conservation biology—perspectives from practitioners in tropical environments. Blackwell-Wiley

Inogwabini BI, Mbende L, Bakanza A. Bokika JC (2013) Crop damage done by elephants in Malebo Region, Democratic Republic of Congo. Pachyderm 54:59–65

Jafy LTM (2014) Externalities and social responsibilities. Int J Acad Res Manag 3(2):1747–2296

Jepson P, Whittekar RJ (2002) Histories of protected areas: internationalisation of conservationist values and their adoption in the Netherlands Indies (Indonesia). Environ Hist 3:129–172

Kangwana KF (1995) Human–elephant conflict: the challenge ahead. Pachyderm 19:11–14

Kangwana FK (1993) Elephants and Maasai: conflict and conservation in Amboseli, Kenya. PhD Thesis, University of Cambridge, UK

Kinzonzi E (2004) Etat de conflict homme-éléphant au nord du parc national d'Odzala-Kokoua, Congo. Presented at the problem animal control methods training, Wildlife Conservation Society. Mid Zambezi Elephant Program and Elephant Paper Trust, Harare-Zimbabwe: Technical Report
Lahm SA (1996) A nationwide survey of crop raiding by elephants and other species in Gabon. Pachyderm 21:69–77
Mebratu D (1998) Sustainability and sustainable development: historical and conceptual review. Environ Impact Assess Rev 18:493–520
Meschkat K (2008) Germany 1968—SDS, Urban Guerrillas and visions of Räterepublik. 1968 revisited 40 years of protest movements. Heinrich Boll Foundation Democracy, vol 7, pp 39–42
Naylor KD (2005) Pinchot, Gifford (1865–1946). In: Taylor B (ed) Encyclopedia of religion and nature. Continuum, pp 1280–1281
Osborn FV (1996) The ecology and deterrence of crop-raiding elephants: research progress. Pachyderm 22:47–49
Osborn FV, Rasmussen LEL (1995) Evidence for the Effectiveness of an Oleo-Resin capsicum aerosol as a repellent against wild elephants in Zimbabwe. Pachyderm 50:55–64
Sachs J (2011) The price of civilization. Random House
Sam MK, Barnes RFW (1998) Elephants and human ecology in northeastern Ghana and northern Togo. Pachyderm 25:43–44
Sam MK, Barnes RFW, Kotchikpa O (1998) Elephants, Human ecology and environmental degradation in northeastern Ghana and northern Togo. Pachyderm 26: 61–68
Tchamba MN (1996) History and present status of the human-elephant conflict in the Waza-Logone Region Cameroon. Biol Conserv 75:3541
Tchamba M (1995) The problem elephants of Kaele: a challenge for elephant conservation in northern Cameroon. Pachyderm 19:23–27
Tchamba MN, Nshombo I (1996) Evaluation préliminaire du conflit homme-éléphants autour du Parc National de Kahuzi-Biega, Zaire. Unpublished Projet Report submitted to IZCN/GTZ - Bukavu, Zaïre
The Economist (2013) The origins of the financial crisis. Crash course: The effects of the financial crisis are still being felt, five years on. The issue of September 7th
Vanleenwe H, Lambrechts C (1999) Human activities on Mount Kenya from an Elephant's perspective. Pachyderm 27:69–73
Ville JL (1996) Man and elephant in the Tsavo area of Kenya: an anthropological perspective. Pachyderm 21:69–72
Waithaka WJM (1999) Monitoring human-elephant conflict through remotely located stations. Pachyderm 27:66–68
Waithaka WJM (1996) Problems and solutions outside of protected areas. Pachyderm 22:91–92
Wasilwa NS (2003) Human-elephant conflict in the Masai Mara dispersal areas of Trasmara District. PhD Thesis, Durrell Institute for Conservation and Ecology, University of Kent, United Kingdom
Werhane PH (1985) Persons, rights, and corporations. Prentice-Hall, Englewood Cliffs, NJ

# Chapter 4
# Landscapes Require New Legal Framework to Conserve Biodiversity

**Abstract** Biodiversity conservation landscape emerged in a legal vacuum because it embraced countries with different juridical history and traditions. Formerly French colonized countries (Cameroon, Central African Republic, Congo Brazzaville, and Gabon) inherited the same juridical history and practices from France but they evolved in different paths on biodiversity conservation issues while Democratic Republic of Congo and Equatorial Guinea have a totally different judiciary history. The idea of landscape was scaled to work at the regional Central Africa level, which lacked the history of joint work on natural resources. The chapter discusses ways in which this critical question was addressed using a mixed cohort of pathways to a juridical framework that was more than just the adaptation of the existing legal framework but rather sound, fully functional, and sensibly regionally harmonized regional framework. The search for this new juridical system offered an opportunity to produce new regulations and laws to democratically govern the natural resources. Process-wise, the chapter describes the integration of international agreements and conventions such as Convention on Biological Diversity, Convention on International Trade in Endangered Species, Convention on the Conservation of Migratory Species, and the International Convention on Wetlands into national laws to ensure that gaps were filled in. The chapter also argues that conservation landscapes offered long awaited opportunities to countries in Central Africa to harmonize their legislations across the region.

**Keywords** Biodiversity conservation laws · International conservation agreements · International conservation treaties · Biodiversity convention · Convention on wetlands · Ramsar site

## 4.1 Introduction

When the landscape process was started in 2002, there were no dedicated legal tools available to support their implementation. First of all, the concept was working at Central Africa as a region, which has never had any experience in working on natural resources within a regional framework. Indeed, the region comprises countries of different juridical history; despite there being a large number of countries (Cameroon,



B.-I. Inogwabini, *Reconciling Human Needs and Conserving Biodiversity: Large Landscapes as a New Conservation Paradigm*, Environmental History 12,
https://doi.org/10.1007/978-3-030-38728-0_4

Central African Republic, Congo Brazzaville, and Gabon), with the French legal system, other countries (Democratic Republic of Congo and Equatorial Guinea) have a different history. Also, despite the fact that former French colonies inherited the same juridical history and practices from French, most of these countries had evolved in different paths on issues of biodiversity conservation. Within these differences, the most important one was the history of biodiversity conservation between the countries: while DRC has had its first national park created in 1925 (Virunga National Park), the first protected areas in Congo Brazzaville and Gabon were created, respectively, in 1993 (Nouabale-Ndoki) and 2002 (13 national parks)—despite the fact that Lopé-Okanda was made into a wildlife reserve since 1946.

## 4.2 Dealing with Obsolete Legal Framework in a New Conservation Paradigm

But, even in countries like the Democratic Republic of Congo where the history of conservation was entering a century of age, the legal framework was, at the time of the conceptualization of the landscape, either obsolete or not fit for the new paradigm that was emerging with the creation of landscapes. In fact, in the early twenty-first century, the legal framework of biodiversity conservation remained centered around obsolete laws that were promulgated during the colonial era. Laws on the management of land and forests can be traced back to land tenure ordinance was passed on 1 July 1885, confirming that lands acquired by Stanley on behalf of King Leopold II would be used by the Belgian Crown but indigenous people would continue to own their properties (Jeal 2008). That ordinance was followed by several other decrees, specifically the decrees of 22 August 1885, 14 September 1886, and that of 3 June 1906, which unilaterally ended the agreements with indigenous people and instituted the registration of all lands, which meant that non-registered land became vacant though indigenous people would continue using lands they collectively owned (Inogwabini 2014). All the laws that would follow had never come to term with the vision of the land ownership; the communities were not seen as a major stakeholder to bring to the discussion table. As argued by Inogwabini (2014), even reforms that came after the Democratic Republic of Congo gained its independence did not deviate so much away from this state-centric vision of the management of natural resources. Inogwabini (2014) argues that Land law reforms were needed in the early 1960s because the 1960 constitution did not clarify land laws but true efforts began with the 1964 constitution. Unfortunately because of the political problems that the country was going through, the legislators remained was vague on land tenure and deferred the settlement of the issues of land tenure to national law to rule on land attributions and concessions acquired before 30 June 1960. Inogwabini (2014) argues, with reason, that the most important reform was the Bakajika law of 1968, which was modified in 1970, 1997, and 1980 (Leisz 1998; Musafiri 2008). This is so because the letter and the spirit of the Bakajika law remained the salient corner-stone for every law on the land and other natural resources management across the country, despite shifts in how these laws would be implemented. The political objectives of the Bakajika law were to challenge the colonial land laws that gave the best cultivable lands to colonists (Leisz 1998; Musafiri 2008) and to provide the land tenure regime instituted by the 1964 constitution. Socially, the Bakajika law aimed to repair the injustices felt by traditional communities. Ironically, the Bakajika law confirmed that 'the soil and anything beneath it belongs to the state'; the 11 April 1949 decree remained unabrogated (Tshikengela 2009), maintaining communities away from becoming true owners of the land where they were born and from which people literally drew most of their livelihoods. Even the forest code of 2002, rather than completely shaking the spirit of the Bakajika law away, reaffirmed it though it left some windows open for communities to acquire land tenure right through a long and cumbersome process, which was not only clearly defined but also was left for some implementation measures to come through later.

In terms of the juridical structures, landscapes in the Democratic Republic of Congo were a new entity not only with the conservation paradigm but also in the management of natural resources. As such and to be fully functional and sensible for different stakeholders, the landscape approach needed more than just the adaptation of the existing legal framework. The coming into being of the landscape as a new way of addressing issues of natural resources was, as such, a new opportunity to help come up with new regulations and laws to democratically govern the natural resources of the country. The process to implement the landscape approach would offer the long awaited opportunities to countries in Central Africa to harmonize their legislations across the region; it would provide the floor for the Democratic Republic of Congo to share its long history being the first country in the region to have a relatively large and well distributed network of protected areas as well as an occasion to learn from the new dynamism of conservation in region and help the Democratic Republic of Congo to come to terms with its own weaknesses in the legal frameworks of the governance of natural resources. Indeed, the history of the country in biodiversity conservation cannot be described only in terms of its successes such as the case of rescuing the northern white rhinoceroses from extinction (Chap. 12) but it has also to be acknowledged that this heroic biodiversity conservation history of the country has been also described to be the one where the state has acted without seeking the informed consent of communities.

## 4.3 Using International Agreements and Treaties to Circumvent the Legal Vacuum

Finally, the landscape planning process as well as the regional movement to harmonize the approaches and the legislations on natural resources would significantly help the Democratic Republic of Congo to align its own juridical arsenal to the numerous international treaties and conventions that the country had signed and had been having difficulties to implement because of the hurdles in its own legal system. Indeed, the Democratic Republic of Congo has become part to many international instruments on biodiversity that it ratified, including the most important such as the Convention on International Trade in Endangered Species of Wild Fauna and Flora (CITES; joined on 20 July 1976), the Convention on Biological Diversity (ratified on 1995-03-03), Convention on the Conservation of Migratory Species of Wild Animals (joined in 1990), the International Convention on Wetlands, and United Nations Convention to Combat Desertification. Of course, the opportunity offered should be also open for integrating other international and regional binding agreements that the Democratic Republic of Congo has ratified to be reflected in the governance of its natural resources. These could include such important international regulations, agreements, and treaties that would have significant impacts on forests and natural resources that are part of different landscapes existing in the country. These would include important treaties and conventions such as the Brazzaville Treaty, which instituted the Economic Community of Central African States (ECCAS), the treaty establishing the Southern African Development Community (SADC), the 2000 Cotonou Agreement between the European Union and the African, and Caribbean and Pacific Group of States commonly known as the ACP countries. Other instruments that should be integrated into the development of the new legal framework to cope with legal questions raised by the new biodiversity conservation paradigm that integrates sustainable development, cultural diversity sustainability, and needs to conserve biodiversity are the questions such as those of power sharing for different stakeholders and the inclusion of minorities. Indeed, to be complete, the new legal framework needs to include items such as the UN Declaration on the rights of indigenous peoples, the 1981-UN Convention of Right to Development and the African Charter on Human and Peoples' Rights, particularly its provision on the right to property or right to own property.

As stated in Chap. 1, the foundational legal reference in the process of drafting and adopting the new legal framework was already agreed upon by the heads of states of Central Africa when, in 2005, they adopted the treaty creating the Commission of Central African Forests that insisted on the inter-country coherence as critical way to help the regional integration. That said, however, the field work on landscapes, particularly in Lake Tumba where there was no significant protected area and no long history of biodiversity conservation, started before all the legal instruments were in place. To circumvent the legal vacuum, different actors started using the regulations that were dispersed and often incoherent between them. Indeed, one of the major problems of the biodiversity conservation in the Democratic Republic of Congo is that each ministry that had something to do with natural resources had its own code or a set of laws; there are codes on mining, agriculture, water, conservation law, etc. In some cases, they have conflicting provisions and bridging these inconsistent requirements is, so often, a very difficult, in some cases impossible, task.

Facing these difficulties often pushed the actors in the Lake Tumba Landscape to use international agreements or treaties to which the country belonged. The recourse to the international regulations was acceptable as the means to solve the dilemma while waiting for new regulations because of the precedence given to the international treaties by the legal architecture of the Democratic Republic of Congo. Indeed, for example, to push the Government to support the regulations on the sustainable use of wetland resources in the Lake Tumba landscape, stakeholders have had to use the Ramsar Convention (Chap. 21). The Ramsar Convention should be bridged by invoking legal instruments that the Democratic Republic of Congo has already signed off. Indeed, the Ramsar guidelines emphasized the importance, among many other items, of developing integrated management plans at wetland sites across the world. To cover the gap in the national regulations of the Democratic Republic of Congo about the demands of management plans for land units found in the Lake Tumba Landscape, stakeholders thought that it is wise to proceed by the classification of the largest World Ramsar Site that covered more than 75% of the landscape in the hope that planning activities would be covered by the provision of the Ramsar Convention. Apart from that, the political will expressed in the COMIFAC Treaty was a very powerful legal instrument even though it was never been included in the national legal framework. Equally important was the support lent by the Convention on the Biological Diversity and the promise made the different Governments of the Democratic Republic of Congo to protect close to 15% of the national territory under a system of protect areas. This was very helpful for the Lake Tumba Landscape particularly because the landscape was a new one and the only one that could be said not to rely on an already existing protected area within it. Apart from that, the planning meeting relied heavily on the new 2002-forestry code, which was adopted partly as a response to numerous demands by the international stakeholders to improve the management of the natural resources in the sector of forest. Different stakeholders felt that despite its limitations, this new forestry code made several advances; chief among them was the decision process making which was clearly calling for the participation of all the major stakeholders.

## 4.4 National Sovereignty Limits the Successes of Regional Planning Processes

The general legal framework of countries involved in the COMIFAC is that of sovereign states that have been trying to create a regional market for several years. Lack of progress in the creation of this community is a major impediment in trying to implement the landscape concept as a conservation paradigm across a region where bi-lateral cooperations are marred with the political incapacities to have serious ambitions in working together to alleviate the poverty of the least abled in the region. Indeed, political leaders of Central Africa have been struggling, with very limited successes, to create a free-movement zone for people and goods using the Brazzaville Treaty, which instituted the ECCAS. The difficulties to make the ECCAS-zone a viable market and political entity whereby regulations and rules apply to communities of different countries have been mostly due to lack of political will and lack of trust among the countries that are members of ECCAS. For example, the two Congos, wherein both Lake Tele and Lake Tumba are located, share similar cultures and, in many cases, similar communities along their common borders but there has never been any sense of being the same communities; Congolese from the Democratic Republic of Congo can be chased as snakes in the neighboring Republic of Congo and be asked to vacate the areas within hours. The same can be said of the Congolese from the two Congos being chased out of Angola with no sensible reasons except that they are not in their countries of origins. Camerounians are said to invade Gabon, etc. With such a bleak neighborhood history, it is rather difficult to see how the COMIFAC Treaty would have more chances to succeed where ECCAS, the umbrella overall embracing treaty had been making only very meager successes. The doubts about the chances of COMIFAC being a successful driver of a new way of managing natural resources would quadruple if one would want to incorporate new developments in the region, which are made of political suspicions that have brought what could be rightly described as former purely politicians' antagonisms to become part of the general community feeling. Indeed, years of wars in Angola, Congo, Democratic Republic of Congo, and Rwanda have pushed people to become more aware of their distinctness for communities that once thought of themselves as being part of a larger network of families and brotherhoods. Congolese from the Democratic Republic of Congo, for example, are now tempted to believe that the natural resources of their country, rather than being a blessing for their communities, are part of the reason why their nation has been cursed and invaded by neighboring countries. Angolans also think that most of the Central African citizens would want to migrate to their country just to make profit of the wealth that their oil has generated after the end of the 3-decade long war. Equally, Congolese from the Republic of Congo believe that Camerounians and Congolese from the Democratic Republic of Congo are the only poachers they can catch in their national parks and other protected areas, etc. In fact, the African Development Bank (2011) is right in stating that from the political standpoint, the attainment of mutual understanding and concord between nations, in the spirit of the African Union, is the biggest challenge to regional integration in Central Africa. Given the history of inter-state cooperations and multiple instruments created by African countries over the last 6 decades, one might be tempted to ask the question whether COMIFAC could not be just another one of these mega-strategies, mega-plans, and mega-ideas that are very common in Africa, and particularly in the context of Central Africa, such as Lagos Plan or even the New Partnership for Development? The main vexing question on the harmonization of the legal frameworks across the region remains: 'why should the issue of biodiversity conservation work where other political attempts to integrate Central Africa as a region failed? The experience of the Lake Tumba Landscape, which officially is the second part of the general Landscape known in the biodiversity conservation circles as 'Lake Tele–Lake Tumba Landscape, is that while the ideal should be to have both parts use the same standards, in reality, these parts (also known as segments) have been operating at different speeds and using different legal frameworks, despite there being an inter-state official document pledging to work jointly. The thing that brings the two segments together, in reality, is the funding mechanisms that were put together by the Congo Basin Forest Partnership, which then led to the creation of the Congo Basin Forest Fund.

## 4.5 Could the Biodiversity Diplomacy Help Where the Normal Diplomacy Failed?

The above makes it that for some people it is easier for a Congolese from the Democratic Republic of Congo to travel to the USA or the United Kingdom than to cross from Kinshasa to Brazzaville. The treatment reserved for communities of neighboring countries in Central Africa is just so awful. Under these circumstances, talking about cross-borders seems to look like a pure logical hallucination. However, it may be that the plight imposed upon countries and communities by the felt and projected effects of climate change could bring countries and communities to think differently. As the African Development Bank (2011) stated, policymakers have also recognized the need to coordinate their policies, in view of climate change and threats to biodiversity conservation across the Basin Congo. This view came as a strong supporting argument for the pledge jointly made by the Secretariat of the Convention on Biological Diversity and the Central African Forests Commission in 2009 that because they recognize the economic and ecological importance of forests to the sub-regional development […], Central Africa Nations have heightened their regional coordination efforts to ensure biodiversity preservation and sustainable forest management in the Congo Basin. But still, saying so does not give the answer to why should this work better than the other treaties and conventions that have known only ochrous fate across this less-politically integrated part of the world? At least two reasons might push this to be a success simply because of the funding currently being channeled by different initiatives to either mitigate the effect of climate change or those that working on how communities could adapt to the same effects. Indeed, the experience of Lake Tumba and Lake Tele described above has had some successes, including the signature of the cross-border country agreement, only because the donors were asking for joint initiatives. In fact, a continuous financial support is likely to forge a new working culture and, if sustained for very long spans of time, could lead to that culture becoming deeply rooted in the practices of both sides of the Ubangui–Congo Rivers. The force of funding to coalesce disparate country-dependent conducts in biodiversity management into a single vision and a community of practice should not be minimized; it has been operational and helped save mountain gorillas in the Eastern Democratic Republic of Congo throughout the 2-decade war. Even at the heights of the conflicts, a moment when official diplomatic links between Rwanda, Uganda, and the Democratic Republic of Congo were, at minimum, bleak, the International Gorilla Conservation Program continued to operate and continued to bring biodiversity conservationists (inclusive of law enforcement teams, researchers, etc.) together to debate about the species survival through the dire tides of the war. This biodiversity diplomacy used several instruments and mechanisms such as convention on migratory species and the United Nations Educational, Scientific, and Cultural Organization as a conduit to ensure that effects of the war on the Virunga ecosystem should be minimal throughout the conflict. Of course, one can say that this was so because of the fact that Virunga National Park in the Democratic Republic of Congo is a world heritage site, which increased its profile on the agendas of international policy and decision makers. This is true but the model can also be applied in different circumstances using other international institutional frameworks. In the case of Lake Tumba Landscape, as indicated above, such an international institutional framework is offered by the Ramsar Convention; the biodiversity diplomacy should be able to work more or less effectively provided that the Secretariat of the Ramsar Convention plays the same pro-active and constructive role as was the case of the UNESCO Commission on World Heritage Sites.

## 4.6 Concrete Steps Toward New Forms of Legal Frameworks

Despite the difficulties of crafting a regional-wide and enforceable legal framework that would be applied throughout Central Africa, some initiatives are under way to at least ensure that convergence in ways natural resources are managed in the region. Convergence is, indeed, the qualification of the regional-wide plan for the management of the natural resources, particularly forests, is known as Plan de Convergence. Indeed, the COMIFAC, which was created in 2005 (see Chap. 1 in this volume), has been operating through its Regional Central Africa Convergence Plans, which with varying degrees had always been centered around its current (2015–2025) objectives, including the harmonization of forest and environmental policies, sustainable valorization of forest resources, sustainable use and conservation of biodiversity, fight against climate change and desertification, ensuring socioeconomic development through the participation of multiple actors, and providing sustainable sources of funding for the management of the forest ecosystems. In the case of the Lake Tumba Landscape, the implementation of the objectives of the COMIFAC's Convergence Plan has been made effective through a bi-lateral agreement signed by the ministers in charge of biodiversity and environment of the Democratic Republic of Congo and the Republic of Congo. This constituted the first concrete step toward the formalization of the notion of the landscape. Indeed, after years of negotiations, which began in 2005; the two countries signed a cooperation agreement on the 5 August 2010 whose main objective was to ensure harmonization of legal practices on both sides of the Congo. Indeed, despite the fact that this accord was a significant milestone in the process of regionalizing the management of natural resources and the environment in particular, the provision of article 5 of the bi-lateral agreement of 5 August 2010 maintained the fact that states would remain sovereign within their territories. Of course, it would have been naïve to believe that limitations imposed by the notion of sovereign states would be washed away over night. Even countries with advanced levels of cooperation, such as in the case of the European Union, have kept most of the sovereignty when it comes to the management of their territories. Nevertheless, agreement to harmonize the legislations, to combine efforts to fight the widespread poaching and illegal bushmeat trade, to work jointly on scientific research and ecological monitoring, to ensure the control over the use and trade of natural resources and the promotion of tourism as a sources of sustainable revenues (see Chap. 23) within the two sections of the same biologically complex zone, as it is stipulated in the article 5 of the cooperation agreement in question, is in itself a major advancement in the way countries in the region work. Furthermore, other provisions of the same article 6, particularly insisting on joint efforts to build the human capacities to manage the natural resources of the area, to institute a direct bi-lateral communication system and the co-sharing of proceeds of activities related to biodiversity was also of a significant value in terms of the philosophy of the mutualization of not only of efforts but also of the geographic space though these parts of the landscape would remain under the sovereign authority of the country within whose boundaries it is part. Finally, of very particular importance, was the fact that the same agreement instituted the participation of local communities and the private sector as an essential part of the management process. This is, indeed, what the landscape, as a biodiversity conservation paradigm brings of different, as compared to other biodiversity conservation paradigms, in the ways we have to look and approach the biodiversity conservation across the region.

In fact, taken from the above perspective, the landscape as a biodiversity conservation paradigm has to be viewed as system-thinking process. Indeed, biodiversity is globally a system in which human communities are part. In this respect, the definition of the biodiversity needs to be broadened or simply brought to the general public. Indeed, very often there are confusions between the number of species and the broader concept of biodiversity. As Inogwabini (2016) has argued, the definition of biodiversity has changed over time and the ethical implications and responsibilities toward biodiversity change depending on the understanding of the word (Bosworth et al. 2011). Narrowly defined, biodiversity is equated with the number of species or the 'species richness' found in a given location (Morgan 2009). However, as Inogwabini (2016) indicated, this definition has moved from this narrow understanding to include living organisms and the complex interactions between these living organisms and their abiotic environments. Biodiversity is the totality of living organisms and functions that ensure that species and life are maintained on earth. Hence, biodiversity consists of three main components: composition, structure, and function (Bosworth et al. 2011), which implies that biodiversity should not be viewed only as the total number of species; it has to include functions that inter-relate different organisms and sustain life on earth. In this sense, different human activities are part of the functions that ensure the maintenance of human life on earth and, as such, should be looked at differently and should strive to be sustainable in their essence. This fully justifies the inclusion of human activities, community participation as key elements in planning and managing biodiversity.

The second legal formalization of the notion of the biodiversity conservation landscape has come through the new conservation law of the Democratic Republic of Congo, known as the law 14/003 of 11 February 2014 whose articles 2 (32) and 31 (6) officially include the terrestrial and marine landscapes in the official nomenclature of protected areas. The inclusion of the landscape within the categories of protected areas can be debatable because it sounds like inverting the reasons that pleaded over the birth of this loose conservation category, which implies the integration of multiple activities, multiple usages, and multiple functional attributes in a single concept. Despite that conceptual debate, however, the clear inclusion of the landscape as a conservation area is a real advancement that fills the gap that was difficult to deal with during the process of developing the landscapes. Indeed, there was a clear legal vacuum, which brought so many difficulties in trying to ensure that long and widespread community mobilization processes required by the landscape, as a biodiversity conservation paradigm, were in place and accepted by all interested parties.

The third legal support to the biodiversity conservation landscape came with the ministerial decree signed by the Premier Ministre of the Democratic Republic of Congo on 2 August 2014 on the acquisition by communities of the local community forests. Despite the lengthy process and the somewhat stringent technical demands, such as detailed maps, this law has opened a wide opportunity for the reconciliation of the traditional land right and modern state land right, which have always been in opposition since the colonial era. This advancement will, in the long run, contribute to better and tranquil negotiations on the land-use planning and the attribution of functions to each land unit falling within the landscape.

## References


African Development Bank (AfDB) (2011) Central Africa Regional Integration Strategy Paper (RISP) 2011–2015. Regional Department Center (ORCE)

Bosworth A, Chaipraditkul N, Cheng MM, Gupta A, Junmookda K, Kadam P, Macer D, Millet C, Sangaroonthong J, Waller A (2011) Ethics and biodiversity. Asia and Pacific Regional Bureau for Education UNESCO Bangkok

Inogwabini BI (2014) Conserving biodiversity in the Democratic Republic of Congo: a brief history, current trends and insights for the future. PARKS 20(2):101–110.

Inogwabini BI (2016) Congo Basin's shrinking watersheds: potential consequences on local communities. In: Rao P, Patil Y (eds) Reconsidering the impact of climate change on global water supply. Use and Management, IGI Global

Jeal T (2008) Stanley: the impossible life of africa's greatest explorer. Yale University Press

Leisz S (1998) Zaire country profile. In: Bruce J (ed) Country profiles of Land Tenure: Africa 1996. Land Tenure Centre Research Paper, vol 130, pp 131–136

Morgan GJ (2009) The many dimensions of biodiversity. Stud Hist Philos Biol Biomed Sci 40:235–238

Musafiri PN (2008) Land rights and the forest peoples of africa: historical, legal and anthropological perspectives. N°3: The dispossession of indigenous land rights in the DRC: a history and future prospects. Forest Peoples Programme, United Kingdom

Tshikengela BKL (2009) Acteurs et interactions autour des ressources halieutiques du Parc National de la Salonga. Cas de l'exploitation de la rivière Luilaka en RDC. Monographie soumise pour l'obtention du diplôme de Master complémentaire en développement environnement et sociétés. Université Catholique de Louvain

# Part II
# Data to Support the Conservation Action

This part of the book presents types of the data that are needed to support both the strategic thinking on how different landscapes would contribute to the overall biodiversity conservation as well as data that would support the daily management of both biodiversity and human communities' needs. Whereas the first part of the book dealt essentially with the examination of what the concept landscape brings as a paradigm in biodiversity conservation, this part is about what are the concrete data that would ground that conservation paradigm within the myriads of alternative paradigms that are available to help conservation biodiversity within countries that have large expanses of biodiversity but which do also have to face genuine demands for economic development and sustainable livelihood for their communities.

This part of the book contains the largest bulk of the materials that make the whole book; it has 12 chapters and could make a book in itself. But this length is understandable in that the part provides a global overview of most of the data collected throughout the Lake Tumba Landscape with a view to help the advancement of the understanding of a conservation biologist on what it is involved in the landscape approach in conservation biology. This part of the book contains both traditional conservation biology aspects of the data as well as new elements that have been brought into the game by enlargements of the views and necessary data required by the landscape as a conservation paradigm.

As it would have to be expected from traditional conservation biology practice, the second part of the book starts with Chap. 5, which is a broad description of habitats found in the Lake Tumba Landscape. In this chapter, habitat types are not described only in their broad categories but also in ways in which they do interact with human communities. Indeed, the chapter emphasizes the fact that humans are the species that modifies most of the niches and accounts for many changes, whether positive or negative ones. In so doing, the aim of the chapter is to lay the ground for the rest of the chapters of this second part of the book, which all describe the biodiversity of the Lake Tumba Landscape. The importance of describing habitats across the Lake Tumba Landscape, as it is always the case in the practice of conservation biology, was necessary because all the biodiversity in the landscape happens in different niches and habitat characteristics. Ecologically, studies of species would not make any sense if they are not coupled with the study of habitats. Chapter 5, therefore, describes in variable degrees of details habitat niches that would provide a global understanding of habitat uses by each species.

The second part of the book then moves on chapters that describe the landscape species, which are four in total and comprising bonobos, chimpanzees, lions, and elephants. Emphasize must be made here of the fact that the Lake Tumba Landscape is the only conservation priority area where both species of Pan are found although they cannot be said properly sympatric as they continue to happen on different sides of the Congo River. Landscape species, as a concept, is not so far from other conservation biology concepts such as 'charismatic species' and 'flagship species'. Indeed, a clear definition of the concept of landscape species has not been agreed upon by different conservation theorists but it would suffice to say that these species are the umbrellas for the Landscapes where they happen to occur; and, as such, landscape species should be viewed as the selling point for external stakeholders to endorse conservation purposes. Of course, the landscape species identified in the Lake Tumba Landscape are the species that are known to play key ecological functions. Hence beyond being the selling point, they should also be seen as ecological keystone species in their own right. The concept of keystone species refers to a species whose extinction within a given ecological community will cause the loss of many others (Mills et al. 1993). Keystone species are so important in determining the ecological functioning of a community that they warrant special conservation efforts (Mills et al. 1993). Mills et al. (1993) link keystone species with the trophic levels and define five categories of keystones species, which are associated with the respective effects of their removal on their ecosystems.

After describing the habitat types, the two following chapters of the book provide data on the bonobo *Pan paniscus*, the great ape species that resides south of the Congo River and does not share its habitats with other species of great apes. The presence of this species in the Lake Tumba Landscape has massive implications in the way its conservation should be organized as Chap. 6 provides data that show that most of the populations of the species were located outside of any protected areas in the Landscape. Documenting the bonobo distribution and its size in the Lake Tumba Landscape, was, hence, felt to be one the most important set of data to be collected before any conservation action could be started. Also because of the important size of the bonobo population in the Lake Tumba Landscape, it was important to document their genetical genealogy not only for theoretical beauty of the genetics as part of the scientific inquiry but also for practical conservation needs such as providing new gene pools that could be used to help salvage populations of the species that would be in direct need of diversification. Indeed, the Bonobo Sanctuary in Kinshasa has been working on the re-introduction program whereby individuals of bonobos that grew up in the Sanctuary were to be re-introduced in the wild; the bonobo populations in the Southern Lake Tumba would provide such a host community for this type of programs. Besides the knowledge on linkages between the bonobos of the Lake Tumba Landscape and the populations of the species found elsewhere, it was also important to understand how the bonobo population of the Lake Tumba Landscape was able to be and remain where it currently happens to be. That is why it was deemed important to construct a robust model of the species evolution, which would better explain what happened in the evolutionary history of the species. This was particularly important when one considered the fact that the other species of Pan (*Pan troglodytes troglodytes*) was also present on the northern side of the Congo River but still in the same Lake Tumba Landscape. As one gets to some of the areas of the Lake Tumba Landscape, particularly south of the town Mbandaka, one wonders, with all the chains of connected islands, why are the chimpanzees of the Ngiri Triangle different from the bonobos that reside just south of the Congo River? Chapter 8 tries to answer such questions while Chap. 9 addresses another key conservation issue, which is the potential impact of diseases on the bonobos. Chapter 10 addresses the issue of how to measure conservation efficiency through the monitoring of the bonobo by the use of less expensive methods. Both Chaps. 9 and 10 address issues that are broader and the implications of key concepts they deal with cover not only bonobos but other species of wildlife. In fact, human–wildlife transmissible diseases are a general conservation concern and theories discussed in this chapter equally apply to species such as the chimpanzees and other primates; the methods introduced by Chap. 10 on the presence–absence as a tool that can be efficiently used to count wildlife populations.

Chapters 11, 12, 13, and 14 give overviews of the conservation statuses of other species of conservation interests in the Lake Tumba Landscape. While three of the species discussed in these chapters (chimpanzees, lions, and elephants) are known to be conservation charismatic species and have been identified as landscape species (as defined above), the Chap. 13, which is about diurnal primates of the Lake Tumba Landscape has an enormous value of being added here because diurnal primates have been qualified to the indicators of the health of the ecosystems wherein they occur (Oates 1986). The other chapters dealing with the landscape species are Chaps. 11, 12, and 14 where knowledge of the distribution and the sense of abundance of chimpanzees, lions, and elephants are, respectively, given. The abundances of these species are mostly given as relative indices of abundance because of the smallness of the data sizes. But, despite that caveat, in the case of lions, some estimates have been provided. For chimpanzees and elephants, relative abundance have been widely used in other regions and are good enough for population monitoring exercises. Providing data for monitoring has been the very reason why survey studies on the landscape species were conducted; so these data are sufficient and fit for the purposes.

After describing populations of landscape species and the diurnal primates, an important landscape conservation objective was to know how good these species do within their environment. This question has been addressed by establishing a relatively easy way of measuring threats, which is covered by Chap. 15 on how to work out a threat index for the Lake Tumba Landscape. While the ideas developed in this chapter were applied primarily on the wildlife species resident in this landscape, they are of a general perspective and could be put to use in other contexts. Hence, most of the theories and ideas advanced in this part of the book are a contribution to the overall conservation biology even though they are coming out from a much contextualized scene. That should not be so surprising as one of the strengths of conservation biology as an area of inquiry has been that it is often built from localized experiments that contribute to the global knowledge. Also, to give a more global picture of the Landscape, it was felt necessary to go beyond the realm of the terrestrial large mammal species, which always occupies a ubiquitous space in the conservation debates. Of course, the ubiquity of the space given to terrestrial large mammal species is often justifiable in the fact that these are the species we can see, admire and often use. Their contribution to our view of biodiversity is easily quantifiable and the fact of being seen makes it easy for humans to feel some sort of attachment. But the landscape has more than just large mammals, which is why two other Chaps. 16 and 17 were added; they, respectively, give synopsis of freshwaters and the birds of the Lake Tumba Landscape. As synopses, they do not bring a lot new knowledge, which is still to be gathered using appropriate methods (Chap. 24); the Page value of these synopses, however, resides in the fact that they were included in the planning processes (Chap. 20).

# Chapter 5
# Qualitatively Describing Forests of the Landscape

**Abstract** Qualitative descriptions of different habitat types are an important part of ecological research. Hence, the chapter is a single-objective one and looks at the broad terrestrial habitat types and major water bodies of the Lake Tumba Landscape. As a qualitative examination, it is limited to forests and savannahs that are two major habitat types of the Lake Tumba Landscape. Qualitatively, and without covering forest floristic compositions, each major habitats type is decomposed into its broad components. Methodologically, the data were collected from combining estimates from transects and the analysis of satellite images. Seven forest types were identified in the Lake Tumba Landscape of which the most important forest types were permanently swampy forest (34.2%) and seasonally flooded forest (28.1%). *Terra firma* forests of the Lake Tumba Landscape had no single species-dominated large swathes of forest, which is often the case in many areas of the region.

**Keywords** Swampy forest · Forest-savannah mosaic · Forest galleries · Uapaca guineensis · Gilbertiodendron dewevrei · Hymenocardia acida · Annona senegalensis

## 5.1 Introduction

Qualitative descriptions of different habitat types are an important part of ecological research. This chapter only discusses the forests and savannahs because these are two major habitat types of the Lake Tumba Landscape. Freshwater habitats will be discussed separately in the summary of the work that has been done of fishes and their habitats in the lakes and rivers of the landscape.

Firstly, characterization of forest habitat types, in the context of tropical ecology is in many cases done with reference to human activities and watercourses (Magliocca et al. 1999; Hall et al. 1998a; Idani et al. 1994; Fruth and Hohmann 1993; Sabater and Vea 1990; Kano 1983, 1984). Secondly, habitats are characterized using floristic compositions and major vegetation recruitments. Likewise, drainage patterns and the structure of the forest floor provide. Thirdly, forests can be characterized by the type of seasons that occur in the region where they occur. Fourthly, forests can also be characterized by their degrees of floristic luxuriance. When describing habitats in



B.-I. Inogwabini, *Reconciling Human Needs and Conserving Biodiversity: Large Landscapes as a New Conservation Paradigm*, Environmental History 12,
https://doi.org/10.1007/978-3-030-38728-0_5

terms of human activities, simple qualitative attributes are used; these are terms such as primary or mature forest, old secondary forest, secondary forest, and cultivated areas. Forest descriptions in terms of floristic compositions include such nomenclature as mono-dominant forest preceded with the species that dominates the entire area being described; sometimes forests are, in this case simply called by the name of the species that outnumbers all other species. This is, for example, the case of *Marantacea* forests or even *Gilbertiodendron* forests. When forests are being characterized by their floors, they respond to two categories, namely seasonally inundated, swamp forest or permanently inundated forest. Forests that are characterized by seasons would generally be named as rainforest, dry forests, etc. Forest characterization by floristic luxuriance comes in when they are broadly called as ever-green forests, deciduous forests, etc.

The objective of this chapter is unique and consists of looking at the broad terrestrial habitat types and major water bodies of the Lake Tumba Landscape. Broadly, terrestrial habitats of Lake Tumba are forests in most of its northern part while the extreme south is within the savannah. The chapter classifies each of these two major habitats types in their respective broad components. The chapter does not cover the floristic composition though it will be giving some specific names just to give the sense of what the habitats would feel like on ground. Broad characterizations of habitats are generally adequate as an approach though in some cases, there is a need for detailed habitat descriptions, sometimes even with the combination of both floristic patterns and physiognomic structure. Broad characterizations of habitat are, indeed, the only way to describe such a big area; 80,000 $km^2$ is a large area and at this scale, there is no way to pin down small details, even though they might be ecologically important. Of course, in some of the chapters that follow, some details would be provided when the need to do so emerges, particularly when talking about the use of habitats by some of the Landscape species such as the bonobos, elephants, and lions. For the time being, however, suffice it to be said that in other chapters, where habitat descriptions will be needed, references will be made to this chapter as the umbrella chapter for habitats of the landscape. Indeed, most of the chapters are written as scientific articles and come with a section on methods, data analysis, and a discussion section. In most of them, the habitat section will be either absent or very limited given the fact that this chapter would have already covered the habitats of the landscape.

Because the habitats respond to human activities more than any other force, the chapter also includes a section on human ecology in the landscape. It does particularly focus on the aspects of human activities that do impact natural resources such as agriculture, fishing, and the extraction of non-timber forest products, which make the stature of the overall economy of the landscape. Before providing some important details on these activities, the chapter also touches upon the land rights and the land uses in each administrative territory located within the landscape. Indeed, for reasons of simplicity, it was felt that that the best social unit to be used in the discussions about the landscape at this macro level is the administrative territory. In fact, the administration of the Democratic Republic of Congo is divided into villages, administrative sectors, administrative territories, and provinces. Administrative territories are the most appropriate level of authority because they are made of space with some sort of homogeneity in the land layout and land characteristics as well as communities that have been unified over long historical periods. Politically, they are not charged as the level of province while they do not exhibit total lack of political weight as do administrative sectors.

As it is a general characteristic of scientific literature to capture weather patterns and human ecology within the descriptions of habitats, this chapter also presents a broader view on weather, human distributions, and human activities across the landscape.

## 5.2 Methods to Assess Habitats

Segregating the general habitats of the landscape into forest and savannahs has been done using the Geographic Information Systems (GIS), using the software ArcGIS. That has generated percentages of the landscape covered by each type of habitat. The same technique was used to dissociate forests by floor types, i.e. swamps, and *terra firm* forests. Combining this with field surveys' data helped delineate permanently inundated forest versus seasonally inundated ones. Then for areas that were surveyed on ground, qualitative descriptions of the forest were systematically taken at 100 m intervals along straight line transect. Randomly positioned parallel transects were laid out, as indicated previously, following an angle $\alpha = 45°$, perpendicular to the gradient introduced by the Congo River. Data recorded on general forest types followed previous studies (Reinartz et al. 2006; White and Edwards 2000; Hall et al. 1998b). Likewise, the composition of the understory and estimation of the canopy cover also followed previous studies (Inogwabini 2015; Reinartz et al. 2006; White and Edwards 2000; Hall et al. 1998b). Forest types were defined as follows.

Mixed mature forest: forest with large trees, high and unbroken canopy, and no single species forming more than a small proportion of the whole community. Secondary forest: areas recently cleared for various kinds of human activity, with some signs of human activity still remaining, including standing crop plants, felled trees, cleared land, dwelling infrastructures such as camping sites, and so on. Old secondary forest: areas that have been used by humans a long time ago, but still retaining plant species indicative of their activities. Mono-dominant forest: mature forest wherein one species is noticeably dominant.

Understory categories were defined as follows. First, there was woody understory, the type of the understory composed essentially of tree saplings. Secondly, mMarantaceae understory was described as forest understory essentially composed of Marantaceae plants, either *Haumania liebrechtsiana* or *Megaphrynium macrostachyum* or both at the same location. There were instances when Marantaceae strata occupied large portions of forests; these Marantaceae strata are characterized by dense ground vegetation made up of free standing and liana-like species of herbaceous plants, mostly wild gingers and Marantaceae (White 2001; White and Abernethy 1997; Fay 1997). Ordinarily, Marantaceae strata form nearly impenetrable thickets about

3–4 m tall. However, liana-like branches can sometimes climb up trees and reach high canopy levels. Where they occur, Marantaceae species can play an important ecological role in pulling down some trees, and opening up gaps in the middle of the forest. Thirdly, a type of understory named liana understory was also present; it was the understory essentially composed of different species of lianas. Fourthly, there was mixed understory, a category of understory composed of plants of different life forms (saplings, lianas, *Marantaceae,* and so on). Fifthly and finally, open understory, the understory with very few plant species and good visibility, was also encountered.

Weather data has come from two different sources. The data for the northern part have come from a long time series of data collected from the Mabali Scientific Research Center while the set from the southern part is made of data collected since the start of the Landscape activities in that zone in 2006 through 2010. Human ecology data were collected throughout different socioeconomic studies (Inogwabini et al. 2013; Colom et al. 2006).

For convenience of the presentation, habitat description (as it is the case of all other elements) for the Lake Tumba Landscape has been always presented by the division between the northern part and the southern part. The northern part of the Landscape was defined not based on the Equator line, which indeed, crosses the landscape. This division was drawn from the distinction of the habitat types and the two landscape ape species (bonobos and chimpanzees); the northern part is mostly a very wet and flat region with forest being sempervirent while the southern part is the beginning of the Bateke Plateaux and is mostly a savannah area. It should also be noted that globally, the geology of the landscape can also characterized generally into two separate zones, the first one being the equatorial flat forest zone at the edges of Lake Tumba and Lake Maindombe, which are composed of extensive swamps and whose soil is predominantly alluvial (Colyn et al. 1991). For all these reasons, this chapter will be presenting habitat types that will be used to describe each specific study and will be using the same nomenclature but items on weather, human characteristics would be presented as specific items if the study requires so, particularly in case the data need to describe the microenvironments in which they were collected.

## 5.3 Forest Types in the Lake Tumba Landscape

Seven forest types were identified in the Lake Tumba Landscape (Table 5.1). Combining estimates from transects and the analysis of satellite images generated the following proportions of swampy forests: (a) 34.2% cover was permanently swampy forest and (b) 28.1% cover consisted of seasonally flooded forest. Permanently swampy forest and seasonally flooded forest are two hydromorphous forests; combined they covered 62.3% of the landscape. The true *terra firma* forests covered only 37.7% of the landscape. A striking finding was that the *terra firma* forests of the Lake Tumba Landscape are not dominated by large swathes of forest comprising a single

**Table 5.1** Proportional cover of different forest types in Lake Tumba landscape

| Forest types | N (Transect) | % cover | % *terra firma* | % Hydromorphous |
|---|---|---|---|---|
| Permanent swamp forest | 39 | 34.2 | | 34.2 |
| Mature seasonally flooded forest | 32 | 28.1 | – | 28.1 |
| Mixed mature closed forest | 20 | 17.5 | 17.5 | – |
| Mixed mature open forest | 10 | 8.8 | 8.8 | – |
| Old secondary forest | 6 | 5.3 | 5.3 | – |
| Young secondary forest | 5 | 4.4 | 4.4 | – |
| Fallow areas | 2 | 1.8 | 1.8 | – |
| Totals | 114 | 100 | 37.7 | 62.3 |

species. While *Uapaca guineensis* and *Gilbertiodendron dewevrei* do occur in mono-dominant stands in some portions of the Lake Tumba Landscape, these stands do not cover such large areas as in some other forests of the Congo basin, where *Gilbertiodendron dewevrei* occurs in extensive swathes. Most of the terra firm areas are in the southern part of the Lake Tumba Landscape, which is due to the fact that this part is the northernmost part of the Batake Plateaux, as indicated above. However, the *terra firma* zone is generally flat and its altitude never exceeds 340 m above the sea level, undulating across large swamps and surrounded by savannah islands. Apart from the swamps that are the major feature of the entire landscape, successions of savannahs (both flooded or on dry land) and forests are also a permanent characteristic of the region. The hydrology of this part of the Lake Tumba Landscape is characterized by an extensive network of rivers and creeks whose most domineering features are the two lakes, which are remnants of a huge interior lake that once occupied the entire basin prior to the breach of the basin's edge by the Congo River and the subsequent drainage of the interior.

## 5.4 Habitats in the Northern Part of the Lake Tumba Landscape

As described by Inogwabini (2013), some 60–65% of the landscape north of 1° 30′ are either seasonally inundated or permanent swamps. A vast zone of seasonally inundated and permanent swamp forest covers is located between the Congo River and Ngiri. By contrast, in the zone south of that latitude, comprising some 35–40% of the Lake Télé–Lake Tumba biome, are found mixed *terra firma* forest types, including mature and secondary forests along valleys and on small hills. These ombrophile (shade-tolerant) and semi-deciduous forests occur in areas between the major river systems here. Both types are composed of stands of leguminous trees, characterized by hardwood timber species such as *Staudtia stipitata, Polyalthia suavaeoleus,*

*Scorodophloeus zenkeri, Anonidium mannii,* and *Parinari glaberrimum,* which serve as important fruit and seed sources for great apes and other wildlife.

The weather (of the northern part of the landscape; see above) was recorded continuously at the Mabali scientific Reserve at Lake Tumba since 1970s. A serial analysis conducted on this data set indicated that there were changes in weather patterns near the Lake Tumba over the last thirty years (Inogwabini et al. 2006). According to that study, the overall, temperatures decreased from mean monthly fluctuations around 25 °C (1970) to 19.5 °C (mid 1990s) and had not retrieved their 1970 levels. Equally, the rainfall regime (intensities and numbers of rains) decreased following the regression equation $y = -20.508x + 1723.5$. However, the mean humidity was constant (72 and 85%), over longer periods, though with striking decreases in 1987, 1991, 1993, 1997, and 1999. As a consequence of decreased rainfalls, the Lake Tumba water depth generally declined. If these long term patterns of weather occur throughout the region, they may have an impact on fruiting phenology in the region. This will certainly impact the abundance and distribution of large mammals in the landscape in the future. These changes in weather patterns have been explained by, at least partially, the overall climatic change phenomenon and may bring severe effects on the biodiversity distribution in the region.

## 5.5 Habitats in the Southern Part of the Lake Tumba Landscape

The Southern Lake Tumba Landscape is *ca* 36,270 km$^2$; it is located between 01°:00′:00″ and 03°:00′:00″ South and 17°00′00 and East 18°:45′:00. Its south-western limit is the Bateke Plateau located at mild elevations, with heights within the range 280–380 m a.s.l. The soil description has not been conducted thoroughly throughout the region but it is inferred from study on the Plateau de Bateke on the other side of the Congo River, where conditions are similar, that the soil of the region is principally sandy and composed of the ferralic Arenosols juxtaposed with the podzols made mostly of continental sediments (Schwartz 1988). The humus in both geological strata of the study area is 1 m thin. Alluvial soils, sands, and podzols form a polymorphous soil in combination with gray sands of the Kalahari's type (SGECN 1999), remnants, and indicators of a large internal water body that swept the central Congo depression in the Tertiary (in the Pliocene: 7,000,000 years ago), which drained out to the Atlantic ocean after the emergence of the Congo River due to geological shift in the crust that once sustained the mountains in the south-western coastal region (Goodie 2005; Schwartz 1988). Of interest is the land layout in the landscape, particularly when its floor smoothly rises up toward the south: small hills are in a succession intercepted by valleys and sometime flat savannahs, which constitute valleys that are often flooded during rainy seasons.

The Malebo region is in the Lake Tumba–Lake Maindombe hinterland (Inogwabini et al. 2007a, b), straddling the provinces of Bandundu and Equateur in their administrative territories of Bolobo, Yumbi, and Lukolela. This area covers ~40,000 $km^2$ (Inogwabini et al. 2011). At its southern edge, the region is located on the Bateke Plateau and descends toward the Congo Central Basin, known as the Cuvette Centrale (Inogwabini et al. 2006, 2011).

Malebo, in the southern Lake Tumba Landscape, deserves a particular point in this section on the habitat because of its importance on bonobos, the landscape species par excellence. Malebo is located in an ecosystem that divides the northern swampy forests and the southern savannahs. Swampy forests in the southern part of the landscape cover essentially the territory of Lukolela and are essentially composed of mixed mature forest with open understory whose main emergent trees are *Uapaca guineensis*, *Uapaca heuloditii*, *Guibourtia demeusei,* etc. (Inogwabini et al. 2011; Inogwabini et al. 2006). The region is also characterized by episodes of flooding, during which water covers approximately 65% of the forest. In the complex of forest-savannah mosaic, forest galleries are composed of *terra firma* mixed mature forest, with species such as *Gilbertiodendron dewevrei* and *Entandrophragma* spp. The understory is comprised of 45–50% of the Marantaceae family and species such as *Humania liebrechtsiana* and *Megaphrynium macrostachyum*. Some of these areas were logged in the last 25–30 years to extract the wenge (*Millettia laurentii*), a high-priced black hardwood. The savannahs of the region are woody, dominated by *Hymenocardia acida* and *Annona senegalensis*, a result of long exposition to fire (Inogwabini et al. 2011; Inogwabini 2010).

At the southern part of the Landscape (see above), meteorological data on the rainfall were collected over a one-year period (February 2006–February 2007), in agreement with the description provided by Bultot and Griffith (1972). These data indicate two rain seasons over the year, one between Mach and May followed by a long dry season (June–August). The peak rain season comes between September and November, followed by a minor dry season starting from the end of December through February (Fig. 2.4). It rains even at dry seasons, though the rainfall decreases at that time of the year. Data on temperature indicate that there is a cooler period over the dry season but the mean annual temperature remains around 24 °C, a little less than in description provided by Vancutsem et al. (2006) and Bultot and Griffith (1972).

A major feature of the Lake Tumba Landscape is its open waters, principally made of the Congo River (the mid Congo River, which is the part of the Congo River between Kisangani and Kinshasa) and two sufficiently large lakes (Lake Tumba and Lake Maindombe). The mid Congo is sometimes called the Central Basin or the Cuvette Central (Banister 1986; Chapman 2001). Several tributaries flow into the rivers throughout the Cuvette Centrale on both of its sides. These major tributaries are Ubangi on the Northern–Western side and Lomami, Lulonga, Ruki-Tshuapa on the southern side. The Kasai–Lukenie–Kwango also flows into this part of the Congo but it is the southern limit of the Lake Tumba Landscape. This is a vast flat zone (altitude: 300-m–700-m a.s.l). In its course, through this area, the Congo crosses major swamps, including Ngiri swamp as well as those located at the junctions of the Congo and its tributaries, and expansive wetlands (greater than or equal to

10–15 km width on both sides) constituted of areas of seasonally flooded and permanently inundated areas (Inogwabini et al. 2012; Gauthier-Hion et al. 1999; Bailey 1986; Evrard 1968). Characteristics forest types have open understory, with the most important plant communities including *Raphia sese*, *Pandanus*, *Guibortia demeusi*, *Uapaca guineensis*, *Uapaca heudelotii,* etc. (Evrard 1968). Indeed, the first of these two lakes has given its name to the entire landscape. Both Tumba and Maindombe are shallow lakes (Corsi 1984a; Bailey 1986). Lake Tumba is 765 $km^2$ large while Lake Maindombe has a surface area of 2,300 $km^2$. The water in the Lake Tumba is dark, acidic (pH = 4–4.5) (Corsi 1984b) and chemical impoverished (Toham et al. 2006; Bailey 1986). Planktonic availability is sparse and the water contains a lot of vegetable debris (Bailey 1986). The region around the lake has been logged for timber and human population is rapidly growing. The fishing rates are very high and fuel markets both in Mbandaka and Kinshasa (Toham et al. 2006; Corsi 1984b). The habitat around the Lake Maindombe is a mosaic of dunes of bare sand, herbaceous savannas, and degraded rain forest, with large expanses of *Raphia* swamps. As its name indicates, the water is colored black (Corsi 1984b), which indicate a low level of mineral content. The town of Inongo is located on the central peninsula of the lake while Kutu is at the southernmost tip of the lake. Large towns are known to discharge their wastes into lakes, threatening thus the biological diversity of the lake water. High levels of fishing are also reported from the lake to feed both the local and the national market. There are four main rivers discharging their water into the Lake Tumba, which are Loko, Bituka, Lobambo, and Nganga. Two main rivers drain into Lake Maindombe: Lokoro and Lotoi (Corsi 1984a). The two lakes are connected to the Congo River via the Irebu channel and Fimi–Kasai–Lukenie–Kwa, respectively (Corsi 1984a).

## 5.6 Biological Diversity of the Landscape

Globally (northern part and southern part), the Landscape encompasses a large assemblage of species comprising the bonobos (south of the Congo River), chimpanzees (north of the Congo River), and several other primates including the Angolan pied colobus *Colobus angolensis*, Allen's swamp monkey *Allenopithecus nigriviridis*, black mangabey *Lophocebus aterrimus*, the Salonga red colobus *Piliocolobus tholonii*, the red-tailed monkey *Cercopithecus ascanius*, the De Brazza's monkey *Cercopithecus cephus*. There are also several other species of conservation concern such as forest elephants *Loxodonta africana cyclotis*, forest buffaloes *Syncerus cafer nanus*, leopards *Panthera pardus,* and lions *Panthera leo*. The fish diversity is rich; the Lake Tumba has 86 fish species, of which the most common species are *Auchenoglianis occidentalis*, *Clarotes laticeps*, *Gephyloglianis congicus*, *Claria buthopogon*, *Distichodus*, *Channa obscurus,* etc. (Corsi 1984b). Other freshwater species include crocodile species: *Crocodylus cataphractus* and *Crocodylus niloticus*; the first species having been reported only in Maindombe. Hippopotamuses (*Hippopotamus amphibius*) have been observed in Maindombe in the recent past but were reported extinct in Lake Tumba. The work initiated in 2005 in the landscape documented some species that were not previously described in the site, including the bushbuck *Tragelaphus scriptus*, golden cat *Felis (Profelis) aurata* and nominate Lesser Black-backed Gulls (*Larus fuscus fuscus*) (Kylin et al. 2010).

## 5.7 Plant Species

Plant specimens were collected from both the Botuali area near Lake Tumba and from the Malebo area by a team of qualified botanists (Brncic et al. 2007). Specimens were collected from different qualitative habitat types consisting of swamp forest, seasonally inundated forest, riparian areas, savannah, savannah-forest ecotones, *terra firma* forest, and village sites. Whenever possible, fertile specimens were collected to aid in identification. In total, approximately 900 specimens were collected, respectively, at Botuali (500 specimens) and at Malebo (400 specimens) (Brncic et al. 2007), including lianas. The aim of this research component is that the final product is a separate storage cabinet in the herbarium dedicated to future research in the zone, containing the fully labeled specimens and a list of species for each area. Currently, the botanical inventory has identified over 420 plant species that have been identified and 4 unknown (Brncic et al. 2007) belonging to 79 families. The twenty most diverse families of plant species include *Rubiaceae*, *Euphorbiaceae*, *Caesalpinioideae*, *Lecythidaceae*, *Papilionoideae*, *Pteridophyta*, *Commelinaceae*, *Poaceae*, *Olacaceae*, and *Melastomataceae*. It is noteworthy that Families that have been cited to be of paramount important as bonobo food such as *Annonaceae* (with species such as *Annonidium manii*, *Greenwayodendron* (*Polyalthia*) *suavolens*) and Marantaceae (with species such as *Haumania liebrechtsiana*, *Megaphrynium macrostachyum*) both as food and as nesting trees (Fruth and Hohmann 1994; Badrian and Malenky 1984) are among the most diverse in the region of the study site.

These vegetation patterns are known to go through cycles of vegetation changes historically; these cycles are related to the overall climatic changes history in the region. As it is now agreed, the world went through major climatic changes due to glaciations in the south and north hemispheres. This resulted in a severely arid period in the tropics resulting in the shrinking of the moist tropical forest cover in most tropical African forested areas leaving only pockets of forest refuges (Maley 1991; Preuss 1990). Most of these forest refuges were located on mountains regions but there is evidence the area along major Congo tributaries and the Congo River itself kept some forests as well (Maley 1991; Colyn et al. 1991). These changes may be not only the cause of the actual vegetation layouts in the region and the succession of forest-savannah (which may picture simply past forest dynamics) but also the causes of genetic diversity observed in some wildlife species. Indeed, as it will be seen below, the genetical variations among bonobo populations across the Congo have been often discussed in terms of the past forest variations (Kawamoto et al. 2013; Inogwabini 2010).

## 5.8 Human Ecology in the Lake Tumba Landscape

The Landscape straddles between two administrative provinces of Equateur (part of the larger former province of Equateur) and Maindombe (which is part of the former larger province of Bandundu). Ethnographically, the region is characterized by its cultural diversity. Bomongo is the first Administrative Territory of the Northern Lake Tumba Landscape and borders with the Congo Brazzaville along the Ubangi River; its populations are made of four major tribes (Dzamba, Lobala, Baloi, and Mabinza); the other territory of the northern Lake Tumba Landscape is Bolomba where the Lulonga River that constitutes the eastern limit of the landscape in the north is located. Bolomba is essentially populated by Ngombe and Mongo; Mongo of Bolomba are divided into 2 groups (Eleku and Baenga), and there are also pygmies (known as Balumbe) who speak either Lomongo or Lingombe depending on where they occur. The Administrative Territory of Makanza is sandwiched between Bolomba and Bomongo; with its 15 tribes (Mabale, Balobo, Libinza, Bamwe, Boloki, Iboko, Mbenga, Ndobo, Motembo, Mbonzi, Mpunza, Bozaba, Likoka, Ngombe, and Nkinga), it has more tribes than any of the other territories in the Landscape. As it should be expected from its geographic position, some of the tribes are those that are found either in Bolomba or Bomongo. Most importantly, a great portion of the populations in Makanza is made of migrants from other provinces and territories of the Democratic Republic of Congo. The reason for this is that Makanza is located along the Congo River, which has been used for millennia as a highway for movements of people and goods.

In the southern part, there are 13 groups, the six first groups being of the larger ensemble of the Mongo: (1) Basengele, (2) Bolia, (3) Bokote, (4) Ekonda, (5) Ntomba, and (6) Losakanyi. These groups cohabit with the minority ethnic group of pygmy named (7) Batswa. The Ntomba dominate the territory of Bikoro (in the province of Equateur), which is shared by Bokote, Losakanyi, and Batswa (Hulstaert 1993a, b). In the southernmost area of the landscape, the study site is occupied by (1) the Bateke, (2) the Batende, (3) the Baboma, (4) the Banunu, and (5) the Basengele. The human densities vary across the site, with the highest density being 23.9 inhabitants/km$^2$ in the region surrounding the lake Tumba (administrative territory of Bikoro) while decreasing to the mean of 6.2 inhabitants/km$^2$ as one moves south the territories Lukolela, Bolobo, and Kutu (PNUD/UNOPS 1998).

Results of the socioeconomic studies carried out by Colom et al. (2006) indicated high birth rates, with the average annual increase of the population 3.8% per year, which is above the national average of 3.4% per year (INS 1984, 2015). The same socio-economic study by Colom et al. (2006) indicates that agriculture, fishing, and collection of non-timber forest products constitute major occupations, and are economic activities that generate most of the income for local communities. Multicultures fields of Cassava, maize, and banana in combination with sweet potato and sugar cane constitute the base of the agriculture but peanuts and rice are also cultivated. Rice culture has been only recently introduced. The culture of Manioc is done through shifting agricultural practice through which cultivated land is left to fallow for one or two years before being reinvested again.

According to the most recent statistics from the National Institute of Statistics (INS 2015), about 90% of households in Bomongo, cultivate 0.45 ha and produces around 3–3.5 tons of maniocs on average annually. Bomongo is known for its production of palm oil, which is its principal commercial product; recent statistics indicate that different communities hold about 2587 ha of plam tree plantations and some of these hectares are made of unplanted naturally growing plam trees, which produce 2069 tons of oil of which 98% is sold to the market in Mbandaka. In addition to palm oil, communities of Bomongo have been also cultivating coacoa but this does not influence so much in their economy because of the evacuation circuit is so poor. The situation described for Bomongo is similar to the one in the administrative territory of Bolomba and, in some extents, to that of Bikoro and Lukolela (included in the present context in the southern part of the landscape) with the exception that the proportions of the production of plam oil differ significantly and there are more bananas planted in Bolomba than it is the case in Bomongo. The differences between Bomongo and Makanza and the rest of the administrative territories (Bolomba, Bikoro and Lukolela) are also in the nature of agricultural trade products: for example, in these latter territories, there is a history of cultivating coffee (*Coffea robusta* sometimes known as *Coffea canephora*) and rubber (*Hevea brasiliensis*). It should be noted that none of these commercial plant species, with the exception of palm oil tree, has originated from Congo or even Africa. Most of them were introduced in the relatively recent history of the people in the Congo Basin to support the commercial balance of the colony. Indeed, the northern part of the Lake Tumba is the area where Congolese were forced to collect rubber saps as part of their contribution to the overall economy of the Royal Crown in Belgium and latter for the colony. The story for this exploitation has been brilliantly described by Adam Hochschild in his famous book that appeared in 1999 entitled King Leopold's Ghost: A Story of Greed, Terror and Heroism in Colonial Africa. Despite their nolvety in this context that is complex both in human ecology as well as in the general ecology, these plants have acquired the notoriety that makes them part of what is known to symbolize either hardwork or even richness and higher social status. Of course, the fact that a higher social status is conferred to owners of large plantations of palms, cocoas, rubbers, and coffees should not be that surprising because higher status characterized the former colonial masters. Indeed, some of the largest plantations of cocoas, rubbers, and coffees were acquired by politicians after independence; they replaced the colonial masters in both the true and symbolic sense.

While people combined many productive activities to cope with different economic fluctuation, fishing remained the principal activity in some villages, as it was the case in the nineteenth century (Harms 1989). Fishing constitutes the most important agricultural activity in territories with massive water bodies such as Bomongo, Lukolela, and Makanza. In most of the territories of the landscape, the dry seasons offer an occasion whereby more than 50% of villagers migrate to river and forest camps which have to be located near their fishing activities. Fishing uses gears that are both traditional and modern; the current situation is, however, that of the predominance of modern fishing gears, notably fishing nets and hooks. These are used in diverse manners and have contributed to the observed depletion of fish stocks over the last decades (Inogwabini 2013). In territories such as Makanza, close to 40% of the total population is said to live exclusively on fishing while in other territories such as Bomongo, 80% of people classify fishing as one the most important activities in terms of lucrative outcomes for their own livelihoods. Hunting by the local communities is not their most important activity; it is basically done for subsistence. However, though local communities clearly indicated that their forests were under high hunting pressure from people they felt were non-resident in their territories. Permanent employment occupies about 15% of the sampled population and is limited to activities such as teaching, territory administration, police, and provide very little income. Some important portions (25–40%) of populations in these territories also reported having permanent employments with the administration and public services such as education (teachers), health (nurses), and housing (masons, carpenters). There are also people employed by the private sector in the logging industry, the transport sector, etc. However, in many cases, and regardless of who the employer is, in nearly all the locations, the petty commerce of all sorts of products is widespread and people clearly indicated that they do not rely on their formal employments to make ends of their calendars meet given the low levels of the salaries they receive from the state. Fishing, hunting, and petty commerce constitute the back up for the sustainable livelihoods of all communities in the landscape.

Apart from the above use of natural resources (fishes, bushmeat, plantations), the petty commerce is also made of non-timber forest products. Indeed, the communities in the Lake Tumba Landscape, as it is the case for many communities across the Democratic Republic of Congo, use a wide range of natural resources for direct subsistence, including mushrooms, caterpillars, stems, leaves, and young shoots of *Megaphrynium macrostachii*, *Haumania libriechtsiana* (Marantaceae), fruits of *Dracryodes edulis*, *Cola edulis*, *Canarium schweinfurthi*, *Annonidium manii*, and roots different *Aframomum* spp species (zinger), barks *Scorodophloeus zenkeri*, copal latex from *Guibortia demeusi*, and fire woods. The list above not being exhaustive, it should be noted that leaves of wild palm trees (*Raphia taedigera)* are used to construct thatches that help cover the houses, which are built of trees and lianas from Ancistrophiluim africanum. Ancistrophiluim africanum and other species of lianas are also widely used in constructing baskets, chairs, beds, etc. This multiple usages of Ancystrophilium are a testimony to the fact that most of the non-timber forest products are used in different ways and can serve purposes so diversified and multiple that there is no way one could come up with a single usage for each of the product. That said, the bottom line is that non-timber forest products are more used than wildlife, fishes, and timber products. They are, in a sense, the rockbed of the daily economy of each village in the landscape.

The land tenure is that of the general property laws passed in 1970, as it was modified in 1980 (Liesz 1998). The provisions of that law on land tenure were recently confirmed by the new forestry code of 2002. In general, this law declares that all land and natural resources belong to the state (Liesz 1998) implying that officially the government does not recognize any other mechanism of accessing the land by local communities. This provision, which is a colonial relic (Inogwabini 2015), remains into force even after the forestry law has been revised in 2002 and even though this new law has introduced the provisions on the land tenancy by local communities. Indeed, despite the fact that communities can currently obtain the land rights over forests in their communities after following a process that seems very cumbersome and would be difficult for most communities, the realities are that the state interests have been kept safe and the rights communities would acquire are in substance the rights of tenancy; they hold until the state finds any interest that is judged to be more important for the public use. Of course, in practice, however, 97% of the land function under traditional land tenure systems, which leads to conflicts and encourages land tenure insecurity for local communities.

The northern part of the landscape is poorly occupied by industrial activities such as logging and the use of land for pasture. This is due to the presence of swamps, which render any mechanized exploitation a very hazardous exercise. Nevertheless, in the limited *terra firma* land that is available, some logging concessions have been allocated by the government but their exploitation is still inoperational in most cases. This does not mean, however, that there is logging in that portion of the landscape. On the contrary, artisanal logging is very extensive and eats very significant expanses of forests. Artisanal logging is mostly illegal and it is difficult for the national laws on logging to be enforced. In the southern part of the Lake Tumba Landscape, the land-use patterns are dominated by logging concessions within the landscape, which have either extraction or prospecting permits. The Logging concessions map covers about 55% of the Lake Tumba Landscape south of the Lake Tumba. Two factors explain the concentration of concessions in this region of the DRC. The first important factor is the presence of highly commercial species such as *Enandrophragma angolense* (Tiama), *Enandrophragma utile* (Sipo), *Piptadeniastrum africanum* (Dabeme), *Strombosia tetranda* (Afina), and specially the *Meletia laurenti* (Wenge). The Wenge constitutes about 75% of the wood extracted in the zone between Lake Tumba and Lake Maindombe. The second factor is the relatively easy access to the zone from Kinshasa either by boat or vehicles in the dry seasons. The forest has been logged in the last 20 years principally to extract timber of *Millettia laurentii* and *Entandrophragma* sp.

At the south-western area of the study site, i.e. in the Malebo region where the study of the bonobos has been conducted, the land-use patterns are characterized by a 15,000 ha ranch subjugated to cattle raising activities since the early 1950s, which usage continued uninterruptedly until now, when the ranch supports an estimate of more than 11,000 heads of cows and bulls. Apart from that large-scale cattle raising enterprise, local communities also raise small numbers of cows, goats, etc., on savannah patches. Local communities also cultivate manioc, maize, peanuts, and the recently introduced rice (SGECN 1999).

An important human ecology feature in the very area of the Lake Tumba Landscape is the use of fire as a management tool in cattle raising concession. Indeed, as indicated above, the *terra firma* habitats in the savannahs are woody, with characteristic trees being Hymenocardia acida, Crossoperyx febrifuga, and Annona senegalensis. The flooded savannahs are characterized by families of Cyperaceae, Gramineae, and Hydrocharitaceaa. Dry savannahs represent 55.7% of the savannah habitats in the southern part of the landscape. These savannahs harbor an interesting ecotone displaying higher species diversity. To repeat what is immediately above, savannahs in this region have been exploited for cattle rising since late 1950s and fire has been used extensively as a management tool. This is far from being unique in Central Africa; fire has been used in similar habitats in the region to maintain landscape mosaics intact in Lopé (Republic of Gabon) and Odzala-Kokoua (Republic of Congo). Two things, however, distinguish these two sites with the case of the Southern Lake Tumba Landscape. First of all, in both Lopé and Odzala-Kokoua, human population densities are very sparse and humans in these areas do not use fire to cattle husbandry. Secondly, in comparison to Lopé and Odzala-Kokoua, there is some scientific knowledge to back up the exercise of using fire as a management tool. There was nothing comparable in terms of knowledge in the Lake Tumba Landscape.

So, clearly, saying that the use of fire as a management tool is far from being unique in Central Africa does not do justice to the specificity of the Lake Tumba Landscape where this practice is expanding and taking high tolls on other ecosystems. That is why it was agreed to include the data on the fire in the data gathering exercise to see the impact of the long use of fire and its expansion with expanding and increasing human populations would have on the diverse terra firm ecosystems of the Landscape. Hence, data were gathered from different sources: The first set of data used in this study is satellite data set acquired from the Department of Geography, University of Maryland (UMD) and Observatoire Satellitale des Forêts d'Afrique Centrale (OSFAC). This raw data set provides number of fire per date and their locations on the ground. The second data set came from fire studbooks kept by ORGAMAN who owns 150,000 ha of the forest-savannahs mosaics in the southern part of the landscape and has been using fire since its arrival in the region back early 50s. Data recorded by ORGAMAN comprised date, time of the day when the fire was lit and location of fire. Beyond these data, the WWF teams collected data on fire incidents within and outside of the ORGAMAN concessions. Data collected by WWF teams also included date and location of fire. We added qualitative data the intensity of fire by incorporating the measure of size for burnt areas in a single fire incident. Categories were (1) very large, (2) large, and (3) small. Very large area consisted of area greater than 10 ha, large equaled area 5–10 ha, and small consisted of area less extended than 5 ha. We also distinguished fire in their causes: (1) hunting, (2) slash–burn agriculture, (3) grass regeneration for cattle, and (4) no apparent reason other than setting fire for pleasure of burning forest. Data from satellite imageries were analyzed using regression equations to track the fire evolution over the time span covered by the data set. Monthly frequencies of fire were produced and examined to see if there was any correlation between fire and seasons. The data set from ORGAMAN was also analyzed to see whether patterns produced by the satellite images corresponded to ground data. Estimated sizes were grouped by categories and simple frequencies were drawn from that classification to produce an index of fire effect on the savannahs.

Data were summed up and divided by number of recorded incidents by the total of recorded incidents to identify the most important cause of fire.

The analysis of all three sets of data indicated that fire is most intensive in the region from June through early September, which corresponds to the long dry season. Using only the data from satellite images, the numbers of the fire has significantly ($R^2 = 0.868$) increased between 2000 and 2005, following the regression equation $y = 1336.7x + 76.7$. Field data collected by WWF teams indicated that 75% ($N = 112$) of fire incidents occurred outside the ORGAMAN concessions. However, the 25% of remaining fire incidents that occurred inside the ORGAMAN concession were the most severe, with 100% being of sizes less than 10 ha, with 68% burning zones greater than 50 ha at one time.

Fire incidents that occurred outside of the ORGAMAN concessions were in their majority for slash-and-burn agriculture (35.7%, $N = 84$), followed by the regeneration of grasses for cattle raising and no obvious reasons (23.8% each respectively) and hunting (16.6%). The analysis of the satellite data indicated also that the geographical spread of the fire incidents expanded in the years covered by the data, moving northward, which means an incursion into the equatorial forest. The geographic expansion of fire incidents into the forests corroborates the results that most fire incidents are produced by slash-and-burn agriculture, which is practiced in forests rather than savannahs. Local people believe that forests are more fertile than savannahs and therefore prefer clearing large areas of forests to grow maniocs and other agricultural products. The increase in fire incidents may also be due to increased cattle raising activities by local communities and other small farmers from other towns in the region, even though the contribution to the total fire incident is greater than 25%. This is likely to be true because it is only in recent years that local communities in the region have adopted raising cattle as part of their culture. The dynamics of different economic activities in the region seem that cattle rising by small farmers will become an important pattern of the region in the near future. As such, there is a need to study fire patterns and fire regime as part of the ongoing ecological studies because this may affect populations of wildlife as shown in other sites (Barclay et al. 2006), and particularly those of bonobos inhabiting in forest-savannah mosaics in the area, and may render fragile forest galleries in their regeneration processes as it is case elsewhere (Cochrane and Laurence 2002). The need to help craft a sound fire management system along with a gallery forest conservation program that takes into account different elements will ultimately include water quality and watershed management, populations' needs in agricultural lands, logging activities, and cattle raising necessities.

Fire as a hunting technique is widespread across savannah people and cultures in Africa and fire has been even documented to be a fishing technique in certain types of habitats during the dry seasons (e.g. Inogwabini 2005), therefore, should not be that surprising in this region. A more complex finding of this study, however, was that 23.8% of fire incidents had no obvious reasons and that many interviewed people simply acknowledged the fact that most often fire was lit for no reason other than the wish for people to clear the grass along their routes in different habitat types. In this sense, fire was used just to see farther away, which may be translated as fire as security device.

Four consequences can be drawn from these results and should provide insights for the management of fire. The first consequence of the fire as a management tool in this region is that agriculture, that uses the fire most often than any other activity, should receive more attention in order to stop the geographic spread of the fire in the region. The best option would be to intensify the agriculture in some portions of the landscape. Secondly, within the ORGAMAN concessions, because their fire incidents are often the most expansive, they should be concentrated in areas where there is higher potential to rejuvenate the grasses and maintain higher numbers of cattle. This should be accompanied by a limitation over time. The plan should also aim at alternating burning sessions and longer periods between burning rather than maintaining the actual frequencies of burning. Thirdly, there is a need to monitor the long term effects of fire on soil dynamics, including erosions, river siltation in order to have a better map of areas to be burnt and those that should be kept out of burning cycles. Finally, Cultural burning will remain a big issue because there is no apparent reason for its usage. There is a need to sensitize local communities about the effect of burning savannah on their own life (pollution, etc.).

## 5.9 Habitats of the Lake Tumba Placed in the Perspectives of the Congo Basin

The proportional cover of hydromorphous forest (63.3–65%) for the cover of swamp reported from the Lake Tumba Landscape (both from the analysis of satellite images and transects) is very high and has not been reported in other areas where bonobos have been studied previously. For comparison reasons, the Salonga National Park, which was created to safeguard the species, has greater than or equal to 75% of its habitat as *terra firma* forest, although Gautier-Hion et al. (1999) and Inogwabini (2005, 2006a, b) reported that close to 50% of the northern sector of the park becomes inundated during major flooding events. However, such events are somewhat sporadic in the Salonga National Park, and should not be considered as part of the normal pattern, as is the case in Lake Tumba Landscape. A land cover assessment carried out by Remote Sensing Solution GmbH, mostly in the southern part of the Salonga National Park, showed that all wetlands in the Salonga National Park region comprised only about 6% cover, excluding open rivers (RSS 2003). Similarly, the Lomako study site supports mostly *terra firma* habitats of mixed mature forest, and only 2.3% of its landscape comprises forest strata that are subject to inundation (CARPE/USAID 2006; Malenky and Stiles 1991). Wamba has a similar landform to Lomako, while the Lukuru area lies at the edges of the dry southern savannahs. Therefore, the Lake Tumba Landscape is the only site where bonobos are known to occur on islands of *terra firma* encircled by swamps throughout most of the year.

Comparisons across Central Africa also give the same specificity of the Lake Tumba Landscape as being the largest wetland area in Central Africa, with the exception of the Lake Tele Landscape, located in the Republic of Congo. Indeed, Lake Tumba and Lake Tele constitute the same swampy zone which extends from the Congo–Ngiri–Ubangu swamps on the side of the Democratic Republic of Congo and continues to the swamps of Ubangui–Likouala aux Herbes on the side of the Republic of Congo. Put together, this complex area represents the largest wetland in the continent. It is because of this that the Landscape in its nomenclature imposed by the political processes is often called as a unique unit. Despite their resemblances in many of their ecology, there are also major differences that make them very distinct entities. Obviously, the sizes of the lakes Tumba (765 $km^2$) and Maindombe (2300 $km^2$ at minimum), which are the essential freshwater feature of the Lake Tumba Landscape, are by no means comparable with the Lake Tele itself (32 $km^2$, at most). Of course, such open environments would induce a different ecology as compared to a compacted body of water such as the Lake Tele. However, in my view, the most important of these aspects is the human occupation, which is rather sparse in the Republic of Congo while in some areas of the Lake Tumba Landscape human densities are sometimes 4–5 times more than would be on the side of the Lake Tele. As a corollary of this situation, the impact of human activities is quite different, levels of offtakes of fishes, for example, are not comparable. Another major difference induced by differences in human densities is the degradation of lands, forests, and other habitat types. A simple comparison of satellite images of the two sides of the Ubangi River clearly shows a more mature biota and less open lands on the side of the Lake Tele. This single fact is, ecologically speaking, a significant dividing factor in that land and forest degradations induce different patterns of habits in species; it acts as a driving force for differences even in the ecology of same species that happen on both sides. The other major difference lays in the fact that the Lake Tele side does have western chimpanzees and lowland gorillas; these great apes species have been reported to occur in large numbers (Blake et al. 1995) On the Lake Tumba Landscape side, great apes (western chimpanzees and bonobos) do occur; chimpanzees occur in the forest blocs between the Congo River and the Ngiri River and the Ngiri and Ubangui, both located north of the Congo River. But these chimpanzee populations happen in very limited numbers (Inogwabini et al. 2012); some of these populations are isolated and are very small populations. Finally, the management regimes of the two sides differ significantly; the Likouala aux Herbes and the Lake Tele have been the focus of conservation efforts and significant international conservation investment years before the idea of creating the landscapes was under way; as indicated above and as it will be the case in chapters that follow, the Lake Tumba Landscape has become a part of the overall regional conservation focus only after the Yaoundé process.

## References

Badrian NL, Malenky RK (1984) Feeding Ecology of **Pan paniscus** in the Lomako forest, Zaire. In: Susman RL (ed) The Pygmy chimpanzee: evolutionary biology and behavior. Plenum Press, pp 325–346

Bailey RG (1986) The Zaïre River system. In: Davies BR, Walker KF (eds) The ecology of river systems. Junk Publishers, pp 201–214

Banister KE (1986) Fish of the Zaïre system. In: Davies BR, Walker KF (eds) The ecology of river systems. Junk Publishers, pp 215–224

Barclay HJ, Li C, Hawkes B, Benson L (2006) Effects of fire size and frequency and habitat heterogeneity on forest age distribution. Ecol Model 197(1–2):207–220

Blake S, Rogers E, Fay JM, Ngangoué M, Ebéké G (1995) Swamp gorillas in northern Congo. Afr J Ecol 33(3):285–290

Brncic T, Menga P, Lejoly J (2007) Preliminary report of botanical activities carried out in the Lac Tumba Landscape. Typescript report submitted to World Wide Fund for Nature

Bultot F, Griffith JF (1972) The equatorial wet zone. In: Griffiths JF (ed) Climates of Africa: world survey of climatology, vol 10. Elsevier, pp 259–312

Chapman LJ (2001) Fishes of African rain forests: diverse adaptations to environmental challenge. In: Weber W, White LJT, Vedder A, Naughton-Treves L (eds) African rain forest—ecology and conservation: an interdisciplinary perspective. Yale University Press, pp 263–290

Cochrane MA, Laurence WF (2002) Fire as a large-scale edge effect in Amazonian forests. J Trop Ecol 18:311–325

Colom A, Bakanza A, Mundeka J, Hamza T, Ntumbandzondo B (2006) The socio-economic dimensions of the management of biological resources, in the Lac Télé—Lac Tumba Landscape, DRC segment: a segment-wide baseline Socio-Economic study's report. Submitted to the World Wide Fund for Nature

Colyn MM, Gautier-Hion A, Verheyen W (1991) A re-appraisal of the palaeoenvironmental history in Central Africa: evidence for a major fluvial refuge in the Zaire Basin. J Biogeogr 18:403–407

Corsi F (1984a) Développement des pêches dans le fleuve Zaïre. Rapport préparé pour pour le projet d'Assistance à l'Office National des Pêches et Amélioration de la Pêche Artisanale. Food Aid Organization, Rome

Corsi F (1984b) Zaïre: Développement et aménagement des pêches du lac Mai-Ndombe et ses affluents. Technical report submitted to the Food Aid Organization, Rome

Evrard C (1968) Recherches écologiques sur le peuplement forestier des sols hydromorphes de la Cuvette Centrale congolaise. Série scientifique 110. Office National de la recherche Scientifique et du Développement du Ministère Belge de l'Education Nationale et de la Culture

Fay JM (1997) The ecology, social organization, population, habitat and history of the western lowland gorilla (**Gorilla gorilla gorilla** Savage & Wyman, 1947). PhD thesis. Washington University. United States of America

Fruth B, Hohmann G (1993) Ecological and behavioral aspects of nest building in wild bonobos (**Pan paniscus**). Ethology 94:113–126

Fruth B, Hohmann G (1994) Comparative analysis of nest building behavior in bonobos and chimpanzees. In: Wrangham RW, McGrew WC, De Wall FBM, Heltne PG (eds) Chimpanzee cultures. Harvard University Press, pp 109–128

Gautier-Hion A, Colyn M, Gauthier JP (1999) Histoire naturelle des primates d'Afrique Centrale. Multipress, Ecosystèmes d'Afrique Centrale. Union Douanière et Economique d'Afrique Centrale

Goodie AS (2005) The drainage of Africa since the Cretaceous. Geomorphology 67:437–456

Hall JS, Saltonstall K, Inogwabini BI, Omari I (1998a) Distribution, abundance and conservation status of Grauer's gorilla. Oryx 32:122–130

Hall JS, White LJT, Inogwabin BI, Omar I, Morland HS, Williamson EA, Saltonstall K, Walsh P, Sikubwabo C, Bonny D, Kiswele KP, Vedder A, Freeman K (1998b) A survey of gorillas (**Gorilla gorilla graueri**) and chimpanzees (**Pan troglodytes schweinfurthi**) in the Kahuzi-Biega National Park lowland sector and adjacent forest in eastern Democratic Republic of Congo. Int J Primatol 19:207–235

Harms R (1989) Fishing and systems of production: the precolonial Nunu of the middle Zaïre. Cah Sci Hum 25:147–158

Hulstaert G (1993a) Etudes dialectologiques Mongo. Etudes Aequat 12: 15–4006

Hulstaert G. (1993b) Liste et carte des dialectes Mongo. Ann Aequat 14: 401–406

Idani G, Kuroda S, Kano T, Asato R (1994) Flora and vegetation of Wamba forest, central Zaire with reference to bonobo (**Pan paniscus**) food. Tropics 3:309–332

Inogwabini BI (2005) Common fishes of the Salonga National Park, Democratic Republic of Congo: preliminary survey and conservation issues. Oryx 39(1):78–81

Inogwabini BI (2006a) A preliminary checklist of mammals and plants: conservation status of some species in the Salonga National Park, Democratic Republic of Congo. Endanger Species Update 23(3): 104–117

Inogwabini BI (2006b) Fire as management tool in cattle raising concession: A preliminary assessment to help develop a system that sustains biological diversity in the southern in the Lac Télé–Lac Tumba Swampy forest, DRC Segment. Field report submitted to WWF and CARPE-USAID

Inogwabini BI (2010) Conserving great apes living outside protected areas: the distribution of bonobos in the Lake Tumba landscape, Democratic Republic of Congo. PhD Thesis, University of Kent at Canterbury, United Kingdom

Inogwabini BI (2013) Bushmeat, over-fishing and covariates explaining fish abundance declines in the Central Congo Basin. Environ Biol Fishes 97(7):787–796

Inogwabini BI (2015) The Bonobo Pan paniscus (Mammalia: Primates: Hominidae) nesting patterns and forest canopy layers in the Lake Tumba forests and Salonga National Park, Democratic Republic of Congo. J Threat 7(13):7853–7861

Inogwabini BI, Leader-Williams N (2013) Conservation paradigms seen through the lenses of bonobos in the Democratic Republic of Congo. In: Navjot S, Raven P (eds) Essays in conservation biology–perspectives from practitioners in tropical environments. Blackwell-Wiley

Inogwabini BI, Sandokan MB, Ndunda M (2006) A dramatic decline in rainfall regime in the Congo Basin: evidence from a thirty-four year data set from the Mabali Scientific Research Centre, Democratic Republic of Congo. Int J Meteorol 31(312):278–285

Inogwabini BI, Matungila B, Mbende L, Abokome M, Tshimanga WT (2007a) Great apes in the Lake Tumba landscape, Democratic Republic of Congo: newly described populations. Oryx 41(4): 532–538

Inogwabini BI, Bewa M, Longwango M, Abokome M, Vuvu M (2007b) The Bonobos of the Lake Tumba–Lake Maindombe Hinterland: threats and opportunities for population conservation. In: Furuichi T, Thompson J (eds) The bonobos behavior, ecology, and conservation. Springer, pp 273–290

Inogwabini BI, Mbende L, Abokome M (2011) The relic population of forest elephants near Lac Tumba, Democratic Republic of Congo: Abundance, dung decay, food items and movements. Pachyderm 49:40–47

Inogwabini BI, Abokome M, Kamenge T, Mbende L, Mboka L (2012) Preliminary bonobo and chimpanzee nesting by habitat type in the northern Lac Tumba Landscape, Democratic Republic of Congo. Afr J Ecol https://doi.org/10.1111/j.1365-2028.2012.01323.x

Institut National de la Statistique (INS) (2015) Annuaire Statistique 2014. Ministère du Plan et Révolution de la Modernité, République Démocratique du Congo

Kano T (1983) An ecological study of the pygmy chimpanzees (**Pan paniscus**) of Yalosidi, Republic of Zaire. Int J Primatol 1:1–31

Kano T (1984) Distribution of pygmy chimpanzees (**Pan paniscus**) in the central Zaire basin. Folia Primatol 43:36–52

Kawamoto Y, Takemoto H, Higuchi S, Sakamaki T, Hart JA, Hart TB, Tokuyama N, Reinartz GE, Guislain P, Dupain J, Cobden AK, Mulavwa MN, Yangozene K, Darroze S, Devos C, Furuichi T (2013) Genetic structure of Wild Bonobo populations: diversity of Mitochondrial DNA and geographical distribution. PLoS ONE 8(3):e59660. https://doi.org/10.1371/journal.pone.0059660

Kylin H, Louette M, Herroelen P, Bouwman H (2010) Nominate Lesser Black-backed Gulls (Larus fuscus fuscus) winter in the Congo basin. Ornis Fennica 87:106–113

Leisz S (1998) Zaire Country profile. In: Bruce J (ed) Country profiles of land tenure: Africa 1996. Land Tenure Centre Research Paper 130: 131–136

Magliocca F, Querouil S, Gautier-Hion A (1999) Population structure and group composition of western lowland gorillas in north-western Republic of Congo. Am J Primatol 48:1–14

Maley J (1991) The African rain forest vegetation and palaeoenvironments during late quaternary. Clim Change 19:79–98

Preuss J (1990) L'évolution des paysages du bassin intérieur du Zaïre pendant les quarante derniers millénaires. In: Lafranchi R, Schwartz D (eds) Paysages Quartenaires de l'Afrique Centrale Atlantique. Office de la recherche scientifique et technique outre-mer, pp 260–270

Programme des Nations Unies pour le Développement (PNUD) and United Nations Office of Project Services (UNOPS) (1998) Programme de national de relance du secteur agricole rural (PNSAR) 1997–2001 : Monographie de la Province de l'Equateur. Ministères de l'Agriculture et de l'Elevage, du Plan, de l'Education Nationale et de l'Environnement, Conservation de la Nature, Forêts et Pêche. République Démocratique du Congo

Reinartz G, Inogwabini BI, Mafuta N, Lisalama WW (2006) Effects of forest type and human presence on bonobo (Pan paniscus) density in the Salonga National Park. Int J Primatol 27(2):603–634

Sabater PJ, Vea JJ (1990) Nest building and population estimates of the bonobo from the Lofeke-Lilungu-Ikomaloki region of Zaire. Primate Conserv 11:43–48

Schwartz D (1988) Some podzols on the Bateke sands and their origins, people's Republic of Congo. Geoderma 43:229–247

Secrétariat Général à l'Environnement, Conservation de la Nature, Pêche et Forêts (SGECN PF), 1999 Plans provinciaux de la biodiversité—appendice au plan d'action national. Ministère de l'Environnement, Conservation de la Nature, Pêche et Forêts, Kinshasa Gombe, République Démocratique du Congo

Toham AK, D'Amico J, Olson D, Blom A, Trowbridge L, Burgess N, Thieme M, Abell R, Carroll RW, Gartlan S, Langrand O, Mussavu RM, O'Hara D, Strand H (2006) A vision for biodiversity conservation in Central Africa: biological priorities for conservation in the Guinean: Congolian forest and freshwater region. World Wide Fund

Vancutsem C, Pekel JF, Kibambe LJP, Blaes X, De Wasseige C, Defourny P (2006) Republique Democratique du Congo: occupation du sol. Carte geographique. Presses Universitaires de Louvain

White LJT (2001) The African Rain forest: climate and vegetation. In: Weber W, White LJT, Vedder A, Naughton-Treves L (eds) African Rain forest: ecology and conservation. Yale University Press, pp 3–29

White LJT, Abernethy K (1997) A guide to the vegetation of the Lopé Reserve. Multipress, Gabon

White LJT, Edwards A (2000) Conservation research in the Central African rain forests: a handbook. Wildlife Conservation Society

# Chapter 6
# Bonobos in the Lake Tumba: Describing the Landscape Species

**Abstract** The chapter describes an 8-month survey of large mammals in the Lake Tumba landscape with the focus on the Bonobo (*Pan paniscus*). After an epic walk of about 86 km in straight lines (transects) and 324 km of forest reconnaissance, research teams documented the distribution and estimated the abundance of bonobos. Five separate bonobo groups were located in the southern part of the landscape. Mean bonobo densities ranged from 0.27 individuals km$^{-2}$ to 2.2 individuals km$^{-2,}$ respectively, in the vicinity of Lake Tumba and in the Malebo–Nguomi area.

**Keywords** Line transect · Forest reconnaissance · Bonobo · Pan paniscus · Population density · Forest-savannah mosaic · Terra firma habitat

## 6.1 Introduction

In the years 1990, scientists were convinced that the world population of the bonobo (*Pan paniscus*) was down to less than 5000 individuals, which meant that the species was reaching the limits of risk of extinction (Thompson-Handler et al. 1995). The range of the animal was restricted to the dark heart of the Democratic Republic of Congo. Destruction of the forest, its main habitat, the increasing trade in bushmeat, and epidemic diseases were thought to put pressure on the species.

From 1996 onward, the DRC was hit by successive waves of a violent civil war that hugely affected conservation efforts in the country (Draulans and Van Krunkelsven 2002), although it remained unclear to what extent bonobo populations were affected, as political instability does not necessarily translate directly into an elevated threats to animal populations (Draulans and Van Krunkelsven 2002). However, toward the end of the generalized political nightmares, estimates of the total number of bonobos surviving were increased to 20.000 animals, using suitable habitat models in combination with the mean bonobo density from different sites (Reinartz and Inogwabini 2001). The difference between the 1990 estimate and that of 2001 may have been a reflection of differences in methodologies and what each study has thought to be the most suitable habitat for the species. Though the species was described in 1929,

---

*Written with contribution from Dirk Draulans*

 
B.-I. Inogwabini, *Reconciling Human Needs and Conserving Biodiversity: Large Landscapes as a New Conservation Paradigm*, Environmental History 12,
https://doi.org/10.1007/978-3-030-38728-0_6

paucity of the field-based knowledge on the species abundance, its distribution, and its ecology remained the major characteristic when trying to craft a sound conservation strategy to prevent the felt decline in the species population (Thompson-Handler et al. 1995).

The news of a newly discovered bonobo population in the area of the Lake Tumba Landscape came as a total surprise for conservationists and great ape researchers, especially as the population was situated way out of what most researchers regarded as the classic bonobo living area (Inogwabini et al. 2007a). Furthermore, the area in which the animals were discovered was never in any formal way protected. Field observations and measurements (see Chap. 9 in this volume) suggested that, on average, up to 3550 bonobos live in the area, which implies that the bonobos of the Lake Tumba landscape could represent up to 18% of the estimated world population (Inogwabini 2010) It is probably the largest single population of bonobos remaining in the wild. Here, we explore some of the reasons why the bonobo population in the Lake Tumba Landscape area remained undetected for so long, and what this can teach us from the point of view of conservation efforts. The chapter will mainly be descriptive and does not involve calculations and statistics in the assessment of its goals. It, broadly, describes the new habitat where the species has been documented present. From this broad habitat description, it proceeds to presenting a summary of the presence of the species in the Malebo region and then discusses the findings in relationship to habitat, primate research history in the Democratic Republic of Congo. Finally, the chapter draws conservation biology conclusions that is the finding of bonobos in such an areas. The chapter is based on field trips by the authors in the area and field observations therein made. These field hard evidence data are complemented by conversations with key people in the region.

## 6.2 Bonobo Presence in the Lake Tumba Landscape

Bonobos have always been present in the area, since the description of *Pan paniscus* as a new species by Schwartz in1929 (Thompson 1997). Local elderly have always known their presence. They called the species '*ebobo*' and learned the name bonobo only in the 2000s onward, when a local non-governmental organization known as Mbou-Mon-Tour began talking to local communities about the importance of the presence of this particular species in their area. The name became more popularized when, after the initial scientific exploration was undertaken in the area and confirmed the presence of the bonobos, scientists started to organize larger expeditions, preparing for long-term field studies and brought in international media to document this '*newly discovered*' population in early 2006. The most surprising fact that scientists first found was that observed animals were never shy, and often build their nests close to villages. As indicated in other studies (Estes 1991), this single trait of behavior provided a clear indication that bonobos were less persecuted by humans residing in the area. As a proof that bonobos had always been in this area, local chiefs indicated that during an official visit of former DRC-leader Mobutu Sese Seko to the area in the late 1960s, traditional chiefs offered him a young bonobo as a present, but what happened to this animal is not known.

From a scientific point of view, it is especially striking that the population has never been scientifically described. First, there are many villages in the area; the 90 km long stretch of the road between Malebo and Lediba on the Kwa River is inhabited by ±15,000 people. Henry Morton Stanley visited the area in May 1882, spent the month of June 1882 at Bolobo before positioning Edward Glave, his Lieutenant at Lukolela in September of the same year (Jeal 2008). This very same part of the road used to be part of the main road (National Route 2) that linked Congo's capital Kinshasa with the town of Lisala in the Equateur Province. So, many people, including many western explorers, must have passed the area in the two past centuries without realizing that they were crossing the habitat of a rare great ape. Only Fenart and Deblock (1973) and Kano (1984) inferred from anecdotic information that bonobos were present in the area. Unfortunately, though, these anecdotes inspire less people to go and check that information, particularly science-driven spirits.

This becomes the more surprising because the southernmost bonobo population in the area is situated not far from the town of Bolobo on the Congo River. Bolobo is the very same place where Henry Morton Stanley left close to a hundred of his collaborators (Jeal 2008), and its distance from Malebo is a mere 40 km, straight line distance. Bolobo is also supposed to be the name that was misread to become bonobo today. The main hypothesis about the origin of the name bonobo is that the name is a misspelling of the name of the town of Bolobo, which was written on a crate in which some individuals of a great ape species, later to be called *pygmy chimpanzees*, were shipped to Europe in 1936 (Thompson 1997). The director of the zoo who received the cargo must have thought that the name was a local name given to the species. It is likely that the animals in the cargo were caught in what is known today as the Lake Tumba Landscape.

In fact, the discovery of this population is an initiative of local traditional Batéké chiefs, led by the highest of them Mr. François Lebo from the large village Mbee (Democratic Republic of Congo), who got worried about the activities of the forest company Société Africaine des Bois (SAFBOIS) in the land of their ancestors, without them knowing for what avail. SAFBOIS arrived in this region under the aegis of the logging laws that preceded the 2002 Forestry Code, which left local communities with little power to bargain with companies extracting timbers in their forests. Local communities and their traditional chiefs perceived that the arrival of the company would, one way or the other and in significant magnitudes, impact their living conditions. The plea for the Batéké was double-facets: (1) they wanted the timber company to exclude from their logging activities in the forests wherein their ancestors and themselves have been hunting, growing staples, and undertaking the traditional ceremonies for millennia and (2) for those areas where logging activities would be carried out, local communities should make tangible benefices. The first of these pleas was, of course, what is known as micro zoning of large areas, which is an essential part of sustainable forestry. The second plea conforms with the stipulations of the resolutions of the Biodiversity Convention from the Rio de Janeiro, which was endorsed by the DRC since 1993 or simply some months after its adoption.

Unfortunately, the juridical instruments included in the Convention on the Biological Diversity were not to be implemented under the regime of the *Bakajika Law*, a carbon copy of the land law that was crafted under the *Leopoldian* era. In fact, the Bakajika law, promulgated under Mobutu, copied word by word the royal decree of 22 August 1885 (Chap. 4 above). And was clear that the land (and forest ipso facto) and what it contained was owned the government. With that background, the plea of the Batéké to a company that felt empowered by the central government licence could only be ignored or, at the very best, partially honored. Whether the pleas of the local communities were honoured or not is beyond the scope of this narrative but what is certain is that in reaction to what was felt as some sort of injustice, the traditional chiefs strongly backed the creation of Mbou-Mon-Tour, a local ONG in 2001. The main objective of this non-governmental organization is to try and reduce the impact of increased forestry in the area through the umbrella of the bonobo as flagship species. Local communities, through Mbou-Mon-Tour increased their efforts to lobby provincial and national authorities as well as international conservation organizations when the Batéké chiefs evidenced the fact that the logging companies were unable to keep their promises. Presentations were given at the national level on the plight of the forests of the Malebo area and a trip to the offices of the World Wide Fund for Nature (WWF) in Kinshasa, with claims that their living area was threatened by plans for large-scale forestry, and that this was particularly painful as it housed large numbers of bonobos. It was this visit that triggered the first visit to the area by researchers from WWF who, to their surprise, immediately confirmed with signs and direct sightings of bonobos.

The bonobo population in the Lake Tumba Landscape is isolated from other populations located at the very heart of the forest habitats, but very dense (Inogwabini 2010). It has four subpopulations, of which the most southern one has the highest density of bonobo populations ever recorded (Inogwabini 2010). The Salonga closest protected area to this zone was National Park, situated at a distance of 180 km (straight line measured from GIS). The Salonga with a relatively lower density of bonobos was created principally to preserve the species. The population lives both within a forest-savannah mosaic and within swampy forests, both habitat types that until recently were thought to be marginal for the species.

Crucial to the success of the survival of this population was the fact that the war never reached the area. There were hardly any soldiers or militia passing through, which implies that there have always been less weapons in the area than in many other regions of the DRC. There was also less pressure on the population with regard to looting of fields and food resources, which can force people into a living of gathering and harvesting wild plants and animals. The area is also partly used by a large cattle company (ORGAMAN), which has thousands of animals roaming in the landscape, that is protected by well-armed and disciplined guards who have interest in keeping poachers out of the area. Interesting was that the guards knew there were bonobos in the area, but they considered it to be 'normal', hence they never reported the presence of the animals to their chiefs, who were as surprised by the discovery as the scientists of WWF, with whom they cooperate. The owners of the company (the Belgian family Damseaux) do not see any potential conflict between the bonobos and their cattle.

Crucial was also the fact that the local population protected the bonobos. Bonobos were never eaten in the area. They are considered to be 'another human' that should not be hunted or even bothered. The taboos went so far as to indicate that men harming bonobos would never be father to children. Social control was important, in that people tend to denounce villagers that would have hunted bonobos, in an attempt to protect their village from bad luck. Bonobos have been protected by people out of respect for their traditional taboos and for their traditional chiefs, rather than through complying with national and international laws and regulations. The traditional knowledge of bonobos shows considerable overlap with scientific knowledge of bonobo ecology. Research has shown that bonobos occurred at higher densities in areas where local communities held positive views of them (Inogwabini 2010).

It was also striking that the animals never showed any clear fear for people. They do not tend to run away as people enter their living area, as they did in areas such as the huge Salonga National park that was created in 1970 in the heart of bonobo living area to protect the species (Van Krunkelsven et al. 2000). In Salonga, former soldiers from different armies and militia turned into poachers that with machineguns attacked any animal species in the forest, obviously forcing bonobos into hiding when people approach their area. It is in this view also not surprising that bonobo density in Salonga National Park was on average only half of that in the Lake Tumba Landscape area (Inogwabini 2010; Van Krunkelsven 2001).

There seemed to be one exception on this 'natural habituation to humans', as was assessed from anecdotic observations made during the first visits of foreign (read: white) researchers to the area: the bonobos seemed to be afraid of white people (Draulans 2007). To the surprise of as well guides as Congolese researchers, they ran away when they saw white people approach them. They clearly have to get used to the presence of these 'other' visitors—a fact that should be taken into consideration when efforts are made to organize tours to the area. Also, local people have to get used to the presence of foreign visitors. They consider the presence of bonobos in their living area as absolutely normal, which makes them suspicious as to the real reasons for the visits. They are convinced that the foreigners are not interested in the bonobos, but that they have discovered gold or copper or another precious metal in the area.

Finally, it has to be mentioned that bonobos were not the only surprise discovery in the area, which also houses a good number of monkeys and buffalo (*Syncerus caffer*), and a large population of elephants, surprising, as the only mention of elephants (*Loxodonta africana*) in the area so far made seems to date back to the expeditions of Henry Morton Stanley, who reported in his diary that he shot four animals in the area to feed his troops, a story that has become ingrained in the memory of the helders of the communities and has been told to whomever wants to believe it. To everyone's surprise, there were even lions *Pantera leo* in the area, although reportedly the last animal in the region was shot several decades prior to the arrival of researchers in the region (see Chap. 8). The Lkae Tumba Landscape area hence probably is the only region in the world in which bonobos and lions coexist.

The discovery of the bonobo as a new species was a, proportionally to the description of other great apes, late and conflictual situation (Thys Van den Audenaerde 1984). As usual, the animal was first described based on skulls in museum collections. The first observations of living animals are related to individuals in captivity, mainly in European zoos. The first field studies started only late in the years 1970, first in the Lomako area, later at Wamba, both in the northern range of the general bonobo distribution area (Susman 1984). The first field expedition to reach the notorious Salonga National Park took place in the winter 1996–1997 (Van Krunkelsven et al. 2000). As a consequence of the insecurity in the area, it took many more years for scientific research in the park to get going (Inogwabini and Omari 2005).

Regarding the logistic difficulties to reach the main study areas for bonobos, it is surprising that the Lake Tumba Landscape area had never been identified as probably the best—in casu: easiest—place to study the animals. However, a few elements of its whereabouts might be part of the explanation of why this should have been the case. Thys Van den Audenaerde (1984) mentions that the area was largely neglected by the collectors that were sent out by the Royal Museum of Mid-Africa in Belgiums Tervuren, from where many collection expeditions in the former Belgian colony were organized (Fenart and Deblock 1973). Perhaps it was considered to be too close to the capital Kinshasa to be of special interest. It would be useful, however, to re-evaluate the recordings of the origins of specimen in the museum in view of the new insights into bonobo distribution, as the museum did have at least one regular local collector in the area (Thys Van den Audenaerde, *personal communication*).

It seems also likely that the first field people that started to explore the possibilities to study bonobos in the field, inspired by the successes in studying chimpanzees (*Pan troglodytes*) by Jane Goodall and mountain gorillas (*Gorilla beringei*) by Dian Fossey from the years 1960 onward, would have considered it best to look for large tracts of dense and virgin rainforest to start their studies, which would have excluded the Lake Tumba Landscape area from their options, as it is atypical of what was considered to be the ideal bonobo habitat. It is significant in this view that the Lake Tumba Landscape area so far has the highest density of bonobos on record.

Many people still have difficulties distinguishing between bonobos and chimpanzees, and as the chimpanzee is the most common and well-known of the two *Pan*-species, it is likely that not every traveler in the area would have realized that he saw another species when he was confronted with bonobos along the National Route Number 2, especially as the chimpanzee is also present in the DRC, be it on the other side of the river Congo. Most 'chimpanzees' that were brought into the capital Kinshasa must have been bonobos, though, as is witnessed by the orphaned baby's that are confiscated in the town the past decades. The bonobo has definitely suffered from its similarities with the chimpanzee.

It is also sad that the knowledge of local people in issues of biodiversity and conservation is often neglected, as it is considered to be not accurate enough, and not having any solid relevance to the issues involved, as it seemed that there was no solid link between the traditional knowledge and scientific facts (Colding and Folke 2001; Berkes et al. 2000). This is a pity, as new insights indicate that traditional knowledge can be extremely useful in understanding what is happening in a region, and in laying the foundations of new conservation areas and scientific enterprises (Inogwabini 2010; Redford and Fearn 2007). Reliance on traditional knowledge could have led to the discovery of the bonobo population of the Lake Tumba Landscape much sooner than has been the case.

The question that needs to be raised,is if this would have been a good thing, especially regarding the turmoil that hit the Congo the past decades. The Lake Tumba area is close enough to Kinshasa to have the opportunity of becoming an economically favorable source of wood and bushmeat for the capital, especially in periods of large instability, in which classic economical logistics are limited. Draulans and Van Krunkelsven (2002) stressed, however, that instability does not necessarily imply large-scale damage to ecosystems, as, for example, forestry activities could be temporarily halted, which implies a diminished threat to vulnerable rainforests. More animals might be shot, but large-scale destruction of habitat could be stopped.

However, the fact that the Lake Tumba Landscape area is home to thousands of pieces of cattle, that serve as a major food source for Congo's capital Kinshasa, and that is claimed to have helped saving the city regularly from starvation during periods of instability and famine (Draulans 2007), could have helped in safeguarding the area from major assaults on its wildlife and forests, especially as well-paid armed guards provided protection for cattle, and for everything else in the region, by discouraging armed poachers and thieves to enter the area. The discretion on the presence of elephants and other animals could have helped in preventing the area from being raided by armed poachers.

On the other hand, the Lake Tumba Landscape area faces the same cultural changes in its societies that affect other areas of the country, and the continent. Due to human migration people from tribes without taboos against hunting bonobos, and with established trading elements, enter the area, attracted by its potential as a productive area—which some see as one of the reasons why bonobos were never hunted, there was never a strong incentive in doing so (Inogwabini 2010). Younger people also lose their interest in traditional activities, so they might lose as well taboos against hunting bonobos. The introduction of money as an alternative to other forms of trade has also increased pressure on wildlife, including bonobos (Inogwabini et al. 2007b). Even mythical species like bonobos are now being hunted by ethnic groups in areas where the species had not been hunted before. Local communities believe on-going poaching of bonobos is mostly driven by trade to make money. Loss of traditional knowledge often does not mean good news for conservation efforts (Hardin and Remis 2006). This erosion of protection from taboos goes hand in hand with the steady increase in the availability of automatic weapons in the country (Draulans 2003; Draulans and Van Krunkelsven 2002).

So despite the fact that the bonobos of Lake Tumba Landscape have survived a long time without protection, it is clear that it would be too risky to let the situation continue as it has been in the past. Times are changing, even in bonobo areas. In dealing with the Lake Tumba Landscape area, care has to be taken as not to upset the major players in the area. As well the cattle owners, all the local chiefs have already expressed worries that conservation efforts could interfere with their activities, despite them having been instrumental in the protection of the area in the past. Removing and banning cattle is not a realistic option, anyway, as large tracts of the area are owned by the Orgaman-company. Odadi et al. (2011) recently showed that cattle should not necessarily compete with wild herbivores, on the contrary, that by eating mainly plants that are less palatable to cattle native herbivores may increase access to 'cattle-friendly' food plants.

Local people should benefit from the presence of bonobos and other key species for conservation, in order to ensure their protection. People in the Lake Tumba Landscape area have already benefited from a school being built in the area, and the road between Kinshasa and the area being improved substantially, including the rehabilitation of a bridge on one of the main rivers crossing the area. People have also found employment as rangers and local trackers (locally known as *pisteurs*) that go into the forest to census, guard, and habituate bonobos and other animals. This fits well with the trend of improving conservation efforts by including local people in the programs, hence providing them with the options for improved living circumstances.

A crucial development for the region will be the implementation of plans for bonobo tourism. Tourism is increasingly seen as a major asset to conservation programs all over the world. Many African countries benefit to a large extent from tourism as a major source of income. Many populations of great apes prevail as a consequence of them generating enormous amounts of money from foreign visitors. It is not farfetched to claim that the mountain gorilla has not gone extinct during the civil wars that raged across the three countries (Congo, Rwanda, Uganda) that house populations of the species, because of its unlimited potential to generate money for the countries that house them.

So far hardly any bonobo tourism has been established. It is arduous to reach the classic areas where bonobos have been habituated and studied (Lomako and Wamba), and even more difficult to reach the Salonga National Park, unless small airplanes can be chartered, which does not solve the problems of transport on the ground. It is logic that the Lake Tumba Landscape area, which is situated at about less than one hour flight from Kinshasa, and which in the dry season can be reached after a drive that can be done in one day, would be the best option to launch bonobo tourism. The logistics would be far easier to organize than in the other areas. The first facilities for housing and feeding visitors have already been built in the area .

The organization of a profitable bonobo tourism industry would of course mean that the DRC shows more stability than it has done in the past decades, so that travel agents and interested visitors can be reassured about the safety traveling in the area. Democratic Republic of Congo has never been an easy country to travel in, so agents should be able to guide visitors without too much difficulty through the procedures for visiting the country. Even the Virunga National Park in East-Congo, one of the best African parks to visit, still suffers heavily from the negative image that Democratic Republic of Congo is carrying after all the turmoil. The reestablishment of enough confidence in the Congolese authorities as to guarantee safety and sure logistics will be crucial in determining the success of bonobo tourism as a means of generating income that can be beneficial to both conservation and the lives of the local communities.

# References


Berkes F, Colding J, Folke C (2000) Rediscovery of traditional ecological knowledge as adaptive management. Ecol Appl 10:1251–1262

Colding J, Folke C (2001) Social taboos: invisible systems of local resource management and biological conservation. Ecol Appl 11:584–600

Draulans D (2003) Handelaar in Oorlog ("Dealer in War"). Atlas

Draulans D (2007) Congo blijft verrassen. Soms ook in positieve zin. Congolese biologen ontdekten meer dan zevenduizend nieuwe bonobo's. Knack (2007-09-12). https://www.knack.be/nieuws/magazine/leven-met-een-andere-mens/article-normal-1009445.html?cookie_check=1578314370

Draulans D, Van Krunkelsven E (2002) The impact of war on forest areas in the Democratic Republic of Congo. Oryx 36(1):35–40

Estes RD (1991) The behavior guide to African mammals: including hoofed mammals, carnivores, primates. The University of California Press, Berkley

Fenart R, Deblock R (1973) Pan paniscus et Pan troglodytes: craniométrie – étude comparative et ontogénique selon les méthodes classiques et vestibulaires. Tome 1, Musée Royal de l'Afrique Centrale, Tervuren, Belgique

Hardin R, Remis MJ (2006) Biological and cultural anthropology of a changing tropical forest: a fruitful collaboration across subfields. Am Anthropol 108:273–285

Inogwabini BI (2010) Conserving great apes living outside protected areas: the distribution of bonobos in the Lake Tumba landscape, Democratic Republic of Congo. PhD Thesis, University of Kent at Canterbury, United Kingdom

Inogwabini BI, Omari I (2005) A landscape-wide distribution of **Pan-paniscus** in the Salonga National Park, Democratic Republic of Congo. Endangered Species Update 22:116–123

Inogwabini BI, Matungila B, Mbende L, Mbenze AV, Tshimanga WT (2007) Great apes in the Lake Tumba landscape, Democratic Republic of Congo: newly described populations. Oyx 41(4):532–538

Inogwabini BI, Matungila B, Mbende L, Abokome M, Tshimanga WT (2007a) Great apes in the Lake Tumba landscape, Democratic Republic of Congo: newly described populations. Oryx 41 (4): 532–538

Inogwabini BI, Bewa M, Longwango M, Abokome M, Vuvu M (2007b) The Bonobos of the Lake Tumba–Lake Maindombe Hinterland: threats and opportunities for population conservation. In: Furuichi T, Thompson J (eds) The Bonobos behavior, ecology, and conservation. Springer, pp 273–290

Jeal T (2008) Stanley: the impossible life of Africa's greatest explorer. Yale University Press

Kano T (1984) Distribution of pygmy chimpanzees (**Pan paniscus**) in the central Zaire basin. Folia Primatol 43:36–52

Odadi WO, Abdulrazak SA, Karachi MM, Young TP (2011) African wild ungulates compete with or facilitate cattle depending on season. Science 333:1573–1755

Redford KH, Fearn E (2007) Protected areas and human displacement: a conservation perspective. The Wildlife Conservation Society. Working Papers 29. New York, United States of America

Reinartz G, Inogwabini BI (2001) Bonobo survival and the wartime mandate. In: The great apes: challenges for the 21st Century. Conference proceedings, Brookfield Zoo 52–56

Susman RL (1984) The locomotor behavior of Pan paniscus in the Lomako Forest. In: Susman RL (ed) The pygmy chimpanzee, Evol Hum Behav, Plenum Press 369–396

Thompson JAM (1997) The history, taxonomy and ecology of the bonobo (Pan paniscus Schwarz 1929) with a first description of a wild population living in a forest/savanna mosaic habitat. Ph.D. Thesis. Oxford University, Oxford, United Kingdom

Thompson-Handler N, Malenky RK, Reinartz G (1995) Action plan for Pan paniscus: report on free ranging populations and proposals for their preservation, Zoological Society of Milwaukee County

Thys van den Audenaerde DFE (1984) The Tervuren Museum and the Pygmy Chimpanzee. In: Susman RL (1984) The Pygmy Chimpanzee, Evol Hum Behav, Plenum Press
Van Krunkelsven E, Inogwabini BI, Draulans D (2000) A survey of bonobos and other large mammals in the Salonga National Park, Democratic Republic of Congo. Oryx 34(3):180–187
Van Krunkelsven E (2001) Density estimation of bonobos (Pan paniscus) in Salonga National Park, Congo. Biol Conserv 99:387–391

# Chapter 7
# Genetics of Bonobos in the Lake Tumba Landscape

**Abstract** Documenting an important population of bonobos west of Lake Maindombe had critical implications for elucidating the biogeographical history of great apes, and for the conservation of bonobos. The chapter assesses the genetic diversity of the isolated western population of bonobos, and identified patterns of gene flow that have occurred between the bonobo population in the Lake Tumba landscape and previously known populations. The haplotype diversity (Hp = 6) of the bonobos in the Lake Tumba Landscape was lower than Hp = 16 found in bonobos in the central population but was similar to those found in eastern, southern, and northern populations. This difference seems to reflect real changes between different bonobo populations because it is supported by the significant differentiation ($p < 0.01$) of haplotypes across the entire range. The combination of these two factors suggests that the genetic variation seen within different populations also varies across the extent of occupancy of species. Biogeographical contextualization of this finding led to infer that the more generic *Pan* (comprising chimpanzees and bonobos) would have emerged from the center of the lowland forests and dispersed outwards, in support of the Wrangham's hypothesis that great apes were living in the rainforest in Central Africa before they moved toward forest-savanna edges where they would evolve to other forms (2009).

**Keywords** Genetic diversity · DNA · Biogeography · Pleistocene forest refugia theory · Species evolutionary path

## 7.1 Introduction

The extent of genetic variation among wild populations of bonobos is gradually becoming understood. Field samples have recently been collected from sites located to the northeast and north of the extent of occupancy of bonobos, including Wamba and Lomako in the Democratic Republic of Congo respectively. Recently published genetic studies have confirmed that the eastern population of bonobos, located between the Lomami and Lualaba rivers, the central population located in the Salonga National Park, and the southern population located between the Lukenie and Sankuru rivers only diverged some years ago (Inogwabini et al. 2007a; Eriksson et al. 2004).



B.-I. Inogwabini, *Reconciling Human Needs and Conserving Biodiversity: Large Landscapes as a New Conservation Paradigm*, Environmental History 12,
https://doi.org/10.1007/978-3-030-38728-0_7

Most excitingly, the documentation for science of the western population of bonobos (Inogwabini et al. 2007a has extended the confirmed extent of occupancy of bonobos by a straight-line distance of ~100 km from their previous known western limit (Inogwabini et al. 2007a; Eriksson et al. 2006) and by a straight-line distance of 400 km from the main southern population (Fig. 7.1). Furthermore, even among the recently discovered western population of bonobos, topographic features include potential ecological barriers that suggest that there maybe five isolated sub-populations of bonobos within the southern area of the Southern Lake Tumba Landscape (Inogwabini et al. 2007a).

The newly documented western population represents ~ 18% of bonobos estimated to remain in the wild (Inogwabini et al. 2007a, b). Therefore, it is critical for their future conservation and management to investigate the levels of genetic variation within this large, recently documented and relatively isolated population of bonobos. Furthermore, it was important to assess the gene flow between all the genetically described wild populations. Consequently, this paper aimed, using non-invasive sampling techniques, at extracting and sequence DNA from the recently documented western population of bonobos, calculating the level of gene flow between the western population and previously documented populations throughout their extent of

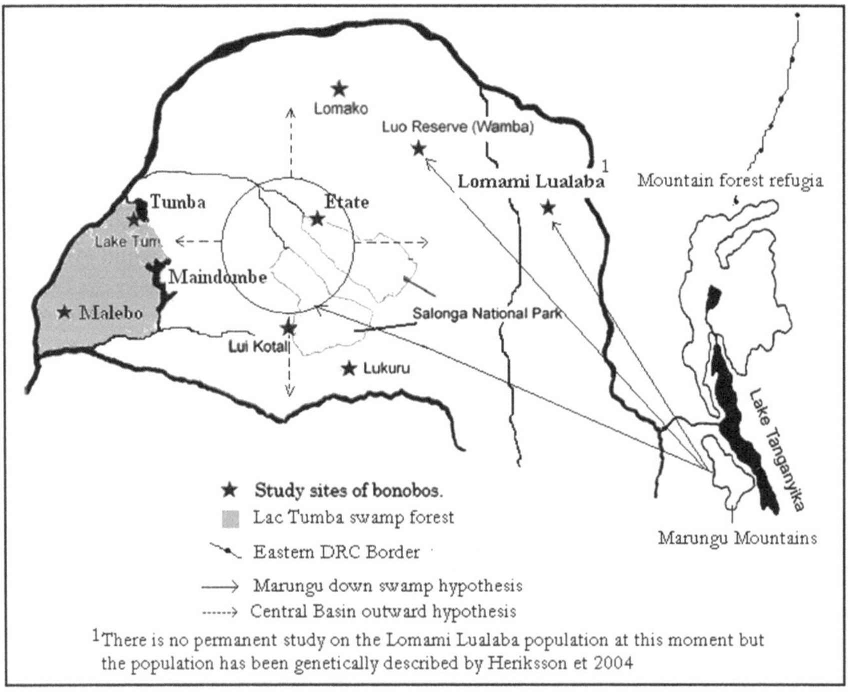


**Fig. 7.1** Suggested direction of the migration of bonobos after drought periods

occupancy, discussing the findings within the general context of the species evolution and, for conservation purposes, enabling sound management decisions, and appropriate actions to conserve bonobos.

## 7.2 Materials and Methods

The research presented here is a summary of work carried out collaboratively. Inogwabini (the editor of this volume) designed the sampling methods, collected samples, and contributed to the interpretation of results. As part of a landscape-wide assessment of genetic variation among bonobos, Inogwabini designed the sampling and collected a total of 56 fecal samples from Nkala (S02.54070, E16.47156) and Mpelu (S02.59127, E16.47433), located within the Malebo-Nguomi sub-population, hereafter known as Malebo sub-population, within the southern Lake Tumba Landscape. Nkala and Mpelu are 6 km apart using a straight-line distance. The western bonobo population in Lake Tumba Landscape is isolated from most other areas within the bonobo range by the River Congo to the north and west, the rivers Kwa and Fimi to the south and Lake Maindombe and Tumba to the east, where there is a small gap of land between the two lakes containing smaller waterways but densely populated by humans since historical times. This continuous human presence has fragmented the forests of the region to the point that corridors for large mammals exist no longer.

Fresh fecal samples were steeped in methanol and stored in sealed bags at room temperature. DNA was extracted approximately 15 days after collection at the Department of Biosciences at Cardiff University, UK. To import the samples into England, a 'general license for the importation of certain animal pathogens and carriers of animal pathogens' was obtained from the Department for Environment, Food and Rural Affairs (DEFRA). Likewise, the export of samples from DRC was authorized by the Ministry of Environment and Conservation.

DNA was extracted from the fecal samples using QIAamp DNA stool minikit (Qiagen, 51504). After sequencing, a genetic analysis was carried out at both microsatellite and mitochondrial levels. Genetic differentiation and gene flow between populations were measured in DNASP 4.20.2. Pairwise $F_{ST}$ values were used to calculate genetic distance between two different groups within the bonobo sub-population in the Malebo region in the Lake Tumba forests, and those populations that have been previously described genetically (Eriksson et al. 2004). The statistical significance of the levels of differentiation between bonobo populations was tested using a non-parametric permutation approach, as implemented by DNASP 4.20.2 with a total of 1000 permutations. Using this test of significance, the null distribution is reached by allocating each individual sample to a random population, while maintaining population sizes as defined by the original data. The mean genetic distance between regions was then compared to geographical distances between the populations, measured as straight lines but diverting the route around geographical barriers to dispersal, using GOOGLE EARTH 4.2 (Google). The significance of the correlation between geographical and genetic distances was measured using a statistical correlation test, the Mantel test, as implemented by ARLEQUIN 3.1. An analysis of molecular variance (AMOVA) test was also carried out in ARLEQUIN 3.1 to further determine the degree of significance between the genetic and geographical distances between sampled populations. AMOVA allows for the components of DNA variation to be analyzed hierarchically, both among groups within the same region and between populations from different regions, independently testing the hypothesis of genetic differentiation based on geographic distance.

Mismatched distributions were plotted using the software NETWORK 4.5.0.0, which allows the comparison of observed pairwise nucleotide site differences between regional bonobo populations, to identify whether or not a population has recently undergone a rapid demographic expansion. Tajima's D test was also used to test for the recent expansion of populations, under the infinite site model used in ARLEQUIN 3.1, an indication of any selection upon the mutation of the HV1 sequence and any inherent heterogeneity of mutation rates among sites is given. Results from the genetic analysis were discussed using the prevailing theories of great apes dispersion in Central Africa through an overall review of the current status of the available knowledge.

## 7.3 Genetic Variations in the Bonobos in the Lake Tumba Forest and Other Populations

Haplotype diversity (Hp), mean pairwise difference (MPD), and nucleotide diversity (phi) were compared within different populations across the extent of occupancy of bonobos (Table 5.1). This comparison shows that the newly documented western population of bonobos, particularly the Malebo sub-population, had a higher haplotype diversity (Hp) than eastern and northern populations (Table 7.1). Furthermore, the same Malebo sub-population had lower Hp than those found for the central and southern populations (Table 7.1). Nevertheless, haplotype diversity ($h \pm$ SD) was not significantly different ($p > 0.05$) between the six populations (Table 5.1). However, the Malebo sub-population tended towards highest haplotype diversity ($h \pm$ SD),

**Table 7.1** Range-wide variations in the mtDNA haploptypes for bonobos

| Population | N | Hp | $h \pm$ SD | $\pi \pm$ SD | MPD |
|---|---|---|---|---|---|
| 1. West | 56 | 6 | 0.880 ± 0.077 | 0.013 ± 0.03 | 6.4 |
| 2. East | 15 | 5 | 0.781 ± 0.064 | 0.023 ± 0.002 | 7.8 |
| 3. South | 26 | 7 | 0.813 ± 0.036 | 0.030 ± 0.002 | 9.5 |
| 4. North | 36 | 5 | 0.819 ± 0.019 | 0.038 ± 0.004 | 12.1 |
| 5. Central | 63 | 16 | 0.923 ± 0.014 | 0.032 ± 0.003 | 10.3 |

Sample size (N), number of haplotypes found (Hp), haplotype diversity and its standard deviation ($h \pm$ SD), nucleotide diversity per site and its standard deviation ($\pi \pm$ SD) and mean pair-wise difference between nucleotides (MPD)

**Table 7.2** Genetic differentiation of bonobo populations based on net number of nucleotide substitutions per site between the populations

| | West | Central | East | South | North |
|---|---|---|---|---|---|
| West | – | | | | |
| Central | 0.0775624 | – | | | |
| East | 0.257709 | 0.248645 | – | | |
| South | 0.0887242 | −0.060763 | 0.271365 | – | |
| North | 0.306164 | 0.0320365 | 0.413517 | 0.0300911 | – |

while those in the east tended the lowest. In contrast, nucleotide diversity ($\pi \pm$ SD) did differ significantly ($p > 0.05$) between individuals from different populations (Table 7.1). The northern population showed the highest nucleotide diversity ($\pi \pm$ SD) between individuals, while the western population showed the lowest nucleotide diversity. Pairwise comparison showed that, with the exception of the eastern population, statistically significant differences were found between the western population and all others. The eastern population also differed significantly from the north. The western population showed the lowest MPD of 6.4 between individuals, while the northern population showed the highest MPD of 12.1 between individuals.

$F_{ST}$ values were based on pairwise differences, and Nei's net number of nucleotide substitutions per site between the populations, $D_A$ (Table 7.2). $F_{ST}$ values were calculated between each pair of populations (Table 7.2) to show the genetic differentiation of haplotypes across the entire extent of occupancy. Permutation tests showed that all were significantly different ($p < 0.01$). The three most differentiated pairs of populations, in order from most to least differentiated, are the north versus east ($F_{ST} = 0.414$), north versus west ($F_{ST} = 0.306$), and south versus west ($F_{ST} = 0.088$). The three least differentiated pairs, from most to least, are north versus central ($F_{ST} = 0.032$), north versus south ($F_{ST} = 0.030$), and south versus central ($F_{ST} = -0.061$). The net number of nucleotide substitutions per site ($D_A$) was also calculated and showed the same level of statistical significance as $F_{ST}$ values ($p < 0.05$).

## 7.4 Genetics of the Bonobos in the Lake Tumba Forests in Perspectives

The documentation of an important population of bonobos west of Lake Maindombe (Inogwabini et al. 2007a, b) has critical implications for elucidating the biogeographical history of great apes, and for the conservation of bonobos. Therefore, it is important to assess the genetic diversity of the isolated western population of bonobos, and to identify any patterns of gene flow that have occurred between this recently documented, and previously known, populations. The western population was shown to have a haplotype diversity (Hp) = 6, which was not as high as that of 16 found in bonobos in the central population, but similar to those found in eastern, southern, and northern populations (Table 7.1). This difference seems to reflect real changes between different bonobo populations because it's supported by the significant differentiation ($p < 0.01$) of haplotypes across the entire range (Table 7.2). The combination of these two factors suggests that the genetic variation seen within different populations also varies across the extent of occupancy of species. The eastern population was the most distinct overall and showed the highest total $F_{ST}$ values relative to the four other populations with which it was compared (Table 7.2). Furthermore, the newly documented western population in Lake Tumba Landscape was quite highly differentiated compared to northern and eastern populations, but less so when compared to southern and central populations. Furthermore, these differences followed the same general pattern as $F_{ST}$ values, with the single exception that only east versus west populations were placed two places higher in the order of differentiation. Consequently, the major conclusion of this genetic study is that both western and northern populations of bonobos are genetically distinct from bonobos within other regions of the extent of occurrence.

The study has shown associations between the haplotype of both western subpopulations, with haplotypes from both the central and eastern regions. Consequently gene flow has occurred to a certain degree between western, central, and eastern populations. Nevertheless, associations between the haplotypes of the western population with those of the eastern population appear somewhat counterintuitive, given the geographical distance between both locations. However, these associations can be more easily understood when viewed from the perspective of the biogeographical history of bonobos. Based on the genetic data, the central population may have dispersed in two opposite directions, one dispersal taking an easterly direction while the other dispersed westward. In this way, both dispersals would have retained some of the genetic similarities that they would have shared before their separation.

## 7.5 Genetics and the Habitat Types Can Help in Identifying the Species Evolutionary Path

This result has implications for the historical biogeography of the species and the way it separated from its cousin common chimpanzees. The newly documented and isolated population of bonobos in the Lake Tumba Landscape possesses genetic traits previously found in populations of bonobos from both western and eastern extremes of their extent of occurrence. This very interesting result raises two questions that go beyond the scope of this study and that need to be addressed in the future. The first of these questions relates to where inferred forest refugia (White 2001; Horn 1976) would have been located in the past. The second of these questions relates to where this bonobo population would have originated. Basically, the genetic findings outlined in this paper pose the question of where the evolutionary journey of the *Pan paniscus* described by Thompson (2003) started.

To answer those two questions, it will be necessary to briefly outline the Pleistocene forest refugia theory, as it is now accepted for African forests (Thompson 2003; Colyn 1990). This theory proposes that actual differences in forest types in Central Africa have arisen as a result of Palaeo-climatic change (White 2001; Fay 1997; Schwartz 1991), and argues that the history of African rain forests was punctuated by major shifts in forest cover over the last 100,000 years (Schwartz 1991). Thus, savannahs would have covered large areas of Africa during an arid climate phase. When more humid conditions returned, the forest re-colonized savannahs from forest refugia. In terms of a time series model, these major climatic changes are thought to have occurred in five phases (White 2001). The first phase occurred during the Middle Stone Age Industry period from 70 to 40,000 years Before Present (BP), when Africa's landscape would have been entirely characterized by major erosion and little vegetation cover (Schwartz 1991). The second phase occurred from 40 to 30,000 years BP and would have been characterized by very heavy rainfall, ranging from 2,000 to 2,500 mm $yr^{-1}$, a climatic regime that would clearly have favored the expansion of an exuberant forest cover (Schwartz 1991). The third phase occurred from 30 to 12,000 years BP and would have been characterized by a cooler and drier climate and forest cover regression. Abundant grass would have covered the savannahs, as evidenced by archeological data available on the period, which contains different pollen species belonging to the family of *Graminaceae*. Isolated forest refugia would have been located across the Congo Basin, mostly on mountainous zones such as Mount Cameroon and the Rift Albertine (Schwartz 1991). Beyond those easily locatable, mountain refugia sites, remnants of gallery forests would have survived along major rivers, and constituted additional fluvial forest refugia from which the current lowland Congo Basin forests and their important biodiversity would have re-colonized the savannahs to occupy their present distribution (Schwartz 1991). Archeological remains of forest species in hydromorphous soils and the actual distribution of primate species (Schwartz 1991; Colyn 1990) provide evidence of these 'lowland' forest refugia' along rivers. The fourth period ranged from 12 to 3,000 years BP was characterized by a much more humid and rainy period (Schwartz 1991). During this period, forest cover greatly expanded, to include even what is now the Sahara Desert. The fifth period 3000 years BP to the present day would have experienced a semiarid period that witnessed the sudden disappearance of the evergreen forest of the Congo coastal region, around 3000 years BP (Schwartz 1992). Archeological data show an increase of pollen from *Graminaceae* that provide evidence of the appearance of savannahs in areas such as Chaillu, Mayombé, and Ogoué (Rietkerk et al. 1995; Schwartz 1991). The sudden disappearance of forest cover has since reversed, as forests have re-colonized savannahs from forest refugia.

Whether their patterns of disappearance are long and slow or sudden and short, losses of forest cover during drier periods have reversed naturally if left without subsequent human intervention (Oslily and Dechamps 1994; Oslily 1998). Thus forests can spring back from *forest refugia* to occupy formerly lost ground through re-colonizing savannahs. For great apes, it is critically important that colonizing and refugia forests support an abundant cover of *Marantaceae* (e.g., *Haumania liebrechtsiana*, *Megaphrynium* spp.) and *Zingiberaceae* (e.g., *Afromamum* spp.) that would support great apes through transitional phases between savannahs and semi-deciduous forest (White et al. 1995). It appears that re-colonization of savannahs by forest may have experienced three successional stages (Tutin et al. 1997a) as follows: (i) 50–100 years: The re-colonization of savannahs by mono-dominant, fast-growing species, including *Lophira alata*, *Acoumea klaineana,* and *Sacoglottis gabonensis*); (ii) 100–200 years: The colonization of the mono-dominant and fast-growing species forests by *Marantaceae* and *Zingiberaceae* spp.; and (iii) 300–500 years: the development of mixed *Marantaceae* under-storey and young mature forest. Tutin et al. (1997a) have suggested a number of implications of forest refugia theory for primates. If recent forest history was punctuated by several reductions of forest cover arising through climatic change during the past 20,000 years, these fluctuations would have played an important role in the evolutionary history of primates in Central African forests (Gautier-Hion et al. 1999; Colyn 1990). Obviously, if individuals of some species took refuge in different forest refugia and did not re-emerge after an arid period, high genetic variability might be predicted.

The genetic variability shown among the western bonobos, and the habitat differentiations documented in this study, raises important questions about the location of the fluvial forest refugia (Colyn et al. 1991; Preuss 1990; Horn 1976). Horn (1976) suggested that fluvial forest refugia might have been located in the area of Lake Tumba because this zone was very low-lying and supported a massive body of stagnant freshwater (Peters and O'Brien 1999). In contrast, Colyn et al. (1991) and Preuss (1990) thought that the fluvial forest refugia would run along major rivers and might probably have even included the current region of Salonga National Park. However, there is no pollen evidence to resolve the debate over the location of lowland fluvial forest refugia. Nevertheless, it seems logical that, if Lake Tumba bonobos share more traits with bonobos both in Salonga National Park and in the far east of their extent of occurrence, bonobos may have emerged from the central zone. The new genetic evidence from Lake Tumba Landscape has important consequences for the fluvial forest refugia theory in that it suggests a reversal of the assumed direction of the emergence of bonobos. Thompson (1997, 2002, 2003) has suggested that bonobos emerged from the Mountain of Marungu to colonize the lowland forests. However, the 'Marungu down to the swamps' hypothesis was based solely on the idea that all *Pan* species and subspecies may have emerged from the Gombe area in Tanzania where they may have taken refuge during the arid periods. Instead, new genetic evidence from Lake Tumba Landscape suggests that *Pan* would instead have emerged from the center of the lowland forests and dispersed outwards. This argument agrees with the hypothesis that great apes were living in the rainforest in Central Africa before they moved toward forest-savanna edges where they would evolve to other forms (Wrangham 2009).

In order to enable them to survive through changes in habitats and food availability induced by long-term climatic fluctuations, another important consequence of the forest refugia theory is that primate species would have to adapt their feeding behavior (Tutin et al. 1997b). Consequently, the re-colonization of savannahs by Marantaceae might have influenced the evolution of the social organization, distribution, and even terrestriality among African apes (White et al. 1995). Consequently, it is unsurprising that Marantaceae species play a key role in the present distribution of bonobos (Tutin et al. 1997a). Thus, the role played by Marantaceae in the distribution of bonobos and other species of great ape is rather a historic consequence of forest dynamics and history.

The western population of bonobos in the Malebo region of Lake Tumba Landscape occur in a complex forest-savannah mosaic and an undulating land layout with a succession of creeks, terra firma, and swampy savannahs, a habitat that can only be compared to some extent, to that found in the Dekese-Lukuru (DL) (Thompson 1997, 2002, 2003), but without swampy savannahs. Until the discovery of the LTSF bonobo population, the habitat type in the southeast zone of the extent of occurrence at DL was the most readily differentiable from other bonobo habitats (Inogwabini et al. 2007a, b). Based on data from the Lake Tumba Landscape, it seems that this type of habitat would be the preferred habitat for bonobos, contrary to long-held beliefs. More importantly, the discovery of bonobos in the Malebo region considerably extended the known extent of occurrence of bonobos. Bonobos of Malebo exhibited ground night nesting behavior that so far only been seen from Etate in the northern Salonga (Inogwabini, unpublished data). Reasons for these differentiated patterns remain unclear but may combine ecological, historic, and evolutionary factors. The forest-savannah mosaic at Lake Tumba Landscape provided year-round sources of fruit in the savannahs, upon which bonobos could fall back when forest resources were scarce (Inogwabini and Matungila 2009). Furthermore, bonobos preferred habitats with higher plant species diversity (Inogwabini 2010). Ecotones harbor high species diversity, comprising a mix of species from adjacent habitats, another explanation of why bonobos find very suitable niches in complex savannah-forest mosaics. Higher plant species diversity provided by savannah-forest ecotone system alone could explain why the average group size of bonobos in Lake Tumba Landscape was larger than those reported from other sites (Inogwabini et al. 2007a, b). Judging by grouping patterns and local densities, forest-savannah mosaics may constitute the preferred habitat for bonobos, while dense forests may represent more marginal habitats. Another important ecological factor is that most savannahs and gallery forests in which bonobos are found in the Malebo region remain soaked in water for most of the year. Previous studies have also shown that some other species of great ape occur at higher densities in swampy lowland areas. Lowland gorillas in the Republic of Congo occur at the highest densities in the Likouala swamps (Fay and Agnagna 1992), adjacent to Lake Tumba Landscape. The same patterns have been also reported in orangutan (Singleton 2000). Although similar findings have not so far been made for mountain gorillas and chimpanzees, the distribution patterns of bonobos in Lake Tumba Landscape appear to reflect a general trend seen among at least two other species of great apes across their respective areas of occupancy.

In terms of management, the most important finding of this genetic analysis is that comparisons between western versus central (the Salonga National Park) and west versus northern populations resulted in greater genetic distances than expected when existing natural barriers such as major rivers that are considered. This implies that the exchange of haplotypes between western and both central and northern populations is being unexpectedly hindered. One probable explanation is that the western population of bonobos in Lake Tumba Landscape (Fig. 7.1) is surrounded by major bodies of water that prevent dispersal, apart from a 60 km gap between Lake Tumba and Lake Maindombe. This gap of land between the lakes is the only significant stretch of terra firma forest in Lake Tumba Landscape. Therefore, it is a hub for human settlements and human agricultural activities. Human activities are relatively intense and cause noticeable deforestation that fragments potential corridors for dispersal of large mammals, and specifically for bonobos. The region has been logged in the recent past, and armed poachers undertaking illegal logging activities have colonized this particular gap, thereby further isolating the western population of bonobos. If this human barrier is sufficient to prevent effective gene flow between the west and central populations, then the passage of western haplotypes to the northern population, via intermittent populations, would also be reduced, despite the low genetic distances between central and northern bonobo populations. This finding further augments the conservation importance of the western population, which accounts for ~20% of the known wild population of the bonobos.

## References


Colyn MM (1990) Distribution of guenons in the Zaire-Lualaba–Lomami river system. In: Gautier-Hion A, Bourlière F, Gautier J-P, Kingdon J (eds) A primate radiation: evolutionary biology of the African guenons. Cambridge University Press 104–124

Colyn MM, Gautier-Hion A, Verheyen W (1991) A re-appraisal of the palaeoenvironmental history in Central Africa: evidence for a major fluvial refuge in the Zaire Basin. J Biogeogr 18:403–407

Eriksson J, Hohmann G, Boesch C, Vigilant L (2004) Rivers influence the population genetic structure of bonobos (**Pan paniscus**). Mol Ecol 13:3425–3435

Eriksson J, Siedel H, Lukas D, Kayser M, Erler A, Hashimoto C, Hohmann G, Boesch C, Vigilant L (2006) Y-chromosome analysis confirms highly sex-biased dispersal and suggests a low male effective population size in bonobos (**Pan paniscus**). Mol Ecol 15:939–949

Fay JM (1997) The ecology, social organization, population, habitat and history of the western lowland gorilla (**Gorilla gorilla gorilla** Savage & Wyman, 1947). PhD thesis. Washington University. United States of America

Fay MJ, Agnagna M (1992) Census of gorillas in northern Republic of Congo. Am J Primatol 27:275–284

Gautier-Hion A, Colyn M, Gauthier JP (1999) Histoire naturelle des primates d'Afrique Centrale. Ecosystèmes d'Afrique Centrale. Union Douanière et Economique d'Afrique Centrale. Multipress

Horn AD (1976) A preliminary report on the Ecology and behavior of the Bonobo chimpanzee (**Pan paniscus** Schwarz 1929) and a reconsideration of the evolution of the chimpanzee. PhD thesis. Yale University. New Haven, United States of America

Inogwabini BI (2010) Conserving great apes living outside protected areas: the distribution of bonobos in the Lake Tumba landscape, Democratic Republic of Congo. PhD thesis, University of Kent at Canterbury, United Kingdom

Inogwabini BI, Matungila B (2009) Bonobo food items, food availability and bonobo distribution in the Lake Tumba Swampy forests, Democratic Republic of Congo. Open Conserv Biol J 3:1–10

Inogwabini BI, Matungila B, Mbende L, Abokome M, Tshimanga WT (2007a) Great apes in the Lake Tumba landscape, Democratic Republic of Congo: newly described populations. Oryx 41(4):532–538

Inogwabini BI, Bewa M, Longwango M, Abokome M, Vuvu M (2007b) The Bonobos of the Lake Tumba – Lake Maindombe Hinterland: threats and opportunities for population conservation. In Furuichi T, Thompson J (eds) The Bonobos behavior, ecology, and conservation. Springer, New York, pp 273–290

Oslily R (1998) L'influence de l'homme sur son milieu naturel: Préhistoire et environnement à la Lopé au Gabon. Canopée 11:10–11

Oslily R, Dechamps R (1994) Découverte d'une zone d'incendie dans la forêt ombrophile du Gabon ca 1500 BP – essai d'une explication anthropique et implications paléo climatiques. Bull du Centre de Recherche Académique et Scientifique de Paris-II 318:555–560

Peters CR, O'Brien E (1999) Palaeo-lake Congo: implications for Africa's late Cenozoic climate: some unanswered questions. In: Proceedings of the XVth INQUA conference, Durban, South Africa, 3–11 August 1999. Palaeoecology of Africa and surrounding Islands, vol 27, pp 11–18

Preuss J (1990) L'évolution des paysages du bassin intérieur du Zaïre pendant les quarante derniers millénaires. In: Lafranchi R, Schwartz D (eds) Paysages Quartenaires de l'Afrique Centrale Atlantique. Office de la recherche scientifique et technique outre-mer, pp 260–270

Rietkerk M, Ketner P, De Wild JJFE (1995) Caesalpinoideae and the study of forest refuge in Gabon: Preliminary results. Bull de Museum d'Histoire Nat Natl de Paris 1 et 2:95–105

Schwartz D (1991) Les paysages de l'Afrique Centrale pendant le quaternaire. In: Lafranci R, Clist B (eds) Aux origines de l'Afrique Centrale. Centres Culturels Français d'Afrique et Centre International des Civilisations Bantu (Libreville), pp 41–45

Schwartz D (1992) Assèchement climatique vers 3000BP et expansion Bantu en Afrique Centrale: quelques réflexions. Bull de la Société Géologique de Fr 163:353–361

Singleton I (2000) Ranging behaviour and seasonal movements of Sumatran orangutans in swamp forests. PhD thesis. Durrell Institute of Conservation and Ecology, University of Kent at Canterbury, Canterbury United Kingdom

Thompson JAM (1997) The history, taxonomy and ecology of the bonobo (Pan paniscus Schwarz 1929) with a first description of a wild population living in a forest/savanna mosaic habitat. PhD thesis. Oxford University, Oxford, United Kingdom

Thompson JAM (2002) Bonobos of the lukuru wildlife research project. In: Boesch C, Hohmann G, Marchant LF (eds) Behavioural diversity in chimpanzees and bonobos. Cambridge University Press, Cambridge, pp 61–70

Thompson JAM (2003) A model of biogeographical journey from Proto-pan to Pan paniscus. Primates 44:191–197

Tutin CEG, Ham RM, White LJT, Harrison MJS (1997a) The primate community of the Lopé Reserve, Gabon: diets, responses to fruit scarcity, and effects on biomass. Am J Primatol 42:1–24

Tutin CEG, White LJT, Abernethy K, Oslily R (1997b) Station d'étude des gorilles et chimpanzés de la Lopé, Gabon – dossier de présentation. Centre International de Recherche Médicale, Gabon. Wildlife Conservation Society et Musée National d'Histoire Naturelles de Paris

White LJT (2001) The African Rain forest: climate and vegetation. In: Weber W, White LJT, Vedder A, Naughton-Treves L (eds) African rain forest: ecology and conservation. Yale University Press, New Haven, pp 3–29

White LJT, Rogers MER, Tutin CEG, Williamson EA, Fernandez M (1995) Herbaceous vegetation in different forest types in the Lopé Reserve, Gabon: implications for keystone food availability. Afr J Ecol 33:124–141

Wrangham R (2009) Catching fire: how cooking made us human. Basic Books

# Chapter 8
# Forest Refugia Theory, Density Dependence and Stress Syndrome and the Proto-Pan

**Abstract** Bonobos and chimpanzees diverged three million years ago, a period during which the world went through major climatic changes due to glaciations in the south and north hemispheres. This phenomenon resulted in the aridification of the tropics, which shrank the tropical forests in Africa and left only pockets of forest refuges. The scientific consensus is that the shrinking of tropical forests likely caused bonobos and chimpanzees to diverge. Conventional logical conjectures locate the center from which the species diverged into two different strands would be around Gombe (Tanzania). Geographically, Gombe is at the edge of the distribution of the *Pan* (both bonobo and chimpanzee). The chapter compares genetic distances between subpopulations and, with the finding that bonobos at the edges of the species distribution maintain the same genetic distance from the populations located at the geographic center argues that the subpopulations were once around the center and inverts the route of Pan migration to move from the center to outer areas of the region. Explaining why glaciations or its subsequent aridification of the tropical forest caused Pan to diverge, the chapter provides density-dependence and stress syndrome as the most probable causes.

**Keywords** Glaciations · Forest refugia · Genetic diversity · Proto-pan · Density-dependence and stress syndrome

## 8.1 Introduction

Bonobos and chimpanzees diverged three million years ago (Doran and McNeilage 1998). At that very period, the world went through major climatic changes due to glaciations in the south and north hemispheres (see Chaps. 5 and 10). The glaciation resulted to the aridification of the tropics resulting in the shrinking of the moist tropical forest cover in most tropical African forested areas leaving only pockets of forest refuges, mostly located on mountains but also along the Congo (White 2001; Rietkerk et al. 1995; Maley 1991; Colyn et al. 1991). As indicated above, researchers (Kawamoto et al. 2013; Chap. 10) believe that this single event was highly likely one of the major causes that made the bonobos and the chimpanzees to diverge. The finding of the genetic studies suggests Pans (which were a single species) became



B.-I. Inogwabini, *Reconciling Human Needs and Conserving Biodiversity: Large Landscapes as a New Conservation Paradigm*, Environmental History 12,
https://doi.org/10.1007/978-3-030-38728-0_8

very dense because the forest cover diminished; this major climatic event made them become very dense in a tiny island of the forest. Discussions have been logical conjectures about where probably these pockets of Pans populations were spread: some have suggested that this was around Gombe (Tanzania) but the current genetic evidence that bonobos at the edges of their natural home range maintain the same genetic distance from the populations located at the geographic center of the area of the species occurrence suggests that the populations were once around the center. Hence, it has moved from the center to the outer areas of the region. It follows that the patches of small fragmented populations that followed the glaciations were spread in the flat zones around the Congo River where there were still some pluvial forest refugia (White 2001; Colyn et al. 1991). How did they then move to new areas where they would be kept separated as to acquire new genetic traits that separate them into two distinct species, *Pan paniscus* and *Pan troglodytes*? This chapter tries to provide a model that would explain the process of that evolution. It is based on density dependence and stress syndrome to explain the facts that followed the glaciations and its effects on the Pans.

## 8.2 Potential Effects of Density Dependence on Proto-Pan During the Glaciation

Density dependence is an important concept in population ecology. In general, density dependence happens metrics to measure growth in populations are caused by fluctuations in population sizes or densities. The metrics in question are factors affecting the growth rate (birth and immigration) and those affecting the decrease rate (death and emigration). There are two types of density dependences: positive and negative density dependences. The positive density dependence happens when population growth is stimulated by the increasing numbers of individuals within a given population while the negative density dependence is the case where the increase in population is restricted or curtailed by the upsurges in densities. If one wants, this could be expressed in terms of feedback loops: the negative and positive feedback loops; one acts positively while the other acts negatively. The negative density dependence restricts the population from growing. In large in mammals such as bonobos, density dependence prevents populations from growing through increased predation and prohibitively higher competition.

According to Brook and Bradshaw (2006), citing several authors, most biologists accept that density-dependent demographic processes, or more generally, negative feedback mechanisms, work to regulate natural populations, at least under some circumstances; and the density dependence plays an important role in natural selection (Gomulkiewicz 1999).

Effects of the changes induced by the glaciations include, but are not limited to, species extinctions and changes in the ranges of taxa that survived (Hewitt 2004). Hewitt (2004) also suggests that the spatial effects of the glaciations depended on latitude and topography; there were extensive extinctions and re-colonization in higher latitudes and altitudinal shifts and complex refugia nearer the tropics. As such, the most impacting effect of the glaciation that happened in the Pleistocene (3M Years BP) on the African forest habitats has been the reduction of forest covers in most of the continent. African forests, in their diversity, were reduced significantly (White 2001) and, according to available records, most of the continent remained baren of the vegetation with a few exceptions. As indicated above (Chaps. 5 and 10; and the introduction to this chapter), forests retracted in some smaller areas where ecological conditions remained somewhat favorable for some tree species to flourish. These areas are generically known as forest refugia and occur in most cases in the mountainous zones of the continent comprising the refugia of the Northern Africa (Marrocco–Algerian–Tunisia and Egypt; see Faleh et al. 2012), the Eastern Rift Valley (stretching from Ethiopia down to the South Eastern South Africa), and the West African complex of refugia starting from the Mayumbe Forest in the Democratic Republic of Congo to Western border of Nigeria with a center around Mount Cameroon (Central West Africa) (Plana 2004). There was, however, pluvial forest refugium in Central Congo Basin, along the Congo River and some of its major tributaries such as the Ruki-Tshuapa-Salonga, Ruki-Tshuapa-Momboyo (Colyn et al. 1991).

It follows that the patches of small fragmented forests that followed the glaciations were spread in the flat zones around the Congo River where there were still some pluvial forest refugia (White 2001; Colyn et al. 1991). This is the region where bonobos occur. As a consequence of forest fragmentation, the populations of bonobos were also parceled in small pouches, that pattern is still current in their distribution to this day. The species is known to be patchily distributed even today (Reinartz et al. 2006; Inogwabini and Omari 2005; Kortlandt 1995; Kano 1984). Needless to say, there are no estimates of the populations of the *Proto-pans* in 4–3 MY BP. However, the shrinkage of the forest cover and the consequences of the forest fragmentation that it shadowed can logically be thought of as having reduced the suitable habitat that would be available to the species. As a consequence of the limitation of the suitable habitat, the populations would have been divided and cramped in higher densities where there would be sufficient food. Indeed, compacted in these small regions (forest patches), their population density became very high, reducing therefore food to scarce levels. *Proto-pans* then became density-dependent, competition over food has then become very high as to act as an evolutionary selecting force.

The competition over food would be the first consequence of the fragmentation and the reduction of suitable forests (niche) on the *Proto-pans*. Current studies of the feeding ecology for bonobos have identified a major pattern from all the sites where they were conducted. That important pattern is that bonobos mostly consum fruits. But, the availability of fruits, as a food resource, varies seasonally (Malenky and Wrangham 1994; Wrangham 1986; Kano 1983). This leads to the second important pattern that has been observed throughout different study sites during periods of fruit shortage, which generally coincide with dry seasons in the species range, and which is that *Marantaceae* species become an important constituent of bonobo diet (Idani et al. 1994; White 1992; Badrian and Malenky 1984; Kano and Malavwa 1984; Kano 1983; Badrian et al. 1982; Horn 1980). This is particularly the case of the consumption of *Haumania liebrechtsiana*. During these periods, *Marantaceae* availability could act as a limiting factor on party (group) sizes and distribution (Malenky and Wrangham 1994; Wrangham 1986). Studies also suggest that bonobos ingest more herbaceous than do other *Pan species*. This is particularly true when it comes to *H. liebrechtsiana*. As stated previously, feeding adaptation is essential in the evolution of mammalian species. Hence, these divergences in the preferences of food types would be expected to differentiate the species in different subspecies and if combined with other factors and in the long-term perspectives to full different species.

## 8.3 Stress Syndrome Theory

Stress syndrome theory says that as population rises, the interactions between individuals of the same population become progressively more frequent. Social pressure such as intraspecific competition and social strife mount. This increasing pressure provides a progressively stronger stimulus to the central nervous system (Vaughan 1978). Because the nervous system is affected, the neuroendocrine's mechanisms are affected, altering the hormonal balance. Experimental studies suggest that stress syndrome can lead to numerous effects at different levels. These can be immunological, gastrointestinal, and cardiovascular changes (Woodford et al. 2002). For example, Rosmond (2005) suggests that continuously changing and sometimes threatening external environment may, when the challenge exceeds a threshold, activate central pathways that stimulate the adrenals to release glucocorticoids. Equally Lupien et al. (2009), found that chronic exposure to stress hormones, whether it occurs during the prenatal period, infancy, childhood, adolescence, adulthood or aging, has an impact on brain structures involved in cognition and mental health with the timing and the duration of the exposure explaining specific effects of stress on the brain.

In general, and as reported by Muller and Wrangham (2004), there are two ways vertebrates respond to stress, with the first response being mediated by the sympathetic nervous system and involving the secretion of catecholamines (epinephrine and norepinephrine) from the adrenal medulla and the second reaction involves an increase in the concentration of circulating glucocorticoids, 21-carbon steroids produced by the adrenal cortex under stimulation of ACTH from the pituitary. The molecules epinephrine and norepinephrine facilitate an arousal and vigilance; they have wide-ranging physiological effects, including increased heartbeat rates and narrow arteriolar systems. Primates, particularly humans, stress leads to hyperlipemia and polycythemia; it affects the growth hormone, and severe stress produces severe metabolic brain effect that subsequently ends by hyperactive stretch reflexes (Wyngaarden and Smith 1982). In wild chimpanzees, studies (e.g. Muller and Wrangham 2004) suggest that stress is an important factor in assessing the cost paid by the males in their fights for ranking. Indeed, contrary to bonobos, chimpanzees are known to show spells of excessive anger and fight for dominance and food. Maybe of the utmost importance, stress-induced changes lead to consequences that impact the normal reproductive function; certain reproductive performance parameters such as the timing of ovulation and relative fecundity were documented to be affected by exposure to stress (Contreras-Sánchez et al. 1998).

## 8.4 All Together: Bonobos, Chimpanzees, Refugia, Density Dependence, and Stress

As a logical consequence of reductions in food and species niches meant that the densities of the **Proto-pans** increased as a purely mechanical effect of the sheer reduction in forest habitats. This has increased the intraspecific competition over the limited food that was available by all the individuals in one given forest fragment. Increased intraspecific competition over food would have acted in the immune system as a decreasing factor for the reproduction of individuals; i.e. the chances for individuals to contribute with offsprings to the next generation would have been reduced. Given the fact that one of the strongest evolutionary forces is the survival of the fittest, which is defined as the possibility or rather the potency to procreate, different strategies would have emerged because of the hardships imposed by the shrinkage of forest habitats would have required different adaptation scenario if the Pans would have to survive through the reduction in their habitats. This would lead one group of ***Proto-pans*** to become aggressive. Aggression is known to be a behavior that mammals acquire when they become density-dependent and when food resources become scarce. Aggression can lead to *stress syndrome* and the stress syndrome would have led to the emigration of the aggressive group of ***Pans***. This could be either a defensive strategy or a way to go elsewhere in the hope to find peace and sufficient resources. Humans demonstrate these very traits; they often leave the room when they bang on the table in the hope that their message has gone through or in the hatred of individuals they feel are just not worthy to stand as their equals.

Following the above trail of arguments, the split between bonobos and chimpanzees would have been caused as a consequence of increased densities due to the fragmentation and the consequent diminution of suitable habitat and the foods it contained because of glaciation. The first group of the ***Pans*** would have been the most infuriatable, which would have left small pockets of habitats to procure foods in remote areas from their initial locations. This group, obviously, was the one that would later give birth to *Pan troglodytes*. If this paradigm works, the model would be that *Pan troglodytes* have emigrated from the Central Congo Basin to colonize the then '*vacant area*' at the edges of the forest and savannahs. Aggressivity is also a common feature documented in the *Pan troglodytes*. The second group did not take the risk to venture outside and remained within different spots of forests where food has become so limited. This second group of ***Pans***, the one that would become *Pan paniscus*, would have remained in the current distribution of the species in the interior basin of the Congo. This pluvial refuge group of Pan, which is now the bonobos, would continue to live mostly on *H. liebrechtsiana*, which plant species is very abundant in the interior of the Congo Basin.

The two groups would not have been able to re-merge because at the meantime, the geographical barrier (Congo River), which re-emerged after the crust of the continent has gone through a shock that moved some geological plates and resulted in the creation of the Albertine Rift and the separation between the Congo and the Nile basin (Horn 1976). The Congo River was a major barrier to any return pathway. The ***Pans*** that out-migrated became aggressive and, once more, divided into subspecies because of the same stress syndrome; becoming hence, the Central chimpanzee, ***Pan troglodytes***, the Nigeria–Cameroon chimpanzee, ***Pan troglodytes vellerosus***, the Eastern chimpanzee, ***Pan troglodytes schweinfurthii***, and the West African chimpanzee, ***Pan troglodytes versus***. The subspecies need not be considered as having appeared at once; a significant amount of time must have elapsed before these subdivisions could coalesce into significant differences to make subspecies. The bonobos, because they were caught in limited areas, continued to exchange genes but when the normalcy of the conditions in the area of their distribution resumed, they started moving out of the smaller patches of forests located within the pluvial forest refuge. As they migrated from one locations to the new ones within their range, they too became somewhat differentiable (Reinartz et al. 2000; Reinartz 1997).

The difference in diets and diets availability have been used to explain both the distribution and the abundance of African great apes; species of African great apes (*Gorillas gorilla* (*berengei*, *gorilla* and *graueri*), *Pan paniscus*, and *Pan troglodytes*) are more abundant in areas with abundant ground vegetation and open canopy, and tend to be in low densities in closed canopy forest (Fay 1997; Williamson 1988; Casimir 1975; Schaller 1963). This trait is attributed to their primary herbivorous feeding behavior (Fay 1997; Tutin and Fernandez 1984; Sabater Pi 1977; Schaller 1963). A magnetic factor attracting all these species to these open canopy forest types is the availability of terrestrial herbaceous vegetation types, which are often felt to be the fallback food species in cases of food shortage. But, in the case of the bonobos, consumption of the terrestrial herbaceous vegetation has been demonstrated to be more than in its nearer cousin, chimpanzees. Discussing why the chimpanzees and bonobos have sensibly different cultures and why the former is more aggressive than the latter species, scientific studies have been conducted in genetics and other areas of inquiry. What they have come to establish was that the abundant contribution of the terrestrial herbaceous vegetation in the diet of the bonobos would be the most important factor separating the species.

The finding on the effects of diets in the behavior of bonobos and chimpanzees is of critical importance in the discussions about their separation as two distinct species. Indeed, if that finding is brought into the perspective developed in this chapter, it has several implications. Firstly, it would help decide about which species is nearer to humans between the two ***Pans***. Genetically, this question cannot be answered by a clear-cut separation as both bonobos and chimpanzees share the same amount of genes with humans. But if we look at the model being presented here and factor in the model the fact that humans have emerged in savannah of the eastern Africa, the model here suggests that chimpanzees are closer to humans than bonobos. This is because while bonobos remained in the Central Congo Basin, the chimpanzees had to move to open areas; it is in these areas that they have started to learn becoming bipedal walkers, which would have been forced on them by fears of being preyed by savannah resident predators such as lions. Indeed, bipedalism has been described as one of the most important traits separating humans from the rest of the species of great apes. The second important implication is about the use of tools: chimpanzees have been described to use tools for fetching their foods and they do this more often than do bonobos. Indeed, this model would suggest that because of food scarcity that was imposed on chimpanzees during the glaciations, they had to learn alternative ways of finding food in the savannahs. A better alternative to seating and uprooting shoots of Marantacea species was to start using tools to dig tubers of yams and other roots. Hence, chimpanzees acquired these skills and internalized them throughout their evolutionary trip until to where we currently stand. The last thing that could be explained by the evolutionary model being presented here is the differences observed in the temperaments of each species of the ***Pans***.

## References

Badrian NL, Malenky RK (1984) Feeding ecology of Pan paniscus in the Lomako forest, Zaire. In: Susman RL (ed) The pygmy chimpanzee: evolutionary biology and behavior. Plenum Press, pp 325–346

Badrian NL, Badrian AJ, Susman LR (1982) Preliminary observations on feeding behaviour of Pan paniscus in the Lomako Forest of Central Zaire. Primates 22:173–181

Brook BW, Bradshaw CJA (2006) Strength of evidence for density dependence in abundance time series of 1198 species. Ecology 87(6):1445–1451

Casimir M (1975) Feeding ecology and nutrition of an eastern gorilla group in the Mount Kahuzi region, République du Zaire. Folia Primatologica 24:136–181

Colyn MM, Gautier-Hion A, Verheyen W (1991) A re-appraisal of the palaeoenvironmental history in Central Africa: evidence for a major fluvial refuge in the Zaire Basin. J Biogeogr 18:403–407

Contreras-Sánchez WM, Schreck CB, Fitzpatrick MS, Pereira CB (1998) Effects of Stress on the Reproductive Performance of Rainbow Trout (Oncorhynchus mykiss). Biol Reprod 58:439–447

Doran DM, McNeilage A (1998) Gorilla ecology and behaviour. Evol Anthropol 6:120–131

Faleh ARB, Granjon Tatard C, Othmen AB, Said K, Cosson JF (2012) Phylogeography of the greater Egyptian Jerboa (Jaculus orientalis) (Rodentia: Dipodidae) in Mediterranean North Africa. J Zool 286:208–220

Fay JM (1997) The ecology, social organization, population, habitat and history of the western lowland gorilla (Gorilla gorilla gorilla Savage & Wyman, 1947). Ph.D. thesis, Washington University, United States of America

Gomulkiewicz R (1999) The effects of density dependence and immigration on local adaptation and niche evolution in a black-hole sink environment. Theor Popul Biol 55:283–296

Hewitt GM (2004) Genetic consequences of climatic oscillations in the quaternary. Philos Trans R Soc Lond B 359:183–195

Horn AD (1976) A preliminary report on the ecology and behavior of the Bonobo chimpanzee (Pan paniscus Schwarz 1929) and a reconsideration of the evolution of the chimpanzee. Ph.D. thesis. Yale University. New Haven, United States of America

Horn AD (1980) Some observations on the ecology of the bonobo chimpanzee (Pan paniscus Schwarz 1929) near Lake Tumba, Zaire. Folia Primatol 34:145–169

Idani G, Kuroda S, Kano T, Asato R (1994) Flora and vegetation of Wamba forest, central Zaire with reference to bonobo (Pan paniscus) food. Tropics 3:309–332

Inogwabini BI, Omari I (2005) A landscape-wide distribution of Pan-paniscus in the Salonga National Park, Democratic Republic of Congo. Endanger Species Update 22:116–123

Kano T (1983) An ecological study of the pygmy chimpanzees (Pan paniscus) of Yalosidi, Republic of Zaire. Int J Primatol 1:1–31

Kano T (1984) Distribution of pygmy chimpanzees (Pan paniscus) in the central Zaire basin. Folia Primatol 43:36–52

Kano T, Mulavwa M (1984) Feeding ecology of the pygmy chimpanzees (Pan paniscus) of Wamba. In: Susman RL (ed) The pygmy chimpanzee: evolutionary biology and behaviour. Plenum Press 233–274

Kawamoto Y, Takemoto H, Higuchi S, Sakamaki T, Hart JA, Hart TB, Tokuyama N, Reinartz GE, Guislain P, Dupain J, Cobden AK, Mulavwa MN, Yangozene K, Darroze S, Devos C, Furuichi T (2013) Genetic structure of wild bonobo populations: diversity of mitochondrial DNA and geographical distribution. PLoS ONE 8(3):e59660. https://doi.org/10.1371/journal.pone.0059660

Kortlandt A (1995) A survey of the geographical range, habitats and conservation of the Pygmy chimpanzee (Pan paniscus): ecological perspective. Primate Conserv 16:21–36

Lupien SJ, McEwen BS, Gunnar MR, Heim C (2009) Effects of stress throughout the lifespan on the brain, behaviour and cognition. Nat Rev Neurosci 10:434–445

Malenky RK, Wrangham RW (1994) The relative importance of terrestrial herbs for bonobos and chimpanzees: comparative data from Lomako and Kibale. Bulletin of Chicago Academy of Science 15:7

Maley J (1991) The African rain forest vegetation and palaeoenvironments during late quaternary. Clim Change 19:79–98

Muller MN, Wrangham RW (2004) Dominance, cortisol and stress in wild chimpanzees (Pan troglodytes schweinfurthii). Behav Ecol Sociobiol 55:332–340

Plana V (2004) Mechanisms and tempo of evolution in the African Guineo-Congolian rainforest. Philos Trans R Soc Lond B 359:1585–1594

Reinartz GE (1997) Patterns of genetic biodiversity in the bonobo (Pan paniscus). Ph.D. thesis, University of Wisconsin. Milwaukee, United States of America

Reinartz GE, Karron JD, Phillips RB, Weber JL (2000) Patterns of microsatellite polymorphism in the range-restricted bonobo (Pan paniscus): considerations for interspecific comparison with chimpanzees (P. troglodytes). Mol Ecol 9:315–328

Reinartz G, Inogwabini BI, Mafuta N, Lisalama WW (2006) Effects of forest type and human presence on bonobo (Pan paniscus) density in the Salonga National Park. Int J Primatol 27(2):603–634

Rietkerk M, Ketner P, De Wild JJFE (1995) Caesalpinoideae and the study of forest refuge in Gabon: preliminary results. Bull de Museum d'Histoire Naturelle National de Paris 1 et 2:95–105

Rosmond R (2005) Role of stress in the pathogenesis of the metabolic syndrome. psychoneuroendocrinology 30(1):1–10

Sabater Pi J (1977) Contribution to the study of alimentation of lowland gorillas in the natural state in Rio Muni, Republic of Equatorial Guinea (west Africa). Primates 1:183–204

Schaller GB (1963) The mountain gorilla: ecology and behaviour. Chicago University Press

Tutin CEG, Fernandez M (1984) Nation-wide census of gorilla and chimpanzee populations in Gabon. Am J Primatol 6:313–336

Vaughan TA (1978) Mammalogy, 2nd edn. Saunders Company

White JF (1992) Activity budgets, feeding behaviour, and habitat use of pygmy chimpanzees at Lomako, Zaire. Am J Primatol 26:215–223

White LJT (2001) The African rain forest: climate and vegetation. In: Weber W, White LJT, Vedder A, Naughton-Treves L (eds) African rain forest: ecology and conservation. Yale University Press, p 3–29

Williamson EA (1988) The behaviour ecology of western lowland gorillas. Ph.D. Thesis. University of Stirling. Stirling, United Kingdom

Woodford MH, Butynski TM, Karesh WB (2002) Habituating the great apes: the disease risks. Oryx 36(2):153–160

Wrangham RW (1986) Ecology and social relationships in two species of chimpanzees. In: Rubenstein DI, Wrangham RW (eds) Ecological aspects of social evolution in birds and mammals. Princeton University Press 352–378

Wyngaarden JB, Smith Jr, LH (1982) Cecil textbook of medicine, 16th edn. Saunders Company

# Chapter 9
# Wild Bonobos and Wild Chimpanzees and Human Diseases

**Abstract** Reported effects of Ebola on western lowland gorillas in the last years of the Twentieth Century brought the underestimated threat that infectious diseases pose to wildlife to the forefront of the international conservation community. Human-borne parasites have since been acknowledged to threaten different species of wildlife and particularly African great apes (Bonobos *Pan paniscus*, *Pan troglodytes,* and gorillas *Gorilla gorilla*) that were documented to contract diseases that affect human communities. The chapter examines 6 diseases, which have been documented to be transmitted between wildlife and humans and provides a short description of each type and its mode of transmission and discusses its potential to affecting wild bonobos within the ecological conditions in the species range and the regional biogeography.

**Keywords** Anthrax · Ebola · Herpes · *Human Immunodeficiency Virus* (HIV) · Simian Immunodeficiency Virus (SIV) · Influenza · Malaria

## 9.1 Introduction

Recently reported effects of Ebola on western lowland gorillas (Walsh et al. 2003, 2007; Bermejo et al. 2006) have brought the underestimated threat that infectious diseases pose to wildlife (Rachowicz et al. 2005; Voyles et al. 2009) to the forefront of the international conservation community. Human-borne parasites have since been acknowledged to threaten different species of wildlife (Inogwabini and Leader-Williams 2012; Cunningham et al. 2003; Vitousek et al. 1996). Studies have shown that although many infectious agents are species-specific, a number of pathogenic organisms cross the species barrier and cause severe clinical diseases in new hosts (Walsh et al. 2007). Specifically, bi-directional zoonotic movements of diseases between humans and wildlife have been documented; but for obvious reasons, efforts have been mostly devoted to look at the potential impacts of these cross-species disease transmission on humans in order to prevent catastrophic effects on human well-being. It is currently acknowledged that mild pathogen in one species may emerge as a new infectious disease when it crosses the natural barriers and, as such, would have an unpredictable and severe impact on its new host community (Daszak et al. 2000; Dobson and Foufopoulos 2001). Because of this unpredictability, human transmitted



B.-I. Inogwabini, *Reconciling Human Needs and Conserving Biodiversity: Large Landscapes as a New Conservation Paradigm*, Environmental History 12,
https://doi.org/10.1007/978-3-030-38728-0_9

diseases should, therefore, be one of the central conservation concerns and justifies recent efforts to investigate diseases that are thought to cross natural barriers. This is particularly of interest for species of great apes because of their genetic proximity, which would increase and ease the bi-directional zoonotic flow.

The African great ape species (Bonobos *Pan paniscus*, chimpanzees *Pan troglodytes* and gorillas *Gorilla gorilla*) have been documented to contract similar diseases as those that affect human communities. Evidence of zoonoses being transmitted between humans and these species of great apes are currently accumulating. However, studies on infectious diseases affecting bonobos are only at their early stages, a fact due to several reasons; chief among these reasons is the fact that the species occurs only within a single national jurisdiction, that of the Democratic Republic of Congo, a country that is both poorly known in general and which has known long times of instability that could not allow long term research. In this chapter, we review some of the published data on sicknesses that can affect both humans and bonobos and discuss them within the global framework of great apes and diseases and their conservation. This chapter also explores hypotheses that can explain why bonobos maybe more fragile for one type of disease but naturally buffered from the spread of other sicknesses based on the island biogeography of their range. The chapter deals mostly with the ecological conditions that can provide favorable environments for the transmissibility and spread of each disease discussed.

Table 9.1 provides 12 types of diseases known to cross between humans and great apes; only 6 diseases that have been documented to be transmitted between wildlife and humans, particularly those that have happened to be transmitted to the African great apes were examined in some lengths in this chapter. A short description of each of the first 6 types is presented, its mode of transmission and the move on discussion its potential to affecting wild bonobos given the human ecological conditions in the species range, as well as the biogeography in that region. It is noteworthy that disease types in Table 9.1 are in alphabetical order and the presentation does not imply any order of importance of the prevalence, the severity or frequencies of occurrence of each disease.

## 9.2 Potential Bi-Directional Human Ape Disease Transmissions

### *9.2.1 Anthrax*

Anthrax is caused by the bacterium *Bacillus anthracis*. According to the World Health Organization (WHO 2008), *Anthrax* is primarily a disease of herbivores; humans almost invariably contract it in its natural form directly or indirectly from animals or animal products (handling carcasses, hides, bones, etc. from animals that died of the disease) while animals (wildlife or livestock) generally acquire it when ingesting spores through grazing or browsing. However, in animals, flies appear to

**Table 9.1** 12 types of disease mostly transmitted between human and wildlife

| Disease type | Evidenced in great apes | Evidenced in bonobos by |
|---|---|---|
| Anthrax | Gorillas[1], Chimpanzees[1] | Calvignac-Spencer et al. (2012) |
| Ebola | Gorillas[2, 3, 4], Chimpanzees[2, 3] | Yoshida et al. (2016) |
| Herpes | Gorillas[5], Chimpanzees[5] | Yoshida et al. (2016) |
| HIV - SIV | Gorillas[5], Chimpanzees[5] | |
| Influenza | Gorillas[6], Chimpanzees[6], Orangutans[6] | |
| Malaria | Gorillas[7], Chimpanzees[7] | Liu et al. (2017) |
| Measles | Gorillas[8], Chimpanzees[9] | Dunay et al. (2018) |
| Monkey pox | Gorillas[10], Orangutans[10], Chimpanzees[11, 12] | Maibach et al. (2017) |
| Poliomyelitis | Chimpanzees[13] | |
| Para-influenza | Gorillas[6], Chimpanzees[6] | |
| Pneumonia | Gorillas[13, 14], Chimpanzees[14] | |
| Salmonella | Gorillas[15], Chimpanzees | |
| Trypanosomiasis | Chimpanzees[16] | Maibach et al. (2017) |

[1]Leendertz et al. (2006), [2]Altizer et al. (2006), [3] Bermejo et al. (2006), [4]Gillespie and Chapman (2008), [5]Seimon et al. (2015), [6]Buitendijk et al. (2014), [7,8]Hastings et al. (1991), [9]Calvignac-Spencer et al. (2012), [10]Cho and Wenner (1973), [11]Mutombo et al. (1983), [12]Maibach et al. (2017), [13]Dunay et al. (2018), [14]Butynski (2001), [15]Nizeyi et al. (2001), [16]Keita et al. (2014)

play an important role in explosive outbreaks in areas where the disease is endemic, inhalation of dust maybe important on occasion and direct animal-to-animal transmission is believed to occur but in insignificant extent, excluding carnivores feeding on other victims of the disease. Human trading activities have been described for long to be one of the most important reasons responsible for the geographic spread of the disease. There is a high age- or sex-related bias in animals general, which contrasts with the situation in humans where this is not apparent. Despite the availability of significant knowledge on this disease and how it is transmitted and spread, however, many questions and holes remain in the current understanding of how animals acquire anthrax and how it is transmitted. Animal—human transmission ratios reflect the economic conditions, quality of surveillance, social traditions, dietary behavior, etc., in specific countries or regions where the disease is present. The infection is influenced by two categories of ecological factors, including (a) factors affecting sporulation and germination (pH, temperature, water activity, and cation level) and (b) seasonal factors (availability of herbs for grazing, the health of the host, insect populations, and human activities.

### 9.2.2 Ebola

*Filoviridae ebolavirus causes Ebola.* It is a virus with five distinct types (Ebola virus *Zaire ebolavirus*, Sudan virus *Sudan ebolavirus*, Taï Forest virus *Taï Forest ebolavirus* also known as *Côte d'Ivoire ebolavirus*, and Bundibugyo virus *Bundibugyo ebolavirus*) which are known to cause disease in humans while the fifth species (Reston virus *Reston ebolavirus*) has not caused disease in humans, yet thought it affects nonhuman primates (Baize et al. 2014). According to the American Center for Disease Control (CDC 2015), Ebola is spread through direct contact with infected blood or body fluids; the virus is not transmitted through the air. However, large droplets (splashes or sprays) of respiratory or other secretions from a person who is sick with Ebola could be infectious. However, in an experiment conducted on Six rhesus monkeys (Macaca mulatta), Johnson et al. (1995) demonstrated that animals whose serological status was negative for filovirus reactive antibody before being exposed to aerosol sprays containing very small droplets of Ebola virus (Zaire) were infected after the experiment. This suggested, therefore, that the probability for the disease to be air-transmitted cannot be entirely ruled out. Ebola frequently happens in ecological conditions of dry conditions following the rainy season whereas fruiting patterns and increased numbers and aggregation of insects and mammals act as mechanisms of its seasonality (Altizer et al. 2006).

### 9.2.3 Herpes

Herpes is a latent disease; and recurring infections of herpes are caused by is a group of viruses known as Herpesviridae. The members of this family are also known as herpesviruses. The infections are latently destructive of cells by disruption of lysis (hence their name as lytic infections). Herpes affects wildlife (including mammals, birds, fish, reptiles, amphibians, and molluscs). At least five of its forms, including 9 herpesviruses that are widespread in humans. These latter are herpes simplex viruses 1 and 2, varicella-zoster virus, EBV (Epstein-Barr virus), human cytomegalovirus, human herpesvirus 6, human herpesvirus 7, and Kaposi's sarcoma-associated herpesvirus). Most people have herpesviruses in latency. Modes of transmission are via aerosol, breast milk, oral-fecal transmission, physical contact, respiratory contact, saliva, sexual intercourse, tissue transplant, transfusions, and urine.

Historically, only Herpes simian B virus also known as *Macacine herpesvirus 1* has been reported to be endemic within macaque monkeys. Because when this form of disease is transmitted to humans, the central nervous system of patients becomes severely affected, leading to permanent neurological dysfunction and, eventually to death, Herpes simian B virus has been for long the focus of different attention. Indeed, if improperly treated, the severity of the disease is conducive to an extremely high rate of fatality, nearing high figures such as 80% in human lives, which (of course) deserves more human attention. Recent research (Seimon et al. 2015) does, however, show high levels of prevalence of Herpesviridae in chimpanzees (23.5%) and gorillas (39.1%) in the Congo Basin.

### 9.2.4 Human Immunodeficiency Virus (HIV) and Simian Immunodeficiency Virus (SIV)

*Human immunodeficiency virus* (HIV) or the acquired immune deficiency syndrome (AIDS) ranks as one of the most important infectious diseases to have emerged in the past century and continues to be one of the most serious public health threats in the twenty-first century. AIDS is caused by two lentiviruses: the human immunodeficiency virus types 1 and 2 (HIV-1 and HIV-2). The closest simian relatives of HIV-1 are SIVcpz in chimpanzees (*Pan troglodytes troglodytes*) and SIVgor in gorillas (*Gorilla gorilla*) from west-central Africa (Keele et al. 2006; Van Heuverswyn et al. 2006). SIV found in sooty mangabeys (*Cercocebus atys*) from West Africa are the closest relatives of HIV-2 (Gao et al 1999; Hirsch et al. 1989). HIV/AIDS is transmitted only through certain fluids—blood, semen, pre-seminal fluid, rectal fluids, vaginal fluids, and breast milk—from an HIV-infected individual. However, these fluids cannot transmit HIV/AIDS unless they come to contact with a mucous membrane or damaged tissue or be directly injected into the bloodstream (from a needle or syringe) for transmission to possibly occur. Mucous membranes can be found inside the rectum, the vagina, the opening of the penis, and the mouth.

### 9.2.5 Influenza

According the World Health Organization (WHO 2014), influenza occurs in three large seasonal variants labeled type-A, type-B, and type-C. Type- A influenza viruses are divided into A(H1N1) and A(H3N2) subtypes whereas the type-B viruses are also separated in two lineages (**Victoria and Yamagata**). The influenza virus continues to evolve and causes annual epidemics and occasional pandemics of disease in humans. According to Webster et al. (1992), the gene pool of influenza A viruses in aquatic birds provides all the genetic diversity required for the emergence of pandemic influenza viruses for humans, lower animals, and birds. But Webster et al. (1992) also say that viruses of influenza type-A shift both antigenically and genetically shifting in humans, pigs, and horses; contrasting, therefore, with the avian influenza viruses that are rather static evolutionarily.

In terms of their respective acuities, influenza type-A and type-B are severer than type-C, which causes milder infections. Throughout human history, it has been noticed that Influenza can become pandemic; with the most notorious influenza pandemic being the 1918–1919 'Spanish Flu' that, with case-fatality rates greater than 2.5%, killed between 20 and 50 million people around the world (Taubenberger and Morens 2006; Reid 2005). Zoonotic pathways to human contamination have been described and humans are susceptible to be infected with influenza viruses that are hosted by animals. Hence, for example, humans can contract avian influenza virus, swine influenza virus, horse influenza virus, dog influenza, etc., although some of these are not easily transmittable between animals and humans. However, when transmissions occasionally occur, infected humans may suffer diseases whose effects would range between mild to severe that can cause even death. Influenza is transmitted by airborne, droplet, and direct contacts (Brankston et al. 2007) and the latter mode of transmission (direct contact) is either contact with infected animals or contaminated environments. Of the utmost importance in the particular context of bonobos is the finding that by Lowen et al. (2007) that the airborne transmission of influenza virus is dependent upon both ambient relative humidity and temperature.

### *9.2.6 Malaria*

Plasmodium parasites cause *Malaria*. Plasmodium parasites are transmitted by infected *Anopheles* mosquitoes, that cause Malaria in humans and have been found to infect other great apes such as chimpanzees and gorillas throughout Africa (Liu et al. 2014). There exist about 400 species of *Anopheles* but 30 serve as the vectors for malaria of which only five types cause malaria in humans. The species that cause malaria in humans are *Plasmodium vivax* for Asia and Latin America and *Plasmodium falciparum* in Africa. Ecologically, *Anopheles* is a water-dependent species because it uses stagnant waters to lay eggs that hatch into larvae before being transformed into mosquitoes (Beebe et al. 2000). Despite the fact that all species of Anopheles are water-dependent, each species does, however, exhibit its own preference of what types of aquatic habitats it would survive in. Some prefer small, shallow collections of freshwater, such as puddles and hoof prints, which are abundant during the rainy season in tropical countries. The transmission of Malaria depends on climatic conditions, which exhibit small-scale ecological variability and temporal changes and, therefore, affect both the geographic spread, as well as the temporal transmission patterns. These small-scale ecological variables are inclusive, principally, of rainfall patterns, temperature, and humidity. According to Minakawa et al. (1999), other ecological micro correlates would include aquatic habitat types (spatial sizes of water bodies, forest canopy coverage, and emergent plants), water biophysical attributes (water pH and turbidity, substrates of water bodies, water surface debris), influence of the neighboring environment (distance to the nearest human settlement and the distances to large water bodies (such a large lakes, large rivers), and the biological diversity in the waters (presence of algae, etc.).

## 9.3 Bonobos, Zoonoses and Human Epidemics in Perspectives

Woodford et al. (2002), citing a long list of studies (e.g. Wolfe et al. 1998, 2001; Sleeman et al. 2000; Homsy 1999) stated that the great apes and humans share a substantial collection of transmittable diseases, ranging from the most commonly thought of ones such as cold and pneumonia, influenza and the most difficult to decipher how they could be transmitted from humans to wildlife and vice versa such as hepatitis, smallpox, chickenpox, bacterial meningitis, tuberculosis, measles, rubella, mumps, yellow fever, paralytic poliomyelitis, encephalomyocarditis, and Ebola fever. Parasitic diseases are also shared, including malaria, schistosomiasis, giardiasis, filariasis, and infection with *Strongyloides* spp., *Entamoeba* spp., *Oesophagostomum* spp., *Acanthocephala* spp., *Cyclospora* sp., *Giardia* sp., and *Sarcoptes* sp. This discussion will focus on diseases whose impact could be very serious on bonobos and chimpanzees of the Lake Tumba Landscape.

Anthrax has been described in gorillas and chimpanzees (Leendertz et al. 2006) but not known report of its evidence on bonobos is available at the moment. However, ecological conditions that enhance the disease transmission and spread prevail in their fullest intensities in the natural range of the bonobos. During rainy seasons, temperatures are rather high, which means around 25° C annually; the precipitations are abundant and the water activity in terms of water runoff is equally high. Hot-dry weather characterize what Hugh-Jones and Blackburn (2009) call Anthrax seasons and are expected to increase the transmissibility of diseases. Hence conditions prevailing at the heart of the bonobos distribution where ambient temperatures and humidity levels are high would provide a good environment to the proliferation of *B. anthracis*. However, other conditions of the central region of the species distribution are made of thick forest covers and large expanses of swamps that confine the distribution of livestock and other grazers in some forest gaps, limiting, therefore, the potential transmission. This is not the case in areas such as Lukuru Study site in the Southern part of the Salonga National Park and Malebo, which are at the edges of the bonobo distribution and have large areas of savannahs where livestock and other grazers can thrive. Indeed, industrial livestock had been present in the area of Malebo since the 1950s. If, as Ganz et al. (2014) have shown, as some of its close relatives, *B. anthracis* has some activity outside of its vertebrate hosts, and principally interactions with grass that could promote anthrax spore transmission to grazing hosts, there is possibility for the bonobos in savannah-forest mosaics to contract Anthrax by way of contact with these plants. Indeed, in Malebo, bonobos have been documented to cross savannahs in search of savannah fruits in dry seasons (Inogwabini and Matungila 2009). Given the possibility that cattle in that zone could contract anthrax and leave it dormant in the grass of the region and despite the fact that Inogwabini and Leader-Williams (2012) have not considered the possibility in their mathematical model on diseases and bonobos, bonobo conservation efforts in the Malebo area should have a wildlife–livestock–human disease alert component to track any potential Anthrax outbreak.

Ebola has been documented to affect African great apes, particularly chimpanzees and western lowland gorillas. So far, it has not been described in bonobos yet. However, ecological conditions that help Ebola develop are prevalent in the range of the bonobos (Peterson et al. 2004). In a mathematical model about the potential for diseases to affect bonobos, Inogwabini and Leader-Williams (2012) found a low correlation coefficient between the extent of the occurrence of bonobos but had argued this finding did not mean that Ebola could not affect bonobos but would reflect the lack available documented geographic data. Inogwabini and Leader-Williams (2012) also suggested, following Walsh et al. (2005), that large rivers may have served as barriers that have kept the ravaging effects of Ebola at a safe distance from bonobos; major rivers in other parts of Central Africa having been felt to have played a key role in containing Ebola, and preventing waves of the disease from emerging outside of its original foci. But Ebola did and does occur in the range of the species; it broke out in 2007 at Dekese and Kole and in 2008 at Boende. Very recently (2014) Ebola also occurred at Isaka near Lokolama on the route between Boende and Watshinkengo. While Dekese and Kole are located at the margins of the bonobo range, Boende and Isaka (Lokolama) are located at what can be considered the center of species range. The recent haemorrhagic episode at Isaka (Lokolama) suggests the potency for Ebola to strike bonobos and humans residing in this part of DRC at any time. This bleak projection is of considerable relevance if one considers the fact that the human populations are increasing in alarming rates in the region in question combined with the advanced levels of poverty that push people to rely exclusively on natural resources, and particularly the human dependency on bushmeat as a source of animal protein.

Although historically, only Herpes simian B virus also known as *Macacine herpesvirus 1* has been reported to be endemic within macaque monkeys, recent research findings by Seimon et al. (2015) that chimpanzees and gorillas have high levels of prevalence of Herpesviridae should alert the wildlife conservation practitioners. The alert is justified because the modes of transmission of herpes (particularly through oral-fecal transmission, physical contact, respiratory contact, saliva, sexual intercourse, and urine) are likely to help the transmission to become a two-way directional action, from wildlife to humans, as well as backward (from humans to wildlife). Indeed, conditions of sanitary installations in the range of the bonobos are very minimal and most people excrete and urinate in open areas in the forest. If infected people excrete and urinate in this way, bonobos stepping on defecations and urines are likely to be contaminated by human forms of herpes, which would have effects that are different from the effects they would have on humans.

The exact conditions and circumstances of cross-species transmission of SIVs from primates to humans remain unknown. But, clearly, human exposure to blood or other secretions of infected primates, through hunting and butchering of primate bushmeat, represents the most plausible source for human infection. Bites and other injuries caused by primates kept as pet animals can favor a possible viral transmission. These conditions are all prevalent in bonobo range and possible zoonotic exchange between the species and humans are still possible.

Indeed, factors associated with single cross-species transmission have to be differentiated from those associated with epidemic spread, the latter being a combination of multiple factors. The epidemiological history for the known SIV cross-species transmissions indicates that there are forms that are restricted to West Africa (Pedersen and Davies 2010; Hahn et al. 2000), others are present in isolated rural areas in Sierra Leone and Liberia and another specific independent cross-species transmission from SIVs from apes to humans has spread across Africa and to all the other continents. One form is endemic in Cameroon and represents 1% of HIV infections (Plantier et al. 2009; Brennan et al. 2008). There has been a diversity of subtypes co-circulating in the Democratic Republic of Congo. Recent studies indicated chimpanzees and gorillas are infected by HIV but very little evidence has been gathered from bonobos. Evidence gathered from wild chimpanzees concluded that HIV-1 group arose from geographically distinct chimpanzee populations in south Cameroon (Keele et al. 2006) whereas the SIV borne by western lowland gorillas is closer to the form found in the geographical ranges of central chimpanzees and western gorillas. These variations and the multitude of types and subtypes show that our knowledge on HIV diversity and possible cross-species transmissions is still incomplete, and illustrates how rapidly new viruses can spread today throughout the home range of the bonobos. Therefore, the paucity of evidence of the presence of HIV-SIV on bonobos does not mean that this type of infection will not affect the species. There is a need to be careful in planning conservation activities for the species.

Malaria has been reported to be prevalent within chimpanzee populations (Duval et al. 2009) though there is no record from the chimpanzees of the Lake Tumba Landscape. Of course, this lack of documentation would be mostly attributable to the fact that the chimpanzees in the Lake Tumba Landscape have never been in the center of research activities before; hence the lack of information should not equate with the absence of malaria or any other diseases. Indeed, the area of Ngiri where the chimpanzees are located within the Lake Tumba Landscape is known to harbor large numbers of mosquitoes; people visiting the area are stunned by how many mosquitoes and how frequent they do bite people even during the daylight with no major seasonal differences. That this is the case is, however, understandable given the fact that *Anopheles* is, ecologically a water-dependent species (Beebe et al. 2000; Minakawa et al. 1999). Indeed, as described in Chap. 5, the region between the Congo River and the Ubangui where the chimpanzees occur in the Lake Tumba Landscape is an area essentially made of swamps; even when the area is dried out during dry seasons there are still significant ponds of stagnant waters. These ponds, obviously, provide favorable ecological conditions for mosquitoes to reproduce and multiply easily. Climatic conditions, under which malaria proliferates, particularly high and constant temperature with higher humidity present in the Lake Tumba Landscape are conducive to an environment that would allow easy multiplications of the Anopheles.

These conditions are relatively the same across the Lake Tumba Landscape; so the likelihood for mosquitoes biting bonobos in the wild is also high. Indeed, Krief et al. (2010) identified *Plasmodium falciparum* and *Plasmodium malariae*-related parasites in bonobos; the *Plasmodium falciparum* they identified was located at the mitochondrial genomes but suggested this species was distinct from the one found in humans. These findings indicate that bonobos in the wild may have been suffering from malaria. Whether this evidence means the species may have known epidemics such as those developed by humans is difficult to assert. What can be said is that given the high lethality of malaria in humans, bonobo conservation schemes should include the possibilities for the diseases to become a significant threat on the species.

Other diseases that have been described to potentially and effectively affect wildlife species include measles, monkeypox, poliomyelitis, parainfluenza, pneumonia, salmonella, and trypanosomiasis. While all of them have not been described in bonobos, most were found in other species of great apes. This means that the probabilities for bonobos contracting these diseases are far from being closer to nil. For example, in a mathematical model predicting effects of different diseases on bonobos, Inogwabini and Leader-Williams (2012) found that wild bonobos were in decreased densities in areas of the home range where *Trypanosomiasis* was rather endemic. This finding, while not to be equated with the prevalence of the disease in wild bonobos, is an indication that some non-documented factors may be playing on the species even when there is no scientific evidence. Indeed, this is likely to be the case not only for bonobos but also for other wildlife species as the study of diseases in wildlife species is relatively recent. Also, even if it had been of a long history, research on diseases would not be able to account for all the possible diseases that can affect wildlife species nor even humans.

This latter observation is a plea for the more general precautionary approach to be applied all the time to guard bonobos and chimpanzees against not only measles, monkeypox, poliomyelitis, parainfluenza, pneumonia, salmonella, and trypanosomiasis but also against any potential diseases. In fact, this precautionary approach should be part of the larger conservation campaign to raise the awareness of communities not only for the wellbeing of wildlife species but also for the benefit of their own good health. Of course, an exaggerated gloomy picture of diseases being dispersed by great apes or any other wildlife species can have counterproductive effects as communities may then target incriminated species as the source of high death penalties for humans. This, in turn, would lead communities to retaliate indiscriminately against wildlife species, as has been reported in several occasions.

## References

Altizer S, Dobson A, Hosseini P, Hudson P, Pascual M, Rohani P (2006) Seasonality and the dynamics of infectious diseases. Ecol Lett 9:467–484

Baize S, Pannetier D, Pharm D, Oestereich L, Rieger T, Koivogui L, Magassouba N, Soropogui B, Sow MS, Keïta S, De Clerck H, Tiffany A, Dominguez G, Loua M, Traoré A, Kolié M, Malano ER, Heleze E, Bocquin A, Mély S, Raoul H, Caro V, Cadar D, Gabriel M, Pahlmann M, Tappe D, Schmidt-Chanasit J, Impouma B, Diallo AK, Formenty P, Van Herp M, Günther S (2014) Emergence of Zaire Ebola virus disease in Guinea. N Engl J Med 371:1418–1425

Beebe NW, Bakote'e B, Ellis JT, Cooper RD (2000) Differential ecology of **Anopheles punctulatus** and three members of **Anopheles farauti** complex of mosquitoes on Guadalcanal, Solomon Islands, identified by PCR-RFLP Analysis. Med Vet Entomol 14:308–312

Bermejo M, Rodriguez-Teijeiro JD, Illera G, Barroso A, Vila C, Walsh PD (2006) Ebola outbreak killed 5000 gorillas. Science 314:1564

Brankston G, Gitterman L, Hirji Z, Lemieux C, Gardam M (2007) Transmission of influenza A in human beings. Lancet Infect Dis 7:257–265

Brennan CA, Bodeller P, Coffey R, Devare SG, Golden A, Hackett J Jr, Harris B, Holzmayer V, Luk KC, Schochetman G, Swanson P, Yamaguchi J, Vallaria A, Ndembi N, Ngansop C, Makamche F, Mbaya D, Gürtler LG, Zekeng L, Kaptuel L (2008) The prevalence of diverse HIV-1 strains was stable in Cameroonian blood donors from 1996 to 2004. J Acquir Immune Defic Syndr 49(4):432–439

Buitendijk H, Fagrouch Z, Niphuis H, Bogers WM, Warren KS, Verschoor EJ (2014) Retrospective serology study of respiratory virus infections in captive greatapes. Viruses 6:1442–1453

Butynski TM (2001) Africa's Great Apes: An Overview of Current Taxonomy, Distribution, Numbers, Conservation Status, and Threats. Smithsonian Institution Press

Calvignac-Spencer S, Leendertz SAJ, Gillespie TR, Leendertz FH (2012) Wild great apes as sentinels and sources of infectious disease. Clin Microbiol Infec 18:521–527

Cho C, Wenner H (1973) Monkeypox virus. Bacteriol Rev 37:1–18

Cunningham AA, Daszak P, Rodriguez JP (2003) Pathogen pollution: defining a parasitological threat to biodiversity conservation. J Parasitol 89:78–83

Daszak P, Cunningham AA, Hyatt AD (2000) Emerging infectious diseases of wildlife - threats to biodiversity and human health. Science 287:443–449

Dobson AP, Foufopoulos J (2001) Emerging infectious pathogens in wildlife. Philos Trans R Soc Lond-Biol 356:1001–1012

Dunay E, Apakupakul K, Leard S, Palmer JL, Deem SL (2018) Pathogen Transmission from Humans to Great Apes is a Growing Threat to Primate Conservation. EcoHealth 15(1):148–162

Duval L, Nerrienet E, Rousset D, Mba SAS, Houze S, Fourment M, Le Bras J, Robert V, Frederic Ariey (2009) Chimpanzee malaria parasites related to plasmodium ovale in Africa. chimpanzee malaria parasites related to plasmodium ovale in Africa. PLoS ONE 4(5):5520. https://doi.org/10.1371/journal.pone.0005520

Ganz HH, Turner WC, Brodie EL, Kusters M, Shi Y, Sibanda H, Torok T, Getz WM (2014) Interactions between Bacillus anthracis and plants may promote Anthrax transmission. PLOS Neg Trop Dis 8(6):e2903

Gao F, Bailes E, Robertson DL, Chen Y, Rodenburg CM, Michael SF, Cummins LB, Arthur LO, Peeters M, Shaw GM, Sharp PM, Hahn BH (1999) Origin of HIV-1 in the chimpanzee **Pan troglodytes troglodytes**. Nature 397:436–441

Gillespie TR, Chapman CA (2008) Forest fragmentation, the decline of an endangered primate, and changes in host-parasite interactions relative to anunfragmented forest. Am J Primatol 70:222–230

Hahn BH, Shaw GM, DeCock KM, Sharp PM (2000) AIDS as a zoonosis: scientific and public health implications. Science 287:607–614

Hastings BE, Kenny D, Lowenstine LJ, Foster JW (1991) Mountain gorillas and measles: ontogeny of a wildlife vaccination program. In: Junge RE (ed) Proceedings of the American Association of Zoo Veterinarians Annual Meeting, pp 198–205

Hirsch VM, Olmsted RA, Murphey-Corb M, Purcell RH, Johnson PR (1989) An African primate lentivirus (SIVsm) closely related to HIV-2. Nature 339:389–392

Homsy J (1999) Ape tourism and human diseases: how close should we get? a critical review of the rules and regulations governing park management and tourism for the wild mountain Gorilla (**Gorilla gorilla beringei**). Unpublished type-scripted report submitted to International Gorilla Conservation Programme Kenya

Hugh-Jones M, Blackburn J (2009) The ecology of Bacillus anthracis. Mol Aspects Med 30:356–367

Inogwabini BI, Leader-Williams N (2012) Effects of epidemic diseases on the distribution of bonobos. PLoS ONE 7(12):e51112:1–8

Inogwabini BI, Matungila B (2009) Bonobo food items, food availability and bonobo distribution in the Lake Tumba Swampy forests, Democratic Republic of Congo. Open Conserv Biol J 3:1–10

Johnson E, Jaax N, White J, Jahrling P (1995) Lethal experimental infections of rhesus monkeys by aerosolized Ebola virus. Int J Exp Pathol 76:227–236

Keele BF, Van Heuverswyn F, Li Y, Bailes E, Takehisa J, Santiago ML, Bibollet-Ruche F, Chen Y, Wain LV, Liegeois F, Loul S, Ngole EM, Bienvenue Y, Delaporte E, Brookfield JFY, Sharp PM, Shaw GM, Peeters M, Hahn BH (2006) Chimpanzee reservoirs of pandemic and non-pandemic HIV-1. Science 313:523–526

Keita MB, Hamad I, Bittar F (2014) Looking in apes as a source of human pathogens. Microbial Pathogenesis 77:149–154

Krief S, Escalante AA, Pacheco MA, Mugisha L, André C, Halbwax M, Fischer A, Krief JM, Kasenene JM, Crandfield M, Cornejo OE, Chavatte JM, Lin C, Letourneur F, Grüner AC, McCutchan TF, Rénia L, Snounou G (2010) On the diversity of malaria parasites in African apes and the origin of plasmodium falciparum from bonobos. PLoS Pathogen 6(2):e1000765. https://doi.org/10.1371/journal.ppat.1000765

Leendertz FH, Lankester F, Guislain P, Néel C, Drori O, Dupain J, Speede S, Reed P, Wolfe N, Loul S, Mpoudi-Ngole E, Peeters M, Boesch C, Pauli G, Ellerbrok H, Leroy EM (2006) Anthrax in Western and Central African great apes. Am J Primatol 68:928–933

Liu W, Li Y, Shaw SK, Learn GH, Plenderleith LJ, Malenke JA, Sundararaman SA, Ramirez MA, Crystal PA, Smith AG, Bibollet-Ruche F, Ayouba A, Locatelli S, Esteban A, Mouacha F, Guichet E, Butel C, Ahuka-Mundeke S, Inogwabini BI, Ndjango JBN, Speede S, Sanz CM, Morgan DB, Gonder MK, Kranzusch PJ, Walsh PD, Georgiev AV, Muller MN, Piel AK, Stewart FA, Wilson ML, Pusey AE, Cui L, Wang Z, Färnert A, Sutherland CJ, Nolder D, Hart JA, Hart TB, Bertolani P, Gillis A, LeBreton M, Tafon B, Kiyang J, Djoko CF, Scheneider BS, Wolfe ND, Mpoudi-Ngole E, Delaporte E, Carter R, Culleton RL, Shaw GM, Rayner JC, Peeters M, Hahn BH, Sharp PM (2014) African origin of the Malaria parasite Plasmodium vivax. Nature Communications

Liu Y, Xu S, Woodruff AL, Xia M, Tan M, Kennedy MA, Jiang X (2017) Structural basis of glycan specificity of P[19] VP8*: Implications for rotaviruszoonosis and evolution. PLoS Pathog 13(11):e100670

Lowen AC, Mubareka S, Steel J, Palese P (2007) Influenza virus Transmission is dependent on relative humidity and temperature. PLoS Pathogene 3(10):1470–1476

Maibach V, Hans JB, Hvilsom C, Marques-Bonet T, Vigilant Linda (2017) MHC class I diversity in chimpanzees and bonobos. Immunogenetics 69(10):661–676

Minakawa N, Mutero CM, Githure JI, Beier JC, Yan G (1999) Spatial distribution and habitat characterization of **Anophalipne** mosquito larvae in Western Kenya. Am J Trop Med Hyg 61(6):1010–1016

Mutombo WM, Arita I. et Jezek Z (1983) Human monkeypox transmitted by a chimpanzee in a tropical rainforest area of Zaire. The Lancet ii: 735–737

Nizeyi JB, Innocent RB, Erume J, Kalema GRNN, Cranfield MR, Graczyk TK (2001) Campylobacteriosis, salmonellosis, and shigellosis in free-ranging humanhabituatedmountain gorillas of Uganda. Journal of Wildlife Diseases 37(2):239–244

Pedersen AB, Davies TJ (2010) Cross-species pathogen transmission and disease emergence in primates. EcoHealth. https://doi.org/10.1007/s10393-010-0284-3

Peterson AT, Bauer JT, Mills JN (2004) Ecologic and geographic distribution of filovirus disease. Emerg Infect Dis 10:40–47

Plantier JC, Leoz M, Dickerso JE, De Oliveira F, Cordonnier F, Lemee V, Damond F, Robertson DL, Simon F (2009) A new human immune deficiency virus derived from gorillas. Nat Med 15(8):871–882

Rachowicz LJ, Hero JM, Alford RA, Taylor JW, Morgan JAT, Vredenburg VT, Collins JP, Briggs CJ (2005) The novel and endemic pathogen hypotheses: competing explanations for the origin of emerging infectious diseases of wildlife. Conserv Biol 19:1441–1448

Reid A (2005) The effects of the 1918–1919 influenza pandemic on infant and child health in Derbyshire. Med Hist 49:29–54

Seimon TA, Olson SH, Lee KJ, Rosen G, Ondzie A, Cameron K, Reed P, Anthony SJ, Joly DO, McAloose D, Lipkin WI (2015) Adenovirus and herpesvirus diversity in free-ranging great apes in the sangha region of the republic of congo. PLoS ONE 10(3):e0118543. https://doi.org/10.1371/journal

Sleeman JM, Meader LL, Mudakikwa AB, Foster JW, Patton S (2000) Gastro-intestinal parasites of mountain gorillas (**Gorilla gorilla beringei**) from the Parc National des Volcans, Rwanda. J Zoo Wildl Med 31:322–328

Taubenberger JK, Morens DM (2006) 1918 influenza: the mother of all pandemics. Emerg Infect Dis 12(1):15–22

Van Heuverswyn F, Li Y, Neel C, Bailes E, Keele BF, Liu W, Loul S, Butel C, Liegeois F, Bienvenue Y, Ngolle EM, Sharp PM, Shaw GM, Delaporte E, Hahn BH, Peeters M (2006) SIV infection in wild gorillas. Nature 444(164). https://doi.org/10.1038/444164a

Vitousek PM, D'Antonio CM, Loope LL, Westbrooks R (1996) Biological invasions as global environmental change. Am Sci 84:468–478

Voyles J, Cashins SD, Rosenblum EB, Puschendorf R (2009) Preserving pathogens for wildlife conservation: a case for action on amphibian declines. Oryx 43(4):527–529

Walsh PD, Abernethy KA, Bermejo M, Beyers R, De Wachter P, Akou ME, Huijbregts B, Mambounga DI, Toham AK, Kilbourn AM, Lahm SA, Latour S, Maisels F, Mbina C, Mihindou Y, Obiang SN, Effa EN, Starkey MP, Telfer P, Thibault M, Tutin CE, White LJ, Wilkie DS (2003) Catastrophic ape decline in western equatorial Africa. Nature 422(6932):611–614

Walsh PD, Biek R, Real LA (2005) Wave-like spread of Ebola Zaire. PLoS Biol 3:1946–1953

Walsh PD, Breuer T, Sanz C, Morgan D, Doran-Sheehy D (2007) Potential for ebola transmission between gorilla and chimpanzee social groups. Am Nat 169:684–689

Webster RG, Bean WJ, Gorman OT, Chambers TM, Kawaoka Y (1992) Evolution and ecology of influenza a viruses. Microbiol Rev 56(1):152–179

Wolfe ND, Escalante AA, Karesh WB, Kilbourn A, Spielman A, Lal AA (1998) Wild primate populations in emerging infectious disease research: the missing link? Emerg Infect Dis 4:149–158

Wolfe ND, Kilbourn AM, Karesh WB, Rahman HA, Bosi EJ, Cropp BC, Andau M, Spielman A, Gruber D (2001) Sylvatic transmission of arboviruses among Bornean orangutans. J Trop Med Hyg 64:310–316

Woodford MH, Butynski TM, Karesh WB (2002) Habituating the great apes: the disease risks. Oryx 36(2):153–160

World Health Organization (WHO) (2014) Influenza virus infections in humans (February 2014). WHO Press (Geneva)

World Health Organization (WHO) (2008) Anthrax in humans and animals. Fourth edition. WHO Press (Geneva)

Yoshida T, Takemoto H, Sakamaki T, Tokuyama N, Hart J, Hart T, Dupain J, Cobden A, Mulavwa M, Kawamoto Y, Kaneko A, Enomoto Y, Sato E, KooriyamaT, Miyabe-Nishiwaki T, Suzuki J, Saito A, Okamoto M, Tomonaga M, Matsuzawa T, Furuichi T and Akari H (2016) Epidemiological Surveillance of Lymphocryptovirus Infection in Wild Bonobos. Front Microbiol. https://doi.org/10.3389/fmicb.2016.01262

# Chapter 10
# Alternative Cheaper Methods to Estimate Bonobos

**Abstract** Data were collected using both line transect and presence–absence methods were compared with a control known density to assess the accuracy of bonobo densities each of them produces. The line transect method provided a varied density estimate of the range = 0.24–3.4 individuals $km^{-2}$ whereas the density estimated by presence–absence data was 1.26 individuals $km^{-2}$, 95% CI = [1.23, 1.29]. Estimates yielded by presence–absence framework were consistently higher than estimates provided by line-transects in zones between 500 and 1500 $km^2$. This picture briskly changes around 2000 $km^2$, estimates from presence–absence decreased sharply while those from transect increased, leading to the conclusion that line-transect provides sensible estimates for smaller zones while presence–absence is more accurate when the geographical survey cover increases beyond 2000 $km^2$. Combining different densities the population estimated for Malebo was adjusted to the mean of 8,352 individuals, 95% CI = [7,952–8,747] individuals.

**Keywords** Line-transect · Presence–absence · Estimated population · Bonobo density · Sampling methods · Estimation sensitivity

## 10.1 Introduction

Monitoring wildlife species is an essential part of conservation activities in that it provides an indicator of how well different conservation schemes operate on ground. Particularly for charismatic threatened species such as bonobos (*Pan paniscus*), monitoring is an obligation for conservationists trying to protect their populations in their natural habitats. The monitoring exercise starts with field baseline data collection whose precision thresholds will influence results in the long-term. Bonobos of the Malebo region, Democratic Republic of Congo, have been described as being one of the largest populations of the species reported from the field (Inogwabini et al. 2007a, b). These field baseline data had inspired a monitoring program in order to document the population trends. However, as in many large forests in Central Africa, that monitoring program has revealed to be a daunting exercise because it can be prohibitively expensive, time-consuming, and physical demanding for teams (Walsh and White 1999; White and Edwards 2000). Because of these field constraints, there has been



B.-I. Inogwabini, *Reconciling Human Needs and Conserving Biodiversity: Large Landscapes as a New Conservation Paradigm*, Environmental History 12,
https://doi.org/10.1007/978-3-030-38728-0_10

a heated debate over approaches that would provide accurate trends. This discussion had to help choose between transects (Newing et al. 2002; Nichols and Conroy 1996; Buckland et al. 1993), forest reconnaissance (White and Edward 2000), and presence–absence (Ibáñez et al. 2005; Jepsen et al. 2005; Manel et al. 2001; Furuichi et al. 2001). The aim of this discussion was to minimize survey efforts and related costs without jeopardizing the quality of the information they generate. Over the debate, spearheaded by the urgent need to provide to decision-making bodies with the current state of bonobos, less strenuous and least expensive approaches such as forest reconnaissance and presence–absence presented serious temptations for field biologists. In order to make a sensible judgment on these three approaches, a need to assess comparable advantages that go beyond the financial costs was deemed necessary. Therefore, a comparison study was initiated with the goal to collect data using standard distance sampling methods and presence–absence to see if differences in methods influenced data accuracy under different circumstances, using a control region where the bonobos density is known from a complete count inventory of habituated bonobos. An ancillary objective to comparing results obtained from line-transect and those of presence–absence was to provide an updated estimate of bonobos based on the results of compared densities.

## 10.2 Materials and Methods

Surveys combined line-transect methods (Newing et al. 2002; Nichols and Conroy 1996; Buckland et al. 1993) and field reconnaissance (White and Edwards 2000) to collect presence/absence data (Ibáñez et al. 2005; Jepsen et al. 2005; Manel et al. 2001; Furuichi et al. 2001), as well as perpendicular measures (Distance sampling) to estimate large mammal abundance. Only data from line-transects are presented in this section; presence–absence data are treated separately later. There then follows a summary providing the extrapolation of densities obtained from the study sites throughout the entire Lake Tumba Landscape's forests.

## 10.3 Line-Transects

A total transect effort of 86 km was preset by the equation of Buckland et al. (1993): $CV (Đ) = [b/L (n_o/l_o)]^{1/2}$, Đ being density Đ, $b = 3$, L equals predicted sampling effort, and a pre-fixed coefficient of variation $CV (Đ) = 10\%$, after we collected preliminary data from two locations where fourteen 1 km-long transects were cut. The survey design used the bonobo nest site encounter rate to generate the systematic segmented track-line. Geographical data were collected using the Garmin® 12 × 1 handheld GPS. Data consisted of nest sites, knuckle prints, feeding remains, calls, and direct observation. Only nest sites were used to estimate population sizes. To estimate bonobo density, we first calculated *D* using DISTANCE (Laake et al. 1994),

**Table 10.1** Density and population estimates of bonobos in the Lake Tumba landscape

| Zone | Densities | | | Population estimates | | |
|---|---|---|---|---|---|---|
| | Lower | Mean | Higher | Lower | Mean | Higher |
| Mbala-Donkese | 0.24 | 0.27 | 0.29 | 15 | 17 | 19 |
| Ngombe-Bonginda | 0.24 | 0.27 | 0.29 | 176 | 198 | 212 |
| Botuali-Ilombe | 0.24 | 0.27 | 0.29 | 92 | 103 | 111 |
| Malebo-Nguomi | 1.8 | 2.4 | 3.4 | 1880 | 2,297 | 3,550 |

The estimates came from using line transects and were calculated through the use of Distance. Densities are individuals/km$^2$ while the populations are in total number of individuals. In order to understand where the zones are located in the Lake Tumba Landscape, the reader should refer to Fig. 10.1, which is the map of surveyed zone for bonobos

a software package widely used to estimate mammalian densities across Central Africa. DISTANCE functions on the assumption that all objects on the transect line are detected (Buckland et al. 1993), and provides estimates of sign densities, using the equation $D = \frac{n}{2lESW}$, where $n$ is the number of sightings, $L$ the total transect length, $ESW$ the effective strip width. We then used the formula of McClanahan (1986) as applied under different circumstances (Barnes et al. 1995; Hall et al. 1997): $E = \frac{D}{rd}$, where $D =$ nest site density, $r =$ daily rate of nest production (=1), and $d =$ nest decay rate (nest decay rate (=90 days, Reinartz et al. 2006). To avoid misleading densities generated by extrapolating point estimates over large areas, we accounted for the patchy distribution by incorporating the niche-breadth and suitable habitat metrics, as in some previous studies (Inogwabini and Omari 2005; Reinartz et al. 2006). Therefore, for groups at the edges of Lake Tumba, where 60% of the 2873.7 km$^2$ surveyed area was inundated or seasonally flooded, we excluded 60% of the surveyed zone from the analysis (Table 10.1). Equally, for the Malebo zone, which can be partitioned into 1,043.9 km$^2$ (52.4%) of mixed mature forests, 127.4 km$^2$ degraded forests (6.4%), and 751.2 km$^2$ (37.7%) of savannahs, we excluded savannas and degraded habitat in calculating the southernmost population. Because of small sample sizes from the edges of Lake Tumba, data from Bonginda–Ngombe and Botuali–Ilombe were pooled to estimate densities. Because habitats in these zones were similar, pooling data arguably provided a logical analytical framework. Geographical coordinates were plotted on the geo-referenced map by the use of ArcView 3.2. The same package was used to differentiate forest types using Landsat satellite images.

## 10.4 Presence–Absence Data and Estimation of Density

Grids of 25 km$^2$ (5 km × 5 km) were overlaid on the map of the study area to define sampling sites. The grid sizes were determined to ensure that they represented the closer size of the bonobo home range. Satellite images were used to ensure that randomly selected grids were in relatively homogenous habitat types. Eighty such grids (totalling 2000 km$^2$, representing 5% of the total area of the study site *ca.* 40,000 km$^2$) were selected at random and visited three times at the interval of three months. The interval of three months between two consecutive temporal replicates was chosen because, as stated above, nests completely decay in 90 days and to avoid counting same nest sites over time, it was deemed logical to temporally space consecutive visits by the span of time it would take a nest to disappear. Center point for each of the grids was pre-positioned using ArcView 3.2 and were located on field by teams using GPS. From those center points, 1 km long transect (0.5 km on both sides of the center of the square) were cut following an angle $\boldsymbol{\alpha} = 45^{\text{o}}$. The value of $\alpha$ was chosen based on the assumed gradient of the effect of the Congo River on the distribution of the species. Bonobo nest sites (detection/non-detection) were counted along without measuring perpendicular distances. Three teams, led by the same biologists as in the line-transect survey, collected data over a period of 1 calendar year. The presence–absence survey was undertaken 14 months after the line-transect survey described above. As in all large mammal surveys, the habitat description data were collected both along transects and on the pathway to the center point. The data were analyzed using single-season models available in the program Presence 45.

To estimate the density of signs from the presence–absence data, I analyzed data from different survey temporal replicates in the framework of mark-recapture methods, combining total time spent and total area sampled. In this framework, if $\Im = f(t, A)$, is the total effort as a function of both time ($t$) and surveyed area ($A$) used to detect a number $\eta$ of bonobo nest sites, the abundance $\alpha$ of nest sites is calculated as $\alpha = \frac{\eta}{\Im}$, which can be rewritten as followed, taking into account the number $\mu = nA$ and defining $\Im = t \times \mu$ or $\Im = t \times nA$: $\alpha = \frac{\eta}{t \times nA}$. The symbol $\alpha$ is relative index abundance, expressed in signs/km$^2$. This relative index can be used to estimate the number after two consecutive temporal replicates at times $t_1$ and $t_{2,}$ whose relative abundance estimates are, respectively, $f(t_1, A), \equiv \alpha_1 = \frac{\eta_1}{t_1 nA}$ and that of the time period, respectively, $f(t_2, A) \equiv \alpha_2 = \frac{\eta_2}{t_2 nA}$, by use of the mark-recapture formula.

$$\Gamma = \frac{\alpha(t_1) + \alpha(t_2)}{\alpha_2} = \frac{\frac{\eta_1}{t_1 nA} + \frac{\eta_2}{t_2 NA}}{\frac{\eta_2}{t_2 nA}}$$

$\boldsymbol{\Gamma}$ is the number of signs, which can be used only under the assumptions that: (a) Rate of immigration and emigration $\approx 0$, (b) $t = t_1 = t_2$ and $nA = \text{Constant} = \pi$, (c) Ecological vital parameters such as production of signs, sign decay rates are known and (d) Signs are clearly identifiable and cannot be mistaken throughout the study period. With that in mind, the above sign number formula can be rewritten as follows:

$$\Gamma = \left(\frac{\eta_1}{t_1 nA} + \frac{\eta_2}{t_2 NA}\right)\left(\frac{t_2 nA}{\eta_2}\right)$$

Using the implications of condition (b), the number of signs $F$ can be rewritten as follows:

$$\Gamma = \left(\frac{\eta_1}{tnA} + \frac{\eta_2}{tNA}\right)\left(\frac{tnA}{\eta_2}\right) \leftrightarrow \Gamma = \left(\frac{\eta_1 + \eta_2}{\eta_2}\right)$$

Because $\mathbf{\Gamma}$ represents a density of signs, it has to be converted into the density of individuals using conversion factors, including the rate of production of signs ($\beta$) and the decay rate ($\omega$). From the presence–absence framework, the density ($\Omega$) is obtained by adapting the Lincoln–Petersen equation in the expression $\Omega = \ln \frac{1}{P}(\Gamma)$. The 95% confidence intervals are calculated based on the formula CI (95%) = $\mu - 1.96\text{SE} \leq \mu \leq \mu + 1.96\text{SE}$. The standard error (SE) is the one modeled by the presence–absence.

## 10.5 Updating Bonobo Estimates in the Southern Lake Tumba

As will be seen below, comparing densities from each method and the control estimate, a decision had shown that line-transect provided more precise estimates for smaller geographic areas (ca. greater than or equal to 1000 km$^2$) whereas presence–absence would give more accurate estimates when used in areas greater than 1000 km$^2$. We, therefore, made a decision to adjust bonobo estimates reported previously (Inogwabini et al. 2007a) based on the geographical spread of the survey. Thus, for areas greater than or equal to 1000 km$^2$ the density estimates from line-transect were used while for those that were greater than 1000 km$^2$ densities obtained from presence–absence were used.

## 10.6 Abundance Estimates from Transect and Presence–Absence Survey

From the line-transect data, bonobo densities varied across the Lake Tumba Swampy Forests landscape, range = 0.24–3.4 individuals km$^{-2}$, respectively, near Lake Tumba and in the Malebo zone (Table 10.1, Inogwabini et al. 2007a) whereas the highest density was 1.26 within the 95% CI = [1.23, 1.29]. This highest estimate came out with the small 'Akaike information criterion" (AIC = 336.6940; Table 10.2), indicating that this was the best modeled estimated (Burnham and Anderson 2002). All estimates are within very tight confidence intervals, indicating that more precise estimates. The highest probability of detection was 0.8690 but with a model that generated the highest AIC (345.6729) and the lowest density estimate of 0.72 individuals km$^{-2}$ (Table 10.2).

**Table 10.2** Probability of detection, naïve estimate and modeled density estimates

| Parameters | (ψ) | SE (ψ) | $\overline{P}$ | AIC | E (naïve) | Γ | Ω ± 95% CI |
|---|---|---|---|---|---|---|---|
| $(\psi)_1$ | 0.0002 | 0.0128 | 0.4958 | 336.6940 | 0.9000 | 1.76 | 1.26 ± 0.03 |
| $(\psi)_2$ | −0.0001 | 0.0140 | 0.5125 | 340.1602 | 0.9000 | 1.46 | 1.05 ± 0.03 |
| $(\psi)_3$ | 0.0000 | 0.0141 | 0.4958 | 340.6940 | 0.9000 | 1.64 | 1.19 ± 0.03 |
| $(\psi)_4$ | 0.0011 | 0.0080 | 0.8690 | 345.6729 | 0.9000 | 1.79 | 0.72 ± 0.02 |

## 10.7 Comparing Line-Transect and Presence–Absence Estimates

Except for 2.4 individuals km$^{-2}$ (Table 10.1), all the estimates yielded by the combination of presence–absence (Table 10.2 and Fig. 4) frameworks are higher than estimates provided by line-transects. The mean estimate generated by the distance sampling (0.27 individuals km$^{-2}$; Table 10.1) for most of the area represented 21% best modeled and highest estimate of 1.26 individuals km$^{-2}$. However, the highest estimate of the presence–absence method (1.26 individuals km$^{-2}$) represented about 53% of the highest estimate (2.4 individual km$^{-2}$) generated by the line-transect approach. A two-sample equal variance T-test showed that there was a significant difference between estimated densities from the two methods ($t = 0.0013 < p = 0.05$; $df = 3$).

## 10.8 Updated Bonobo Population Estimates

Adjusting estimates based on the geographical spread of the survey (Fig. 10.1), as dictated by the comparison of the line-transect and the presence–absence data, the population of Mbala–Donkese and Botuali–Ilombe remained the same, with respectively 17 individuals, 95% CI = [15–19] and 92 individuals, 95% CI = [103–111]. All other populations were readjusted with a mean density of 1.26 individuals km$^{-2}$, the mean estimated total population is adjusted to 8,352 individuals, 95% CI = [7,952–8,747] and that of the Malebo–Nguomi is now 2511 individuals, 95% CI = [2451–2571].

## 10.9 Estimates of Bonobo Population in Perspectives

Bonobo density estimates from line-transect methods in the study area were presented by Inogwabini et al. (2007a) and varied significantly between different subsites, with 0.24 individuals km$^{-2}$ near Lake Tumba and 3.4 individuals km$^{-2}$ (maximum) in the Malebo zone (Table 10.1). While the range 0.24–1.8 individuals km$^{-2}$ is similar to reports from other study sites (Reinartz et al. 2006; Van Krunkelsven et al. 2000; Sabater Pi and Vea 1990), the upper limit of 3.4 reported from the Malebo region is probably the highest estimate ever reported. Inogwabini et al. (2007a) argued that reasons for this high density needed more in-depth and long-term research work to be confirmed and decently understood though they had also provided some preliminary insights of probable causes, the first of them being the contribution of the forest-savannah mosaic system in providing year-round fruit sources; bonobos would fall back on savannah fruits when the forest resources are scarce. This would make the forest-savannah mosaics a preferred habitat for bonobos, while dense forest habitats would be more marginal for this species. The second ecological explanation would be the human history in the region, made of taboos that proscribe the Bateke to eat bonobo meat that would also have influenced the patterns of bonobo distribution in that region (Inogwabini et al. 2007a, b).

Estimates provided by the presence–absence framework fall within the range 0.72–1.26 individuals km$^{-2}$, which is around the central values of the range 0.27–2.4 individuals km$^{-2}$ resulting from the line-transect approach and also in agreement with reports from other study sites (Reinartz et al. 2006; Van Krunkelsven et al. 2000; Sabater Pi and Vea 1990). Indications that estimate from the presence–absence are, overall similar to reports from other study sites within the bonobo range should not, however, conceal major differences between the two approaches. The first statistically significant divergence between the two methods is that presence–absence exhibits consistently lower densities than line-transect ($t = 0.0013 < p = 0.05$; d$f$ = 3). The second significant difference is that the two extreme estimates showed a completely opposite picture (Fig. 10.2). This difference is spatially dependent on the spread of the sampling (Fig. 10.1) and indicates that density estimates from both methods are dependent on the size of the surveyed area. Another striking difference is that results from the presence–absence framework have tighter confidence intervals. Tighter confidence intervals may reflect precision in results but do not necessarily mean that those results are representative of the true population parameters. This caution is supported by field observation. For example, a density of 1 individual km$^{-2}$ would mean a population estimate of 828 individuals in the Mopulunge–Mbanzi area (Fig. 10.1 and Table 10.1), which is unrealistic compared to the field realities.

The bonobo habituation process is undertaken covers an area of approximately 190 km$^2$ with repeated count of 45 that are known to occupy the zone. This zone is within the Malebo–Nguomi zone (Fig. 10.1) where the calculated presence–absence estimated density is 1.26 individuals km$^{-2}$ and the line-transect mean estimate is 2.4 individuals km$^{-2}$. Both these estimates are higher than the control estimate of (0.24 individuals km$^{-2}$) (or 45 individuals over 190 km$^2$). The control estimate reported in this paper is much closer to the mean density of 0.27 individuals km$^{-2}$ reported from other places (Reinartz et al. 2006; Van Krunkelsven et al. 2000, Sabater Pi and Vea 1990).

Even though this comparison may seem somewhat puzzling at the first sight, it also shows clearly that there are instances for which presence–absence would be the best way to estimate population sizes and there are those when the line-transects are the most suitable method. Results from the comparison (Fig. 10.2) clearly indicate

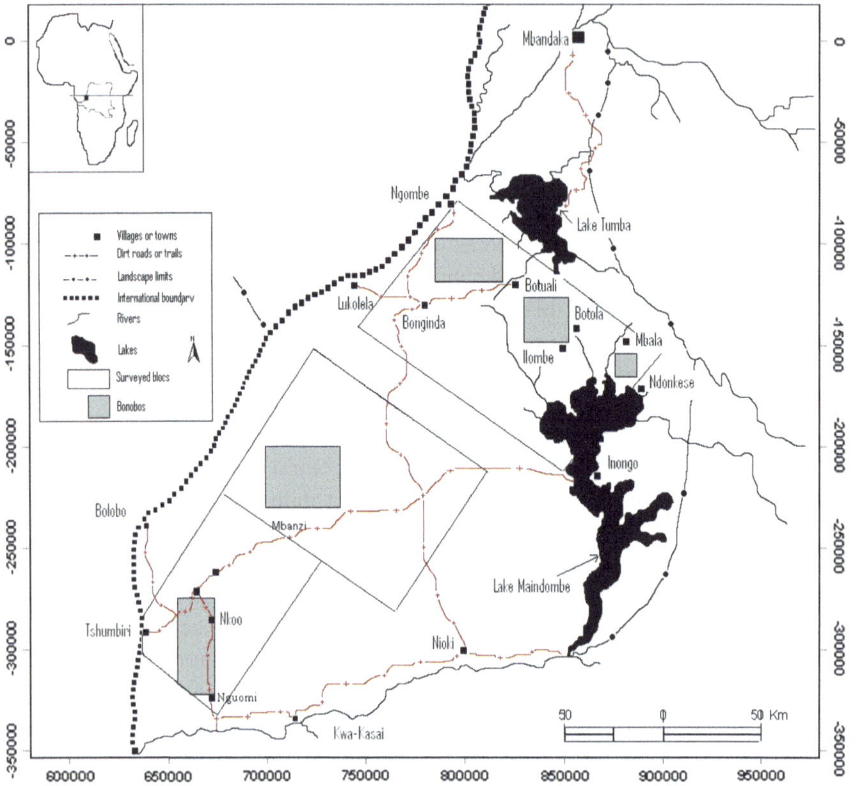


**Fig. 10.1** Bonobo surveyed zone in the Lake Tumba Landscape, DRC

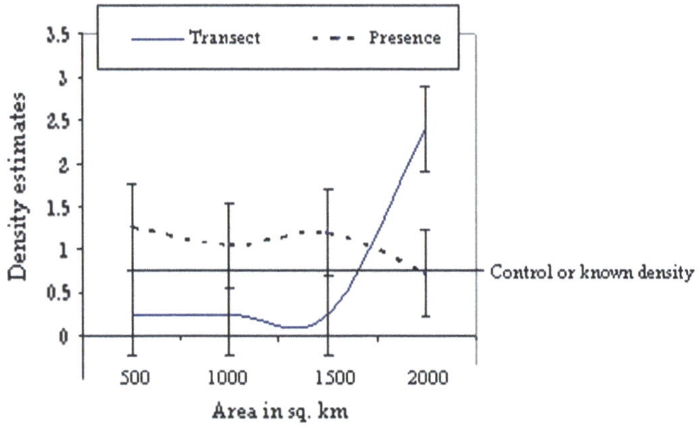


**Fig. 10.2** Comparison of line-transect and presence–absence estimates

that for the monitoring of smaller forest areas (greater than or equal to 1000 $km^2$), line-transect provides estimates that are sensibly closer to true population estimates. It is also clear from these comparisons that once the geographical coverage of the survey is greater than 2000 $km^2$, the presence–absence framework would provide lower estimates, but these seem to give a more realistic picture of the reality on ground.

Adjusting estimates based on the geographical spread of the survey, combining data from line-transect and the presence–absence seems to have increased the population of bonobos in the southern Lake Tumba. This is not exactly so because, first of all, in the actual estimates by groups there is a group that was not included in the previous report, Kenia that adds ca. 400 individuals on average in the picture. Secondly, the population of Mopulenge–Mbanzi that was estimated in the past by simple encounter rate (Inogwabini et al. 2007a) has been converted into a density estimate generated by the presence–absence data set and adds on 1740 individuals on the previously reported population. With an adjusted mean population of 8,352 individuals, the actual estimates indicate that the southern Lake Tumba is one of the strongholds of the bonobo conservation, representing in itself *ca* 42% of the wild population of bonobos, if the often quoted number of 20,000 free-ranging individuals remaining in the forest (Inogwabini et al. 2007a) is close to the reality. The critical conservation effort is desperately needed to help protect that population through an integrated conservation program whereby humans would be part of the envisaged solution rather than being essentially the root cause of declining bonobo numbers.

## References


Barnes RFW, Blom A, Alers MPT, Barnes KL (1995) An estimate of the numbers of forest elephants in Gabon. J Trop Ecol 11:27–37

Buckland ST, Anderson DR, Burnham KP, Laake JL (1993) Distance sampling: estimating abundance of biological populations. Chapman and Hall, Boca Raton

Burnham KP, Anderson DR (2002) Model selection and inference: a practical information-theoretic approach. Springer-Verlag

Furuichi T, Hashimoto C, Tashiro Y (2001) Extended application of a marked-nest census method to examine seasonal changes in habitat use by chimpanzees. Int J Primatol 22:913–928

Hall JS, Inogwabini BI, Williamson EA, Omari I, Sikubwabo C, White LJT (1997) A survey of elephants (Loxodonta africana) in the Kahuzi-Biega National Park lowland sector in eastern Zaire. Afr J Ecol 35:213–223

Ibáñez JJ, Caniego J, Garciá-Álvarez A (2005) Nested subset analysis and taxa-range size distribution of pedological assemblages: implications for biodiversity studies. Ecol Model 182:239–256

Inogwabini BI, Omari I (2005) A landscape-wide distribution of **Pan-paniscus** in the Salonga National Park, Democratic Republic of Congo. Endanger Spec Updat 22:116–123

Inogwabini BI, Matungila B, Mbende L, Abokome M, Tshimanga WT (2007a) Great apes in the Lake Tumba landscape, Democratic Republic of Congo: newly described populations. Oryx 41(4):532–538

Inogwabini BI, Bewa M, Longwango M, Abokome M, Vuvu M (2007b) The bonobos of the Lake Tumba – Lake Maindombe Hinterland: threats and opportunities for population conservation.

In Furuichi T, Thompson J (eds) The bonobos behavior, ecology, and conservation. Springer: 273–290
Jepsen JU, Baveco JM, Topping CJ, Verboom J, Vos CC (2005) Evaluating effect of corridors and landscape heterogeneity on dispersal probability: a comparison of three spatially explicit modeling approaches. Ecol Model 181:445–459
Laake JL, Buckland ST, Anderson DR, Burnham KP (1994) DISTANCE—User's guide V2.1. Colorado Cooperative Fish & Wildlife Unit. Colorado State University
Manel S, Williams HC, Ormerod SJ (2001) Evaluating presence–absence models in ecology: the need to account prevalence. J Appl Ecol 38:921–931
McClanahan TR (1986) Quick population survey method using faecal droppings and the steady state assumption. Afr J Ecol 24:37–39
Newing H, Davies G, Linkie M (2002) Large mammals. In Davies G (ed) African forest biodiversity: a field manual for vertebrates. Earthwatch Institute, pp 69–98
Nichols JD, Conroy MJ (1996) Techniques for estimating abundance and species richness. In Wilson DE, Cole FR, Nichols JD, Rudran R, Foster MS (eds) Measuring and monitoring biological diversity: standard methods for mammals. Smithsonian Institution Press, pp 177–234
Reinartz G, Inogwabini BI, Mafuta N, Lisalama WW (2006) Effects of forest type and human presence on bonobo (Pan paniscus) density in the Salonga National Park. Int J Primatol 27(2):603–634
Sabater Pi J, Vea JJ (1990) Nest building and population estimates of the bonobo from the Lofeke-Lilungu-Ikomaloki region of Zaire. Primate Conserv 11:43–48
Van Krunkelsven E, Inogwabini BI, Draulans D (2000) A survey of bonobos and other large mammals in the Salonga National Park, Democratic Republic of Congo. Oryx 34(3):180–187
Walsh PD, White LJT (1999) What will it take to monitor forest elephant populations? Conserv Biol 13:1194–1202
White LJT, Edwards A (2000) Conservation research in the Central African rain forests: a handbook. Wildlife Conservation Society

# Chapter 11
# Chimpanzees of the Ngiri Triangle

**Abstract** The chapter describes a survey of large mammals in the Lake Tumba landscape with the focus on the chimpanzees (*Pan troglodytes*) using straight lines (transects) and forest reconnaissance to document the species distribution and to estimate its abundance. Unfortunately, after the completion of the surveys, samples were insufficient to conduct a density analysis properly. The encounter rates were used as a sensible indicator of the species abundance and this latter varied with distance from main rivers, with the highest being within the distance 11–15 km; their averages were 0.11 nesting sites $km^{-1}$ and 0.05 nesting sites, respectively, for the Ngiri Triangle and Bosobele, respectively. Swampy forests and seasonally flooded forests combined dominated the overall area; woody swamps occupied 32% of the space and that Ngiri Triangle had 27.6% of mixed mature forest on terra firma, which was higher than most forests in that zone where the average terra firma forest is greater than 20%. Finally, Chimpanzees' nesting sites were found at the highest proportions (70%) in flooded forests and that human signs averaged at 2.3 signs per km in the Ngiri region (ranging between 0.8 and 5.0).

**Keywords** Chimpanzees · Pan troglodytes · Line-transect · Forest reconnaissance · Nesting sites · Swampy forests and seasonally flooded forests

## 11.1 Introduction

The Democratic Republic of Congo is the only country that holds three species of great apes (bonobos, chimpanzees, and gorillas). Of those three species and with the single caveat of the little known western population, gorillas are the most studied great ape species in the Democratic Republic of Congo, while bonobos and chimpanzees have generated less interests from the scientific and conservation communities until recent days. A recent prioritization exercise requested by the Great Apes Survival Project (GRASP) assessed the status of populations of each great species in the Democratic Republic of Congo and came out with clear indications that chimpanzees were the least known species of great apes residing in Congo. Ten populations were identified, all of them residing in protected areas (Inogwabini 2005). However, there has been a bias toward populations residing in protected areas



B.-I. Inogwabini, *Reconciling Human Needs and Conserving Biodiversity: Large Landscapes as a New Conservation Paradigm*, Environmental History 12,
https://doi.org/10.1007/978-3-030-38728-0_11

simply because protected areas are the only zones from which data have been gathered over time and, in most cases, they are sympatric with gorillas. A national field assessment of chimpanzees had never been undertaken and the true status of the species is far from being certain. Threats faced by chimpanzees had been very often aligned to those of other great ape species and they are very frequently treated as simple appendixes to studies whose focuses are either large mammals or other great ape species. Therefore, across their ranges, few have had systematic and extensive surveys and chimpanzees' conservation status even in protected areas is not well understood. The ecology of the species is far from being uncovered and surveys to document that the species abundance are still needed.

The Ngiri and Bosobele chimpanzee populations are among the most recently described populations (Inogwabini et al. 2007). They comprise two distinct populations separated by the Ngiri River: (a) the Bosobele-Lubengo population and (b) the Ngiri Triangle population. These two populations inhabit a zone whose major habitat characteristic is swampy forest (*ca* 52% of the entire zone) and have never been scientifically described prior to the study by Inogwabini et al. (2007). Despite the fact that the study by Inogwabini et al. (2007) has limited general implications for the species conservation, it had revealed that the Lake Tumba Landscape was the only place where both bonobos and chimpanzees could be found, even though still non-sympatric as they are separated by the Congo River.

## 11.2 The Chimpanzees of Lake Tumba Landscape and the Species Conservation

Chimpanzees are the most widespread species of great apes; they comprise four geographic subspecies: the eastern chimpanzees (*Pan troglodytes schweinfurthii*), the Central chimpanzees (*Pan troglodytes troglodytes*), the western chimpanzees (*Pan troglodytes verus*), and the Nigeria-Cameroon chimpanzees (*Pan troglodytes vellerosus*). The Democratic Republic of Congo is home to the eastern and central chimpanzees; the newly described population of Ngiri and Bosobele are part of the central chimpanzee populations. The 2007 International Union for Conservation of Nature (IUCN) Red List of Threatened Species classifies the central chimpanzee as an endangered species while in 1988 they were considered 'vulnerable', which indicates that their conditions in wild deteriorated over the last 20 years. Between 47,000 and 78,000 individuals of chimpanzees are estimated to currently remain in the species natural and historical range. According to the IUCN, the decline in the Central Chimpanzee population is expected to continue for another 30–40 years due to poaching for bushmeat, habitat destruction related to increasing human presence (agriculture, deforestation, and development) and political instability, and hemorrhagic diseases such as Ebola fever.

## 11.3 The Habitats of the Ngiri and Bosobele

Despite the general description of the habitats of the Lake Tumba Landscape provided in Chap. 5, it's worth describing the habitats of this zone where the chimpanzees occur because of its specificities. Indeed, apart from the situation in the Republic of Congo (Brazzaville) where huge numbers of great apes were described in swamps (Blake et al. 1995), the Lake Tumba Landscape is the only other area within Central Africa where great apes are reported in swampy areas. This single fact calls for a detailed description of what constitutes these swamps.

The habitats of the Ngiri and Bosobele regions are characterized by extensive swampy forests whose altitude is very low and hardly surpassed the 300 m above the sea level (Inogwabini et al. 2007). As described by Inogwabini et al. (2012), characteristic tree species of these permanently and seasonally flooded forests include *Raphia sese*, *Pandanus* sp., *Guibortia demeusi*, *Uapaca guineensis*, and *Uapaca heudelotii*. The mean annual rainfall oscillates between 2007 and 2106 mm (Evrard 1968). The remaining 48% of the area is *terra firma* mixed mature forest wherein species such as *Anonidium manii* and *Polyalthia suaveolens* are among the most emergent tree species. Large swaps of forests are dominated by palm trees (*Elais guineensis*), which are either naturally grown or planted by humans. According to the biogeographic study of Gautier-Hion et al. (1999), other diurnal primates of the region include the red-tail monkey (*Cercopithecus ascanius*), the Allen's swamp monkey (*Allenopithecus nigroviridis*), the Wolf's monkey (*Cercopithecus mona wolfi*), and the De Brazza's monkey (*Cercopithecus neglectus*). The zone is also home to other large mammal species such as forest elephant (*Loxodonta africana cyclotis*), leopard (*Panthera pardus*) giant pangolin (*Smutsia gigantea*), African forest buffalo (*Syncerus caffer nanus*), bongo (*Tragelaphus euryceros*), sitatunga (*Tragelaphus spekei*), blue duiker (*Cephalophus monticola*), bay duiker (*Cephalophus dorsalis*), and water chevrotain (*Hyemoschus aquaticus*). Data reported by Inogwabini et al. (2012) showed that chimpanzees in these tow areas residing in this zone use both swampy and terra firma forests. They have been documented to build some significant portions of their night nests on *Raffia sese* and *Pandanus*. Information gathered from local communities indicated that chimpanzees of the Ngiri region tend to concentrate in forest strata with muddy substrates and exhibited the habits of worm-digging and palm nut eating as part of their foraging activities.

Apart from these physical habitat descriptions, it is worth including some notes of the uses human communities in these areas make of these swampy habitats as they would also inform the knowledge of why chimpanzees are present in this type of physical environment. First of all, the region of Ngiri and Bosobele is, in its majority, occupied by communities that are primarily made of fishermen, the hunting culture having been imported in the region in the most recent past. Hence, while people extensively had always used the swampy forests in this zone, their most important activities have been around fishing but not hunting. Also, because of swamps, there are no major human settlements throughout the landscape; human densities range between 5.5 and 5.8 inhabitants km$^{-2}$ (Bomongo and Makanza, respectively) and

18 inhabitants km$^{-2}$ (Bikoro), with Lukolela having a somewhat mid-point density of 14 people/km$^2$. Bomongo, Bikoro, Lukolela, and Makanza represent a little less than 50,000 km$^2$ (62.3% of the total Lake Tumba Landscape (80,000 km$^2$) while Bomongo and Makanza are the swampiest of the area and comprise 34% of the total area of the Lake Tumba Landscape. The highest density (18 inhabitants km$^{-2}$ in Bikoro), is at the point where the terra firm soil begins and goes up smoothly in terms of the altitude and ends up culminating at the northern end of the Bateke Plateaus, located in the territory of Bolobo, just south of Lukolela.

The above populations reside mostly along the major rivers, particularly the Congo River for the case of Makanza, Lukolela, and even Bolobo; the population of Bikoro agglomerates, in its majority, along the shores of the Lake Tumba. Indeed, the rivers and lakes in that region have always served as communication routes, linking people from different regions of the Democratic Republic of Congo and even the wider Congo Basin for centuries. The land layout and the distribution of human populations imply that many areas of the northern part of the Lake Tumba Landscape are rather lightly occupied permanently; agricultural fields are mostly confined within a 2.5 km-wide width from the main passages, leaving extensive swamps unpeopled. Saying that the swamps are mostly unpeopled does not, however, mean that these areas are vacant lands in the juridical interpretation of the term; these widely open yet uninhabited areas belong to communities who know where the limits of their ancestral lands are. These details of human ecology are important to understand why the land use in this region should be treated differently. It does contribute to the current ecological processes happening in the region and provide a strong background against which to read the findings of different research components.

## 11.4 What Interests Were There to Study the Chimpanzees in This Area?

Because the chimpanzees uncovered in the Ngiri Triangle and the Bosobele regions, in the northern Lake Tumba Landscape, were never studied previously, the first and single most important objective of their study was to document their abundance and understand their conservation status. While conducting the survey, the study also assembled other ecological data with the aim to inform the conservation plan of the chimpanzees located in these zones. Of course, the conservation plan had been envisaged to contain sufficient scientific information that would be crucial for the management of these chimpanzee populations. With the presence of these chimpanzees in addition to bonobos, lions, and elephants, the Lake Tumba Landscape can be arguably classified as a site of critical importance for the overall conservation in the Democratic Republic of Congo and, particularly for the chimpanzees that have been documented to decline continent-wide. Indeed, the study was 100%-aligned with the Lake Tumba Landscape's main objectives and goals that, as described below (Chap. 20), were to conserve the Lake Tumba Landscape's swamp forests and ensure that all charismatic species residing therein and their habitats are preserved and their populations monitored through an integrated approach that combined both community conservation and law enforcement. The integrated approach described above was to be based on coherently distributed spatial functions reached through an all stakeholder participative process guided by sound scientific data and should have as a guiding principle to care about ensuring that human wellbeing is compatible with the needs of for the conservation of the biological diversity.

## 11.5 Procedures and Protocols Used to Study Chimpanzees

The survey data were collected using standard survey protocols and statistical analysis used in the Central African Region. These were described in great detail in Chaps. 6, 10, and 14 of this volume. However, just to give a brief summary here, a systematic-random sampling protocol involving line-transects was developed. The design of the survey was set to provide an estimate of chimpanzee abundance with a coefficient of variation (CV) of 5%. As described by Inogwabini et al. (2012), the use of ArcView GIS software helped construct a grid of 5 km × 5 km cells on each study site (Grossmann et al. 2008) to identify the location of each transect. Then, 30% of all these cells were randomly chosen. Coordinates of the center of each randomly selected cell constituted the placement of the starting point of each transect. Chimpanzee's nests provided the basis of animal abundance estimates following standardized methods. Namely, perpendicular distances between centers of clusters of nests and the lines of transects were measured. Data on human activities were collected to document the biological integrity of the zone. Of particular importance were poaching signs (snares, used shotgun cartridges, smoking racks, human fishing, and hunting camps, etc.). Each mammal and human data point was accompanied by a spatial attribute attached to it and a thorough habitat description using the same methods as described above (Chap. 6). The habitat description was both floristic and physiognomic. These habitat descriptions allowed comparisons with the bonobo studies located east and south of the Ngiri and Bosobele. A preliminary botanical inventory, using 10 $km^2$-grids, was conducted to allow a finer scale identification of species diversity in this specific habitat of the chimpanzees and provided an indication of the food availability to the chimpanzee populations.

## 11.6 Chimpanzee Population Estimates in the Lake Tumba Landscape

After the completion of the surveys, it has become obvious that samples were insufficient to run Distance properly, even if boostrapping was tried. Hence, the analysis had to rely on the encounter rates, which as indicated in several instances in this volume (Chaps. 13 and 14) are relative indicators of abundance. As Inogwabini et al. (2012) reported the encounter rates for chimpanzees varied with distance from main rivers, with the highest being within the distance of 11–15 km; their averages were 0.11 nesting sites $km^{-1}$ and 0.05 nesting sites for the Ngiri Triangle and Bosobele, respectively. Inogwabini et al. (2012) also reported that swampy forests and permanently flooded forests combined dominated the overall area; woody swamps occupied 32% of the space and that Ngiri Triangle had 27.6% of mixed mature forest on terra firma, which was higher than most forests in that zone where the average terra firma forest is greater than 20%. Inogwabini et al. (2012), finally, reported Chimpanzees' nesting sites were found at the highest proportions (70%) in flooded forests and that human signs averaged at 2.3 signs per km in the Ngiri region (ranging between 0.8 and 5.0).

## 11.7 Relative Estimates of Chimpanzee Population in the National and Regional Perspectives

As indicated above, in 2005, the United Nations Great Apes Survival Project (GRASP) assessed the status of populations of chimpanzees in the Democratic Republic of Congo and came out with 10 distinct populations of chimpanzees: (1) the Ituri population (located within the Okapi Faunal Reserve), (2) the Maiko population (located within the Maiko National Park), (3) two distinct populations of Kahuzi-Biega (located both in the low land and mountain parts of the Kahuzi-Biega National Park), (4) the Virunga population (located within the Virunga National Park), (5) the Garamba (Kurukwata and Kilwa) population, (6) the Itombwe population, (7) the Luki population, (8) the Rwenzori population, and (9) the Bili-Uele population.

The chimpanzees of the Lake Tumba Landscape were not included; just the fact of describing them was of a significant conservation importance. Apart from that fact, most of the populations (8 over 10) were located within existing protected areas. Hence, describing the populations of Ngiri and Bosobele was also important in that they were populations without particular conservation measures. Indeed, the Lake Tumba Landscape was outside of historically conservation imbedded areas (see Chap. 6); communities dealt in essential with any wildlife species the way they felt was fit. Along the line of the story told (Chap. 6) on the bonobos, these chimpanzees deserved equally important conservation efforts and equally loud media attention. Biases toward chimpanzee populations located in protected areas are easily understandable; protected areas are the only areas from which data on wildlife species have been gathered over long time periods. Furthermore, for long periods, chimpanzees in the Democratic Republic of Congo were studied only through general studies whose main focus was on gorillas. Therefore, with the exception of the Ituri population, it is not surprising that the table shared by GRASP in 2005 had very precise density estimates for populations of Virunga, Kahuzi-Biega, and Maiko; these areas are where chimpanzees are sympatric with gorillas that have been studied frequently.

The encounter rates of nest sites in this zone were lower than compared, for example, with the encounter rates reported by Inogwabini et al. (2000) in the mountain sector Kahuzi-Biega National Park but given the clear indications that chimpanzees were the least known species of great apes residing in Congo and were declining continent-wide, the population in the Lake Tumba was an important population. Furthermore, despite this population being clearly part of the Central chimpanzees (*Pan troglodytes troglodytes*), it is separated by its next neighbors by either major river barrier (the chimpanzees of Likouala Swamps in the Republic of Congo) that the Ubangui constitutes or by long geographic distances intercepted by major human settlement (the chimpanzees of Bili-Uele). Hence, the chimpanzees of the Ngiri might have taken a different genetic direction and warrant conservation efforts to protect their current gene pool. Indeed, as indicated in Chap. 7, rivers act as major barriers for gene flows and species become easily differentiable when separated for a long time by these barriers.

Threats faced by chimpanzees of the Lake Tumba Landscape (see Chap. 15) are very similar to those faced by bonobos in the same landscape. But there are also significant differences. Firstly, the chimpanzees in the Ngiri have been found in a short distance from the main navigable rivers; which exposes them to be targeted by fishermen and hunters. This renders the chimpanzees of the Lake Tumba Landscape particularly vulnerable because this area is swampy and chimpanzees would have no luxury of climbing down to hide when faced with someone decided to hurt them. Of course, this should be expected to be cushioned by the fact that the composition of these swamps is that some areas are difficult to navigate through as their vegetations are essentially made of *Pandanus candelabrum*. Inogwabini et al. (2012) hypothesized that the difficulties to go through this species were among the major reasons why chimpanzees were found in areas with higher encounter rates of human signs.

A second difference between the threats to bonobos and chimpanzees was the very location of each species. While bonobos are also located in areas exposed to human movements, the chimpanzees of the Ngiri Triangle are located in an area that borders the Congo and the Ubangui rivers. These are massive highways connecting major human settlements such as Kinshasa and Kisangani and Brazzaville and Bangui, respectively.

A third big difference between threats to bonobos and chimpanzees in the Lake Tumba Landscape was that while there are wide-spread acknowledgments by the state, the communities and different groups that bonobos are protected, chimpanzees legal status has not been so popularized. In discussing with several stakeholders about this situation, many have been feeling that the potential confusion between the two species would, in the end, serve the conservation awareness on the chimpanzees. However, discussing with local communities, they were blatantly aware of the differences between species; they called chimpanzees using different names as compared to names used for bonobos. Most of the individuals interviewed knew that bonobos were fully protected whereas they were not aware of the legal conservation status of the chimpanzees.

A fourth difference was that, while the majority of bonobo populations were located in areas where they were not protected by traditional taboos, the largest population of bonobos in the southern Lake Tumba Landscape was, indeed, covered by a strong traditional taboo enforced by the strong authority of the traditional Bateke chiefs, which was not the case for chimpanzees in the Landscape. Inogwabini (2010), Inogwabini and Leader-Williams (2012) discussed reasons why bonobos of the southern Lake Tumba Landscape were protected while they were not located within any protected areas; they found out that the strong traditional authority of the Bateke enforced traditional taboos against hunting bonobos, which ended up protecting bonobos better than in the Salonga National Park, for comparison reasons. Hence the importance of traditions in preserving species of great apes is significant in terms of conservation thought and should be put on the balance weighting the conservation status of any given species, and particularly the status of great apes. Broadly speaking, the lesson being drawn from the studies cited above and others worldwide is that where there are supportive traditions, there is a high likelihood for species to be better protected even when there are no official law enforcement mechanisms. But, unfortunately, there are no traditional taboos in the regions of the Ngiri and Bosobele to protect the chimpanzees that reside therein. Hence, alternative legal methods were needed, i.e., the creation of a protected area to ensure the sustainability of the population of the chimpanzees in the Lake Tumba Landscape, as briefly stated below and described in Chaps. 20 and 21 in this volume.

The above differences brought together mean that even though both species are endangered species, landscape-wise, the chimpanzees were more threatened than bonobos and deserved more attention than they were currently receiving. Indeed, the chimpanzees in the Lake Tumba Landscape were not even officially known by the conservation and scientific communities. That is why after the priority setting exercise, which was conducted using the survey data collected from all parts of the landscape, it was proposed to revamp the old idea of creating a natural reserve in the Ngiri area. Indeed, back in 1970s, the then Zairean Government proposed the creation of the reserve in that area to ensure the protection of swamps but the project was never materialized before the program 2011. Hence, the Lake Tumba Landscape proposed the creation of this protected area to provide a secure environment to the chimpanzees while ensuring that human communities in the areas where chimpanzees occurred would not suffer any significant loss in the daily activities to support their sustainable livelihood.

## References

Blake S, Rogers E, Fay JM, Ngangoué M, Ebéké G (1995) Swamp gorillas in northern Congo. Afr J Ecol 33(3):285–290

Evrard C (1968) Recherches écologiques sur le peuplement forestier des sols hydromorphes de la Cuvette Centrale congolaise. Série scientifique 110. Office National de la recherche Scientifique et du Développement du Ministère Belge de l'Education Nationale et de la Culture

Gautier-Hion A, Colyn M, Gauthier JP (1999) Histoire naturelle des primates d'Afrique Centrale. Ecosystèmes d'Afrique Centrale. Union Douanière et Economique d'Afrique Centrale. Multipress

Grossmann F, Hart JA, Vosper A, Ilambu O (2008) Range Occupation and Population Estimates of Bonobos in the Salonga National Park: Application toLarge-scale Surveys of Bonobos in the Democratic Republic of Congo. In: Furuichi T, Thompson J (eds) The Bonobos. Progress and Prospects. Springer, Developments in Primatology, pp 189–216

Inogwabini BI (2005) Common fishes of the Salonga National Park, Democratic Republic of Congo: preliminary survey and conservation issues. Oryx 39(1):78–81

Inogwabini BI (2010) Conserving great apes living outside protected areas: the distribution of bonobos in the Lake Tumba landscape, Democratic Republic of Congo. PhD thesis, University of Kent at Canterbury, United Kingdom

Inogwabini BI, Lingopa Z (2007) Evaluation de la biodiversité de poissons au lac Maindombe, Nord-Est/Août 2007. Paysage Lac Télé – Lac Tumba, Segment RDC. Rapport produit dans le cadre du Programme CARPE

Inogwabini BI, Leader-Williams N (2012). Effects of epidemic diseases on the distribution of bonobos. PLoS ONE 7 (12):e51112:1–8

Inogwabini BI, Hall JS, Vedder A, Curran B, Yamagiwa J, Basabose K (2000) Conservation status of large mammals in the mountain sector of Kahuzi-Biega National Park, Democratic Republic of Congo in 1996. Afr J Ecol 38:269–276

Inogwabini BI, Matungila B, Mbende L, Abokome M, Tshimanga WT (2007) Great apes in the Lake Tumba landscape, Democratic Republic of Congo: newly described populations. Oryx 41(4):532–538

# Chapter 12
# Lions of Malebo: Population and Conflicts with Humans

**Abstract** Lions (*Panthera leo*) have disappeared in large portions of their African range, which historically covered most of the countries. Yet, the species population has never been comprehensibly assessed in Central African though distribution maps indicate lions occur therein fragmentally. The lions of the Bateké Plateaus were absent from scientific records, though local communities report the presence of the species in the region in historical times. However, based on field observations dated between December 2006 and September 2008, 21 lion's calls have been recorded from the research field base at Malebo and footprints were observed along roads and were measured. Combining measurements of footprints and the fact that attacks on cattle herds occurred at different geographic locations, the study estimated that there were, at least 4 groups at Malebo. The mean group size was 3 individuals per group, with at least, 12 lions in the entire region of Malebo. Direct encounter with lions were also recorded four times and one person was killed by lions. The incidents of lions killing livestock were also reported, raising therefore the conflict between human and lions.

**Keywords** Lions · Panthera leo · Cattle · Ranches · Malebo and bateké plateaus

## 12.1 Introduction

Historically, the lion (*Panthera leo*) ranged across most of the African continent, occupying diverse types of habitat, including desert, woodland, dry forest, savannah, steppe, and even the mosaic of savannah-forest system of the edges of tropical forest (Barnett et al. 2006a, b). However, in the last 150 years, lions have disappeared in some African countries where they were present (Barnett et al. 2006a, b) such as Gambia, Mauritania, Sierra Leone, Burundi, Djibouti, and Eritrea. In areas where lions are still present in West or Central Africa, no single population of lions is large enough to be viable (Marchant 2001). In Central African countries, the lion population has never been comprehensibly assessed but different distribution maps (Bauer and Van Der Merwe 2004) indicate that lions occur in a fragmented distribution (Anonymous 2005). As it is the case with other regions of the continent (Barnett et al. 2006a, b), the central African populations may now be facing rapid decreases and local extinctions in the near future caused by human activities.



B.-I. Inogwabini, *Reconciling Human Needs and Conserving Biodiversity: Large Landscapes as a New Conservation Paradigm*, Environmental History 12,
https://doi.org/10.1007/978-3-030-38728-0_12

In the Democratic Republic of Congo, the species has been reported to present mostly in the eastern and the north eastern parts of the country (Barnett et al. 2006a, b; Verschuren 1975). Available estimates indicate that there were 90 and 150 individuals in the Virunga and Garamba National Parks, respectively, within the margins [60–125] and [100–200], respectively (Bauer 2003); 220 individuals were reported to be present in the Bili-Uele complex in 2005 (Anonymous 2005).

The population of the lions of the Batekke Plateaux in the DRC in general, and particularly of the Malebo area, seems to be absent from all scientific records, though local communities report the presence of the species in the region in historical times. The present study is, therefore, the first that describes this small population of the lions, which still reside in the Malebo area. Firstly, the paper presents a rough estimate of the population and, secondly, it presents the aspects of lion–human conflict, which is the most stringent conservation issue for the preservation of the species in the area.

## 12.2 What are the Characteristics of This Area Where Lions Roam in the Middle of the Forests?

This chapter reports on a species that occurs at these geographic coordinates: S02.45690; E16.48332. This is the same area as Malebo frequently cited above. Apart from the general habitats described for the southern part of the landscape above, this part of the landscape needs some more details here to give a more specific understanding of where lions were documented. To repeat what has been said above, this part of the landscape is within southern savannah of the plateau de Bateke, on the side of the Democratic Republic of Congo of the Congo River. The habitat is predominantly a forest-savannah succession; woody savannah predominates but is intercepted by narrow gallery forests that often follow the course of streams. Species of canopy-emerging trees in those galleries are essentially *Plagiostyles africana*, *Garcinia punctata*, *Greenwayodendron suaveolens*, *Diogoa zenkeri*, *Millettia laurentii*, *Chaetocarpus africanus*, *Isolona hexaloba*, *Entandrophragma* sp., and *Strombosiopsis tetrandra* (Brncic et al. 2007). As also stated above, these forest galleries have been logged in the last 20 years to principally extract timber of *Millettia laurentii* and *Entandrophragma* sp.

However, it should be added, as it is of general knowledge in ecology, that the forest-savannah system provides suitable micro-habitats for diverse species of wildlife, including savannah species, as well as forest species. Therefore, the ecotone, the junction of the forest and savannah in this case is richer than other ecosystems that share with them the same landscapes. This is also the case here (Malebo) where forest resident species of large mammals such as sitatungas, black-fronted duikers, blue duikers are niche-sympatric with savannah species of large mammals such as bushbucks and hyenas. Apart from these species, the site is also a home for species that are high conservation importance such as bonobos and elephants. Other species that are present in sizable numbers include forest buffaloes, red river hogs, and several diurnal primate species such as red-tail monkeys, de Brazza's monkeys, Mona's monkeys, etc. (for more details on primates, see Chap. 13).

From the perspectives of the land use ecology, as indicated earlier, the over prevalence of cattle raising ranches of multiple dimensions is the most important characteristic of the land use patterns in the region. Cattle ranches are mostly owned by small farmers who breed limited number of cows but are spread across different villages. However, breeding cows is also a professional activity for an industrial breeder. Altogether, small size local and industrial ranching occupies about 150,000-ha of the land of the region. The 150,000-ha have been subjugated to cattle raising activities since the early 1950s. This activity and land usage continued uninterruptedly since then until now. From the information gathered from local communities, small farmers practice cattle rising as part of the changes infused into the area by the industrial cattle ranching as a means to curb the theft of the cattle from the company. The practice has been introduced as formal sharecropping whereby communities had to receive first individuals from the company and then redistribute the offsprings to the rest of the community. Judged from the information gathered in the region, raising cattle is the most important income generating activity and confers enormous social and political weight to those who practice it with successes. The industrial ranch is divided into different sub-sectors. All these sub-sectors combined to support an estimate of 11,000 heads of cows and bulls. Apart from that large-scale cattle raising enterprise, local communities also raise small numbers of goats, etc. on savannah patches. Local communities also cultivate manioc, maize, peanuts, and the recently introduced rice (SGECN 1999).

## 12.3 What Was Done to Understand the Lions at Malebo?

Since September 2005 three monitoring field teams have been collecting data on bonobos, elephants, forest buffaloes and documenting the presence and absence of other wildlife species including birds, antelopes, and inventory vegetation diversity in the forests and savannahs of the Malebo area. Data on the presence of the species consisted of notes of direct encounters with species, presence of indisputable signs, identification of dead specimens from local markets and sometimes triangulated reports from local communities. When a species was directly encountered or seen in the market, teams do take pictures that are compared *post-hoc* to descriptions of field guides (Kingdon 1987; Stuart and Stuart 1997; Estes 1991). Equally, signs were confronted with available literature (Parnell 2000) while species reported by local trackers are triangulated either using the existing literature or through an argumentative dialogue with other trackers supported by scientific materials such as pictures from field guides.

Lion presence has been confirmed by three different types of data: triangulated reports from local communities, presence of indisputable signs, and direct encounters. Upon the confirmation of the lion's presence in the Malebo region, we sought to determine the species abundance, its distribution, and conservation issues that would arise from its presence in the region. To estimate the population of a lion at Malebo, we have used two different methods: (1) the metrics of footprints (Gusset and Burgener 2005) as follows: Teams were sent where lions were reported having been heard the previous night in order to collect data on the sizes of their footprints. For each footprint, both the larger radius ) and the smaller one ) were measured. These measures helped us identify how many individuals were in one group. Then were compared footprints from one location to these of the next location visited. Presences reported 45 km–straight line apart in less than 48 h were considered to be distinct groups. Then sighting of lion's footprints were mapped out to determine area of the lion presence. Also included in the presence data were records of calling by lions at nights.

Because lions are signaled in the area of Malebo particularly for killing cows, we have also collected data on reported killings of cows and reported encounters with humans. Each time killings of cows were reported, we interviewed the manager of the ranch to get the information on (1) number of cows killed, (2) how far away from the nearest village, and (3) in which month incidents happened, and (4) at what time of the day incidents happened. When possible, we went to the place where the incident took place and pictured killed cows and collected lion hairs. Data on reported kills were used in a model to provide a second estimate of the lion population in the area; this was possible only when two consecutive kills happened at a relatively similar time of the day over a distance that was felt to be difficult for the lion to travel in a short span of time. That distance was set to be beyond 10 km. This meant that when two kills were reported to happen at two different villages located at a 10 km-distance from each other and at relatively the same time (statement like teenish, etc.), these two kills were attributed to two different groups of lions.

## 12.4 Estimating the Lion Population at Malebo

Over the period covered by this chapter (December 2006–September 2008), 21 calls have been recorded from the research field base at Malebo and footprints were observed along roads and were measured (N = 10 for Malebo). Combining measurements of footprints and the facts that attacks (N = 5) on cattle herds occurred at locations 30–40 km apart sometimes in short spans of time (less than 24: 00 apart) and others (N = 2) happened almost simultaneously, it was estimated that there were, at least 4 groups at Malebo (Table 12.1). The mean group size was 3 Individuals per group. If this mean size is applied across the estimated number of groups, there would be at least, 12 lions at Malebo. The largest group of lions was observed at Malebo, which counted at least 3 individuals clearly differentiable from the prints; a young kitten was also clearly identified by its footprints that were way different from the rest of the knuckles recorded. Since December 2006, direct encounters with lions have been recorded four times and there was one person killed by lions (Table 12.2). Cows were reportedly killed in the same period. Humans encountered the lions in

**Table 12.1** Families and estimated population of the lions at Malebo

| Family | (a – b) Radius (cm) | (c – d) Radius (cm) | Mean number of Individuals |
|---|---|---|---|
| Malebo | 18 (±2), N = 10 | 15 (±2), N = 10 | 3 |
| | 15 (±1), N = 10 | 13 (±1), N = 10 | |
| | 8 (±2), N = 10 | 5 (±2), N = 10 | |
| Sali | 18 (±2), N = 8 | 15 (±2), N = 8 | 3 |
| | 15 (±1), N = 8 | 13 (±1), N = 8 | |
| | 8 (±2), N = 4 | 5 (±2), N = 4 | |
| Lebomo | 16 (±2), N = 6 | 14 (±2), N = 6 | 2 |
| | 14 (±1), N = 4 | 10 (±1), N = 4 | |
| Esanga | 12 (±2), N = 5 | 6 (±2), N = 5 | 4 |
| | 17 (±2), N = 4 | 15 (±2), N = 4 | |
| | 12 (±1), N = 3 | 13 (±1), N = 3 | |
| | 8 (±2), N = 4 | 5 (±2), N = 4 | |

**Table 12.2** Reported incidents since between 2006 and September 2008

| Area | Cow killings | Human killings | Human chasings |
|---|---|---|---|
| Malebo | 7 | | – |
| Esanga | 8 | 1 | – |
| Mbanzi | – | – | – |
| Lebomo | 3 | – | 1 |
| Sali | 3 | – | |
| Edzaengo | – | – | 2 |
| Mbee | – | – | 1 |
| Total incidents | 21 | 1 | 4 |

the early afternoon and lion's roarings were heard at the dusks (mostly around 19:00) and at late hours when people ventured to remain outside. Some of the roarings were heard when the day was dawning at the horizons (between 4:00–5: 30 AM).

## 12.5 Lions of Malebo in the Perspectives of the Species Conservation Status

The evidence of the lion presence at the Malebo area has been confirmed both by direct observations (Table 12. 2), calls, and indirect signs including hairs and footprints. Therefore, the question of the existence of the lions at the Malebo is no longer a pertinent question. This agrees with the records from the ranch owners indicated that lions had always been present in the area until the late 1990s what people claimed to be the last individual was signaled as being killed after it assaulted and killed cows. Records from the cattle raising company agreed with that version of fact though their dating pushed the event back in the mid-1980s. Records from the ranch also confirmed that since mid-1980s no single sign of the lion was signaled around. The presence of the lion is embedded in the traditional memories of the local communities that clearly distinguish the lion, locally known as '*tambue*', from other predators such as leopards that also occur in the region, indicating that the species has always been part of their natural ornamentation. In the literature, the Kingdon's (1997) field guide to African mammals says that lions have existed in the area of Malebo but no recent record was available. For comparison reasons, it should be mentioned that lions have been reported on the Plateau de Bateke in the Republic of Congo until recently and they were confirmed in Gabon (Bauer and Van Der Merwe 2004) in similar habitat. Indeed, the Plateau de Bateke, despite its divide by the majestic Congo River, stretches from the Democratic Republic of Congo to the Republic Gabon via the Republic of Congo. On the side of the Democratic Republic of Congo, the Plateau lions were reported to be present at Bombo-Lumene. The distance between Bombo-Lumene to Kinshasa is about 150 km, which is equidistant, if measured as a straight line, to the length of the travel from Bombo-Lumene to Malebo. However, the lions reported from Bombo-Lumene may have disappeared several decades from now (Vermeulen and Lanata 2006). Being unseen over several, nonetheless decades does not equate going extinct; the accepted definition being that a species would be proclaimed extinct only if unseen at least for 50 years after it has been sighted in a site where it used to happen. This rule should be taken as a crude measure because it is used in combination with other more critical biological criteria.

The re-discovery of lions over 20 years after its last manifestation around Malebo is a good news for conservation. However, this good news also poses a serious problem for species conservation. First of all, nothing is known about the species status, its ecology (population dynamics, behavior, and feeding strategies) and there is no specific data on its habitat uses and requirements because there has never been any study carried out at Malebo focusing on lions. Ecological studies of the species exist in a wide variety of its natural range across Africa. Malebo shares some critical characteristics with the larger Bateke Plateau, which includes Republics of Congo and Gabon but the Congo River separated it with the rest of the Bateke Plateau. This separation has been there since 30 MY BP in the geological history. This when a very large water body of the central Congo Basinstarted draining to the coast and cutting through the high ground at the continental margin on the western rim of the basin (Goudie 2005). The lakes Tumba and Maindombe are remnants of this large water basin. The savannah ecosystems of Malebo are also separated from the southern savannahs by the Kwa–Kasaï River, which constitutes a major biological barrier. So, clearly, these studies, however, cannot simply be extrapolated continent-wide and be used everywhere, particularly in Malebo, an area that has been separated by the rest of the lion's range for that long.

The second conservation problem is that the estimate of 12 individuals of lions, albeit with the caveat of not being a systematic survey of the population across what would seem to be its suitable habitats in the area of Malebo, is certainly a small population that is completely isolated from other lion populations. The smallness of the population will ultimately lead to inbreeding followed by the local extinction of species. Undeniably, among the biological criteria referred to the above is the minimum population size of 50 individuals (Frankham et al. 2014; Mace et al. 2008); 12 individuals is a figure way below the 50-individual genetic viability rule. But this statement would also be tempered down by the fact that the data collected showed clearly that there was at least one young kitten; which is an evidence of the population still being procreant and may show a possibility for the species to grow should there be specific conservation efforts invested in the species management. Saying so is far from being ferrytale; in the biodiversity conservation history of the Democratic Republic such an instance where proper species management and strong conservation measures rescued a species at the brink of its extinction was the case of the northern white rhinoceroses that were rescued from numbers as low as 13–20 to higher figures comprising hundreds within several decades of efforts (Hillman-Smith et al. 1986). Of course, the story of the northern white rhinoceroses was to end in a sobering reality but it shows clearly that when efforts are invested and clear species management plans are agreed upon collectively and implemented thoroughly, success can be achieved.

A further conservation problem posed by the evidence of the re-apparition of the lions at Malebo is attacks on livestock and humans, which has immediately onset a conflict between the species, local communities, and the cattle raising company. Late in 2006, guardians of the cattle herds reported the killing of one cow by a mysterious animal. The guardians presented clearly lion-looking hairs that were collected at the location of the incident. Between January and March 2007, at least seven incidents confirmed that the lions were attacking the ranch and other husbandry species such as goats and sheep. Of these incidents, the ranchers reported 5 attacks on cattle herds; the casualties of these incidents were 7 cows killed and a dozen gravely wounded in different ranches. The same incidents also reported two times when lions attacked humans traveling along the road in the early evening time; the casualties of attacks on humans were that one man was killed and some of the body parts were eaten by the lions. Predation by lions on livestock has been reported from other locations in Africa (Stander 1990) and attacks on humans were also documented in other geographical circumstances across the continent (Durrheim and Leggat 1999); they all had a common denominator which is that they instilled similar conflicts that played against the species long term survival (Woodroffe and Frank 2005; Bauer 2003). At Malebo, local communities openly voiced their fears that the resurgence of the species would result in a major conflict between conservation organizations, the Democratic Republic of Congo Government and their own local interests. The presence of the species in the region does, in fact, paralyze whole communities for months; when attacks of lions are reported, whole villages do not move into the forest to do the work they usually do for their livelihoods such as fetching water, collecting woods for energy, weed out their manioc fields, and hunting. For these communities, there are only two options to solve this situation: either to completely eliminate the species or the conservationists and the government should provide alternative sources of energy, new ways of getting food and financial resources, etc. Of course, there must be ways to, at least partially, reconcile these demands without seeing things in black and white as the demands from local communities are sound; but there is also a need to craft a comprehensive and strong conservation action to diffuse this conflict in order to help the population of lions residing at Malebo survive in the future.

Attacks on herds of cattle and the irruptions of lions at Malebo have been exclusively recorded within the period between January and March. This period is very hot and humid in the region and coincides with the shortest rainy season. Discussions as to why these attacks happen at the same period of the year abound in the region; multiple sets of explanations ranging from the rational to the mystical ones are offered. One of the explanations that merit some words here is the fact that lions might be crossing either the Kwa–Kasaï River or even the Congo River at the driest times of the year and come into the Malebo area. At the first glimpse, some recognition maybe given to this interpretation of facts. Some large carnivores have been documented to swim. For examples, tigers (*Panthers tigris*) were seen to swim across wide tidal rivers in the Sundarbans at the mouth of the Ganges (Miththapala et al. 1991); leopards can swim if they feel that they have to do so and it has been reported that the Anatolian leopard *Panthera pardus tulliana* arrived at Samos from the opposite coast of Turkey, swimming across the channel separating the island from western Anatolia (Masseti 2010). Yet, a deep ecological reading of the situation seems to give an opposite view. This is supported by the fact that the Congo River has a permanently strong current for a species like a lion to swim across it; so too is the Kwa–Kasaï river. The water levels of these rivers can, of course, decrease seriously in dry seasons but their respective currents remain very strong. So, despite some plausible validity of the swimming across the rivers hypothesis, I suggested that this phenomenon could be understood if the appearances and disappearances of lions in the region would be explained using the general ecological features of the region in the period when they abruptly emerge in the human landscapes of the region. One important and widespread ecological characteristic of this region between February and March each year is that the region goes through a dry season. Observations made in the region show a clear clustering of herds of cattle in the gallery forest with a thickly closed canopy. Equally, at the same period of the year, other savannah species such as bushbucks and forest buffaloes are literally obliged to retract inside the forest galleries and concentrate around water points. Indeed, in that period of the year, water points located in the savannahs are dried up because of the evaporation caused by higher temperature of the small dry seasons; sources and points of water that are available in this period of the year remain in the forest galleries. Because the lion prey species move into forest galleries, which is a habitat type where lions are more or less incompetent to hunt, they then rely on easy targets constituted of herds of cattle. They become more prone to hunt openly and near villages where cattle and other husbandry species occur.

The above explanation is plausible and coheres with the ecological realities of the zone. Indeed, in the dry seasons communities use this very knowledge for their net hunting exercises. They come together and deploy their handmade hunting nets in areas adjacent to available water pounds; they often capture species of antelopes such as sitatungas, bushbucks, etc. Because lions cannot go deep into the gallery forests to hunts these prey species, they rely on what remains in the open space: cows, sheep, goats, and (unfortunately) humans too. The plausibility of this explanation, which is an alternative to lion swimming across Kwa–Kasai River or across the Congo River, is supported by the fact that there is no documented evidence of lions swimming through the rivers in this part of the world. The swimming behavior being not an essential trait of the lion's use of habitats and hunting techniques, it is very unlikely that the lions in Malebo would want to take the risks of dying or wasting massive amount of energies just to cross the Congo River in search of preys on the Malebo's side of this grandiose river. The prospects of crossing the Kwa–Kasai River are also very unlikely plausible, unless there is something really ecologically and evolutionarily beneficial for the species to cross such a huge savannah followed by the strong current of the Kwa–Kasai River. Such a beneficial item could be imagined to be only larger prey species that could be provided by herds of cows. But, if this would be the sole reason, lions would not find any interest in going through such lengths because the south side of the Kwa–Kasai River from where they would be coming from is located within a webbing of husbandries of cows, goats, sheep, etc. as it is the case in the Malebo area. Indeed, Bombo-Lumene where Vermeulen and Lanata (2006) described a population of decades unseen lions and from where the lions moving into Malebo would have to come from is an extensive woody savannah area where communities, as well as other industrial landlords, have extensive farms and raise cattle and other types of livestock. Lions, if needed, would hunt on these farms rather than taking long trips and swim across massively strong current and risk their lives. This would be a highly prohibitively expensive cost in terms of the energy to pay for the lions and would make very little sense. Hence, there is likely no lions migrating in the area in the dry seasons and moving out during the heavy rainy seasons; they are likely resident in the area and only show different ecological behavior at different periods of the year because of the movements of their prey species dictated by the patterns of water distribution in the region.

## References


Anonymous (2005) Chapter II: Population survey. Available at the following website: http://www.conservationforce.org/pdf/conservationoftheafricanlion_study_2_of_5.pdf. Accessed 1 Aug 2007

Barnett R, Yamaguchi N, Barnes I, Cooper A (2006a) Lost populations and preserving genetic diversity in the lion **Panthera leo**: implications for its ex situ conservation. Conserv Genet. https://doi.org/10.1007/s10592-005-9062-0

Barnett R, Yamaguchi N, Barnes I, Cooper A (2006b) The origin, current diversity and future conservation of the modern lion (**Panthera leo**). Proc R Soc B. https://doi.org/10.1098/rspb.2006.3555

Bauer H (2003) Lion conservation in West and Central Africa: Integrating social and natural science for wildlife conflict resolution around Waza National Park, Cameroon. PhD Thesis, Leiden University, the Netherlands

Bauer H, Van Der Merwe S (2004) Inventory of free-ranging lions **Panthera leo** in Africa. Oryx 38(1):26–31

Brncic T, Menga P, Lejoly J (2007) Preliminary report of botanical activities carried out in the Lac Tumba Landscape. Typescript report submitted to World Wide Fund for Nature
Durrheim DN, Leggat PA (1999) Risks to tourists posed by wild mammals in South Africa. J Travel Med 6(3):172–179
Estes RD (1991) The behavior guide to African mammals including hoofed mammals, carnivores, primates. The University of California Press, Berkeley
Frankham R, Bradshaw CJA, Brook BW (2014) Genetics in conservation management: revised recommendations for the 50/500 rules, red list criteria and population viability analyses. Biol Conserv 170:56–63
Goudie AS (2005) The drainage of Africa since the Cretaceous. Geomorphology 67(3–4):437–456
Gusset M, Burgener N (2005) Estimating larger carnivore numbers from track counts and measurements. Afr J Ecol 43:320–324
Hillman-Smith K, Ma Oyisenzoo M, Smith F (1986) A last chance to save the northern white rhino? Oryx 20(1):20–26
Kingdon J (1987) The kingdon field guide to African mammals. Academic Press
Mace GM, Collar NJ, Gaston KJ, Hilton-Taylor C, Akçakaya HR, Leader-Williams N, Milner-Gulland EJ, Stuart SN (2008) Quantification of extinction risk: IUCN's system for classifying threatened species. Conserv Biol 22(6):1424–1442
Marchant J (2001) Lions face extinction in large parts of Africa. New Scientist, November 2001. http://www.newscientist.com/article.ns?id=dn1500. Accessed 1 Aug 2007
Masseti M (2010) Homeless mammals from the Ionian and Aegean islands. Bonn Zool Bull 57(2):367–373
Miththapala S, Seidensticker J, Phillips LG, Goodrowe KL, Fernando SBU, Forman L, O'Brien SJ (1991) Genetic variation in Sri Lankan Leopards. Zoo Biol 10:139–146
Parnell RJ (2000) Information from animal tracks and trail. In: White LJT, Edwards A (eds) Conservation research in the African rain forests: a technical handbook. Wildlife Conservation Society, pp 153–184
Secrétariat Général à l'Environnement, Conservation de la Nature, Pêche et Forêts (SGECN PF), 1999 Plans provinciaux de la biodiversité—appendice au plan d'action national. Ministère de l'Environnement, Conservation de la Nature, Pêche et Forêts, Kinshasa Gombe, République Démocratique du Congo
Stander PE (1990) A suggested management strategy for stock-raiding lions in Namibia. S Afr J Wildlife Res 20(2):37–43
Stuart C, Stuart T (1997) Field guide to the larger mammals of Africa. Struik Publishers
Vermeulen C, Lanata F (2006) Le domaine de chasse de Bombo Lumene: un espace naturel en péril aux frontières de Kinshasa. Franc Vert 3(2):6–9
Verschuren J (1975) Wildlife in Zaire. Oryx 13(2):149–163
Woodroffe R, Frank LG (2005) Lethal control of African lions (Panthera leo): local and regional population impacts. Anim Conserv 8:91–98

# Chapter 13
# Diurnal Primates: Estimates and Conservation Issues

**Abstract** Data were collected for a period of 38 weeks (9.5 months) covering both rainy and dry seasons to identify diurnal primate species, document group compositions, and to estimate both encounter rates $\delta$ and densities $\overline{\varpi}$ at Malebo region. Three species were observed in six groups: De Brazza's monkey, Mona monkey, and red tail monkey along a 10 km long trail. The mean relative abundance for all three species $\delta = 0.6$ groups/km $\pm$ 0.2 (SD). The mean group sizes for red tail monkeys $\overline{X} = 7.3$ individuals/group $\pm$ 0.12 (SE), larger than any of the three systematically monitored in the region and De Brazza's monkey, with $\overline{X} = 2.5$ individuals/group $\pm$ 0.23 (SE) had the smallest group size. The most encountered species was Mona monkey (sighting probability $\mu = 0.82$ sightings/visit). The overall mean density $\overline{\varpi} = 0.28$ individuals km$^{-2}$, within the range = [0.13–0.38] individuals km$^{-2}$. The red tail monkey was the most abundant ($\overline{\varpi} = 0.38$ individuals km$^{-2}$) followed by Mona monkey ($\overline{\varpi} = 0.32$ individuals km$^{-2}$) and De Brazza's monkey ($\overline{\varpi} = 0.25$ individuals km$^{-2}$). Comparisons with other areas in Tropical Africa indicated that group sizes, relative abundance, and densities in this region were lower. Despite the potential of other ecological processes (e.g. differences in habitat types, food availability, etc.) in the region to deplete primate populations, bushmeat trade was arguably the main reason for lower primate abundance in this region. Immediate conservation actions were called upon to help conserve some of the most important forest galleries from hunting to preserve the primate populations.

**Keywords** Primates · Group size · Densities · Bushmeat · Conservation status

## 13.1 Introduction

Primates are ecological indicators for the ecosystem's health in areas where they occur (Cowlishaw 2000). They play key ecological roles in regenerating African forests by dispersing seeds (Colyn 1988; Oates 1986; Bourlière 1985) and are integral elements in human mythologies, diets, and scientific paradigms (Fuentes and Wolfe 2002). With such a high profile, diurnal primates should become part of any ecological monitoring program in the conservation of Central African forests (Chapman and Peres 2001). In the conservation history, however, studies (with few exceptions)



B.-I. Inogwabini, *Reconciling Human Needs and Conserving Biodiversity: Large Landscapes as a New Conservation Paradigm*, Environmental History 12,
https://doi.org/10.1007/978-3-030-38728-0_13

and conservation of diurnal primates were considered as secondary priorities of conservation organizations, particularly in Central Africa. Small primates are neglected because most of these organizations operating in tropical Africa are focused on charismatic species such as elephants, great apes, and top predators for reasons of fundraising, visibility of their conservation impact, and limited expertise. As a consequence, nearly 1/3rd of primate species are threatened with extinction (Konstant et al. 2002). Forty-three African primates spread across 24 countries are at the brink of extinction (Mittermeier et al. 2006, 2004); 28% of the most endangered primates in the world occur in Africa (Magnuson 2005; Mittermeier et al. 2004; Konstant et al. 2002). In the recent past, however, studies were conducted to evaluate primate conservation statuses (Magnuson 2005) but large areas of Central African forests had not been adequately surveyed. Basic information such as primate distribution and their abundances are still lacking, time-series data on abundance are rarely available and ecological data such as habitat uses and requirements is absent in many cases. This lack of critical basic knowledge, of course, impedes any attempt to produce sound management strategies for diurnal primates facing threats. Worst of all, the lack of conservation action for diurnal primates to the second-order conservation priority has been occurring when most existing data, where available, suggested that primates were among the species that composed a large part of the bushmeat trade in African countries (Chapman et al. 2006). Bushmeat trade depleted forests across Africa of their primates and resulted in the empty and silent forests (Oates 1999).

It was to address this gap in data on diurnal primates and to help design a sensible conservation primate monitoring and conservation plan that data were collected from the Malebo region. Data were, therefore, collected from the Bambou River forest gallery to (1) provide the first primate conservation status for the Malebo region where the bonobo studies have been going for the last three years (Inogwabini et al. 2007a, b; Inogwabini and Matungila 2009) and (2) to discuss their conservation status in light of existing field evidence. In so doing, the ultimate goal was to propose a conservation strategy for all the primates residing in the forest galleries around the Malebo zone.

## 13.2 How Were the Data on Diurnal Primates Collected in Lake Tumba Landscape?

The data for this chapter were collected from the southern part of the Lake Tumba Landscape, particularly around the Malebo study site from February through October 2010. The habitat of the study area was a forest-savannah mosaic. As part of the ecotone system, there is high vegetation diversity (Inogwabini and Matungila 2009; Brnic et al. 2007; Inogwabini et al. 2007a). Data were collected from the forest gallery within the gallery forest along the Bambou River (S2.48302°; E16.50477° and S2.49713°; E16.49314°). This gallery has a mean width of approximately 0.080 km on both sides of the river within the range 0.030–2 km. Data were collected within the flood zone of the Bambou River, with the most important forest type is mixed mature, with species such as *Uapaca guineensis*, *Raphia sp.*, *Irvingia smithii*, etc. The dominant understorey was liana composed essentially of *Ancistrophilium Africana* and *Cyrtosperma senegalense*. Data collected from the field Base at Bambou showed that the humidity was high (mean = 85%) and that the mean annual temperature was 24 °C (Inogwabini and Matungila 2009), the temperature in this zone is cooler than in nearest forests located close to the equator line. However, gallery forests in this zone do share a common trait of having two rainy seasons (March–May and September–December) and two dry seasons (January–February and June–August). The biological diversity of the zone is high even though more work is needed to complement the existing data sets. Apart from bonobos *Pan paniscus*, chimpanzees *Pan troglodytes troglodytes*, forest elephants *Loxodonta africana cyclotis*, lions *Panthera leo,* and forest buffaloes *Syncerus caffer nanus* that are the charismatic species of the zone, there are seven diurnal primates in the larger zone of the southern Lake Tumba Landscape (De Brazza's monkey *Cercopithecus neglectus*, red tail monkey *Cercopithecus ascanius*, Mona monkey *Cercopithecus mona wolfi*, Angolan pied colobus *Colobus angolensis*, black mangabey *Lophocebus aterrimus*, Tshuapa red colobus *Piliocolobus tholloni*, Allen's swamp monkey *Allenopithecus nigroviridis*) to which would be added the enigmatic historical account describing a species similar to the golden-bellied mangabey *Cercocebus chrysogaster*. Potto *Perodicticus potto faustus*, two species of galagoes (Demidoff's galago *Galagoides demidoff phasma* and *Galagoides thomasi*), two species of squirrel (Congo squirrel *Funisciurus congicus* and Thomas's rope squirrel *Funisciurus aneythrus*), and several species of bats are also present.

A 10 km long trail was laid along the gallery forest of the Bambou following approximately the north-south direction. Following other studies (Chapman and Lambert 2000; Brugiere and Fleury 2000; McGraw 1994) the trail was visited fortnightly from 5:30–8:30 AM and 4:00–6:00 PM. Researchers walked at the speed ranging within 1–1.5 km/hour stopping periodically at every 100 m for five minutes to watch and listen to monkey calls (White and Edwards 2000; Uehara and Ihobe 1998; Whitesides et al. 1988). Data collected included species observed, time of observation, visual estimate of the distance between the observer and the center of the observed group of primates, number of observed individuals, and geographical coordinates recorded on a handheld unit of the Global Positioning Systems (Garmin 12XL). Habitat data were coarse resolution descriptions adapted from a protocol that was used across Central Africa by many research groups (Inogwabini et al. 2007a; Reinartz et al. 2006; Van Krunkelsven et al. 2000; Hall et al. 1998a, b, 1997) and consisted of savannah versus forest, divided into their own categories: woody versus grassy savannahs, mature versus secondary forests. Additionally to those coarse categories, forests were also divided based on their canopy categories (open versus closed).

Data were collected for a period of 38 weeks (9.5 months) covering both rainy and dry seasons. Group of monkey was defined as an assemblage of monkeys that were separated from the others by a minimum of 150 m. Therefore groups numbers were simple additions of each group found responding to this criterion of 150 m-separation. Over the study, we discovered that primates of the Malebo were sedentary because groups were restricted within a radius ≥200 m of their initial location. Using this reality, mean group sizes ($\overline{X}$) were obtained by dividing numbers of counted individuals over the number of encounters.

To estimate the primate abundance, we used both relative measures. Relative indexes of abundance consisted of encounter rates ($\delta$), which were ratios of groups over the 10 km of the sampling trail (White and Edwards 2000) and frequencies of sighting monkey groups ($\mu$), which were the likelihood of seeing monkey groups calculated as ratios of numbers of sightings of each group over the times the trail was visited. As applied by Brugière et al. (2002), densities ($\overline{\varpi}$) were calculated using the mean group sizes that were divided by the estimated area of the Bambou gallery forest, which was defined by the maximum spread of the forest multiplied by the length of our research trail ($= 2\text{ km} \times 10\text{ km} = 20\text{ km}^2$). The overall density used the mean of mean group sizes while the minimum density used the lowest mean group size and the maximum density used the highest mean group size. The habitat uses were simple percentages of how many times species were encountered in a given habitat.

## 13.3 Diurnal Primates: Species and Estimated Populations

Three species of monkeys were observed for ~9 months and data collected in the Bambou Gallery Forest (Table 13.1). These were De Brazza's monkey, red tail monkey, and Mona monkey. These species occur in six different groups of which 4 groups were of De Brazza's monkey and 1 group for each of the red tail and Mona monkeys. Overall, the relative abundance estimated $\delta = 0.6$ groups per km ± 0.2 (SD). With $\overline{X} = 7.3$ individuals per group ± 0.12 (SE), the largest mean group size was that of a red tail monkey while the smallest $\overline{X} = 2.5$ individuals/group ± 0.23 (SE) was that of one of the four De Brazza's monkey groups. The mean sighting likelihood $\mu = 0.72$ groups per visit within the range of 0.3–1.1 groups per visit (Table 13.1).

**Table 13.1** Mean group sizes and estimated densities of monkey of the Bambou River

| Group | Species | # Visits | $\overline{X} \pm$ SE | $\mu$ | $\overline{\varpi}$ |
|---|---|---|---|---|---|
| Group 1 | *Cercopithecus neglectus* | 19 | 6.42 ± 0.25 | 0.63 | |
| Group 2 | *Cercopithecus neglectus* | 19 | 5.89 ± 0.25 | 0.82 | |
| Group 3 | *Cercopithecus neglectus* | 19 | 2.50 ± 0.23 | 0.53 | |
| Group 4 | *Cercopithecus neglectus* | 19 | 5.33 ± 0.12 | 0.75 | |
| Total | *Cercopithecus neglectus* | | 5.03 + 0.21 | | 0.25 |
| Group 5 | *Cercopithecus ascanius* | 19 | 7.60 + 0.12 | 0.77 | 0.38 |
| Group 6 | *Cercopithecus mona wolfi* | 19 | 6.33 ± 0.11 | 0.83 | 0.32 |

**Table 13.2** Habitat use by different primate species at Bambou River

| Forest types | *C. neglectus* (%) | *C. ascanius* (%) | *C. m. wolfi* (%) |
|---|---|---|---|
| Mixed mature forest/closed canopy | 28.79 | 100.00 | 100.00 |
| Mixed mature forest/open canopy | 69.70 | 0.00 | 0.00 |
| Savannah | 1.51 | 0.00 | 0.00 |
| Total | 100.00 | 100.00 | 100.00 |

As would be expected, the mean sighting likelihood varied between groups, with the highest being $\mu = 0.82$ sightings/visit (Group 6) and the least likely seen $\mu = 0.53$ (Group 3). The overall mean density $\varpi = 0.28$ individuals km$^{-2}$, within the range 0.13–0.38 individuals km$^{-2}$ (Table 13.1). The highest density was that of red tail monkey ($\varpi = 0.38$ individuals km$^{-2}$), followed by Mona monkey ($\varpi = 0.32$ individuals km$^{-2}$), and De Brazza's monkey ($\varpi = 0.25$ individuals km$^{-2}$). *C. ascanius* and *C. m. wolfi* spent 100% of their time in mixed mature forest with closed canopy while *C. neglectus* had a varied usage of habitat but with 69.70% of its time being spent in mixed mature forest with open canopy, 28.79% in mixed mature forest with closed canopy, and 1.51% in savannahs (Table 13.2).

## 13.4 Diurnal Primates of Lake Tumba in the Perspectives of Primate Conservation

Mean Group sizes vary between the four species found in the region (Table 13.1). The largest mean group size was that of the red tail monkey (Table 13.1). This fact reflects realities in the other sites across Central African forests where the red tail has been described not only as a ubiquitous species but also harboring large groups (Walker et al. 1964). The mean group size for the red tail at the Bambou River was consistent with that reported from Kahuzi-Biega National Park (Hall et al. 2003) but was lower than group size reported from Lomako in 1994 (McGraw 1994). In fact, all mean group sizes were lower than in any other site in tropical Africa such as Uganda, Gabon, and other sites in the Democratic Republic of Congo where data are available (Brugiere 2005; Chapman 2000; McGraw 1994). For example, the 7.6 individuals per group for the red tail monkey was lower than the figure reported for the same species in Uganda (Chapman 2000) and the overall $\varpi = 0.28$ individuals km$^{-2}$ was lower than 45 individuals km$^{-2}$ reported from Lope, Gabon (Brugiere 2005) even though the encounter of the same species was higher in this zone (0.1 group km$^{-1}$) compared to Salonga National Park where it occurred in lower encounter rate (0.08 groups km$^{-1}$; Inogwabini 2006a). Habitat use by each species corresponded with what would be expected of the behavior of species herein described. *C. ascanius* is a cosmopolitan species using a variety of habitats while *C. mona* is a high canopy species. Both species feed largely on fruit, leaves, and flower (Chapman et al. 2002; Kingdon 1997; Thomas 1991). The circumstances of forest galleries at Bambou are that most high canopy trees are within the mixed mature forest with closed canopy and do offer a diversity of flowers, fruits and leaves that are preferred food for those two species. Therefore, it not surprising that these species had been found in that type of habitat (Table 13.2). That *C. neglectus* has been found using three habitat types contrasts with findings from Lomako (McGraw 1994). The use of savannah appears to be new in this species since *C. neglectus* is known to be restricted in swampy forests (M[c]Graw 1994). *C. neglectus* used different types of habitat, including closed and open canopy forests, low trees, and the ground. The use of savannahs maybe explainable by the species feeding behavior, which includes moving away for about 200 m from rivers to visit major food sources (Kingdon 1997) and is likely an adaptation to the physical environment of the site. In fact, mature individuals of *C. neglectus* had been observed crossing the savannah to access to fruits of *Landolfia spp* and *Annona senegalensis*, which reproduce in dry seasons when most tree species in the forest galleries bear no fruits (Inogwabini and Matungila 2009).

Regarding the three species conservation status, data on the species abundance (Table 13.1) alone would lead to the conclusion that all the three species present in this zone are locally critically endangered and deserve more conservation attention. Abundance (densities and biomasses) of primates is thought to depict the ecosystem's health. Studies (Chapman et al. 2006; Tutin and White 1999; Tutin et al. 1997; White 1994a, b) hypothesized several reasons that could lead to severe reduction of primate abundances in their natural ranges. These reasons included ecological catastrophes such as prolonged low fruit production periods, continual extension–contraction of the forest cover in some areas over the past millennia, which might not have provided enough time for primate populations to reach carrying capacity in the new habitats, hunting for bushmeat trade connected to high human densities, and the needs for protein and cash incomes. More recently epidemic diseases have decimated large numbers of wildlife and were included among paradigms that would explain primate densities decreases. The region of Bambou is similar in many aspects to the Lopé region (Gabon), they share the forest-savannah complex, they have people that historical descended from the same cultural matrix and hence would have the same attitude toward their immediate environment. Furthermore, both areas have been using fire as a management tool for long periods (Inogwabini 2006b). Despite these similarities, there are differences most likely dictated by the Congo River, which is a major barrier. There are, for example, differences in the rainfall regime, which would make it difficult to translate causes for depleting primate populations from Lope to the Malebo region where the Bambou River is located. There are no data, albeit not collected yet, to conclude that ecological processes (prolonged dry periods, contraction–extension dynamics) described in the context of forest-savannah mosaics in Gabon had prevailed here and had affected the populations of primates in the region with the same magnitude. There has been a report of an outbreak of diseases in the region that affected wildlife species, but all records indicated that the disease particularly attacked the cane rat *Thryonomys swinderianus* (Inogwabini 2008). Even though there is no data on the extent of that event on other large mammals of the region, socio-economic data indicated that people were only aware of its effects on cane rat and had never seen any case of for other wildlife species, including primates. On the contrary, the same socio-economic data had indicated that diurnal primates were among the 25–30% of all the bushmeat trade in the region, with only red river hog *Potamocherus porcus* exhibiting higher frequencies. These figures alone are conclusive that, at least in part, the bushmeat trade may have played a key role in the actual lower densities of primates in this region. This suggestion is supported by reports from local communities that of herds of forest buffaloes, Angolan pied colobus, black mangabey and Tshuapa red colobus, Allen's swamp monkey were present. However, these species have now become so rare that over the four years spent in the region the Angolan pied colobus was seen only once in the region of Mpoka (S01.40443°; E017.04747°) located at *ca* 132 km from Bambou (S2.48302°; E16.50477°) and *ca.* 135 km and the Mabali Scientific Reserve (S0.88254°; E18.12256°), the Tshuapa red colobus had been seen only in the market. For the Tshuapa red colobus the origin of where the specimen was captured was even dubious. The region having been logged in the past, the rarity of diurnal primates maybe a simple reflection of intensive bushmeat hunting that occurred several years earlier. This conclusion is consistent with the results of Chapman et al. (2000) who found that populations of *C. ascanius* continued to decline decades after industrial logging occurred in the Kibale National Park, Uganda.

The golden-bellied mangabey, a species that local communities describe with the certainty of its characteristics and its known terrestrial ranging habits has been reported to have disappeared since the early 1980s (Inogwabini and Thompson 2013) when the bushmeat crisis reached its climax was collateral to the elephant killings for the ivory trade. In fact, interviews with local communities indicated between 1980 and 1990, massive invasions of elephant poachers came into the region in search of elephants and buffaloes. The influx of ammunition and guns that were brought helped local communities exterminate all large wildlife species to fuel the market both in Kinshasa and Brazzaville (Inogwabini et al. 2007b). Combined with the ivory and bushmeat trade was the fact that the region is located in the most logged zone in the Democratic Republic of Congo, which is a consequence of its easy access from Kinshasa by road or by boat. Logging camps and logging routes opened the area to more hunting and bushmeat activities in the region, decimating, therefore, populations of easy targets such as the Salonga Red colobus.

The conservation context of this study is that the research reported herein had been carried out in a complex of the forest-savannah mosaic belonging to a private sector commercial ranch operational since 1950s (Inogwabini et al. 2007a). While critical law enforcement had been implemented to protect savannah habitats, little attention has been given in protecting forest galleries, albeit with the exception of bonobos that are protected by traditional taboos in this region. Rampant bushmeat hunting targets large-bodied species such as bushbucks, sitatungas and diurnal primates in forest galleries, which are preferred habitat for most African cercopithecine species (Gautier-Hion and Brugière 2005). People armed with 12-gauze machine guns, hunt nights and days in those forests galleries. Conducting a reconnaissance survey in the area, our team collected 15 used shotgun cartridges over 20 km (encounter rate = 0.75 signs $km^{-1}$), higher than all primate's signs combined. The conservation message conveyed by data presented here is that protection of those forest galleries should be given more attention if the species of diurnal primates residing therein are to be protected. This implies incorporating forest galleries in the management of the ranch to benefit from different existing law enforcement schemes, which is aligned with the new forestry code of the Democratic Republic of Congo where private protected areas are legal.

## References


Bourlière F (1985) Primate communities: their structure and role in tropical ecosystems. Int J Primatol 6:1–26

Brncic T, Menga P, Lejoly J (2007) Preliminary report of botanical activities carried out in the Lac Tumba Landscape. Typescript report submitted to World Wide Fund for Nature

Brugiere D (2005) Monkey community structure in the old growth forests of the Lope reserve, Gabon. Afr J Ecol 43:70–72

Brugière D, Fleury MC (2000) Estimating primate densities using line transect and home range methods: comparative test with the black colobus monkey Colobus satanas. Primates 41:373–382

Brugière D, Gautier JP, Moungazi A, Gautier-Hion A (2002) Primate diet and biomass in relation to vegetation composition and fruiting phenology in a rain forest in Gabon. Int J Primatol 23(5):999–1024

Chapman CA (2000) Constraints on group size in red colobus and red-tailed guenons: examining the generality of the ecological constraints model. Int J Primatol 21(4):565–595

Chapman LJ (2001) Fishes of African rain forests: diverse adaptations to environmental challenge. In: Weber W, White LJT, Vedder A, Naughton-Treves L (eds) African rain forest—ecology and conservation: an interdisciplinary perspective. Yale University Press, pp 263–290

Chapman CA, Lambert JE (2000) Habitat alteration and the conservation of African primates: a case study of Kibale National Park, Uganda. Am J Primatol 50:169–186

Chapman CA, Balcomb SR, Gillepsie TR, Skorupa JP, Struhsaker TT (2000) Long-term effects of logging on African primate communities; a 28 –year comparison from Kibale National Park, Uganda. Conserv Biol 14(1):207–217

Chapman CA, Chapman LJ, Cords M, Gathua JM, Gautier-Hion A, Lambert JE, Rode K, Tutin CEG, White LJT (2002) Variation in the diets of Cercopithecus species: differences within forests, among forests, and across species. In: Glenn ME, Cords M (eds) The guenons: diversity and adaptation in African monkeys. Kluwer Academic/Plenum Publishers, pp 325–350

Chapman CA, Lawes MJ, Eeley HAC (2006) What hope for African primate diversity? Afr J Ecol 44:116–133

Colyn MM (1988) Distribution of guenons in the Zaire-Lulaba-Lomani river system. In: Gautier-Hion A, Bourlière F, Gautier JP, Kingdon J (eds) A primate radiation: evolutionary biology of the African guenons. Cambridge University Press, pp 104–124

Cowlishaw G. (2000). Primate Conservation biology. The University of Chicago Press

Fuentes A, Wolf LD (2002) Face to face: the implications of human-nonhuman primate interconnections. Cambridge University Press

Gautier-Hion A, Brugiere D (2005) Significance of riparian forests for the conservation of Central African Primates. Int J Primatol 26(3):515–523

Hall JS, Inogwabini BI, Williamson EA, Omari I, Sikubwabo C, White LJT (1997) A survey of elephants (Loxodonta africana) in the Kahuzi-Biega National Park lowland sector in eastern Zaire. Afr J Ecol 35:213–223

Hall JS, Saltonstall K, Inogwabini BI, Omari I (1998a) Distribution, abundance and conservation status of Grauer's gorilla. Oryx 32:122–130

Hall JS, White LJT, Inogwabin BI, Omar I, Morland HS, Williamson EA, Saltonstall K, Walsh P, Sikubwabo C, Bonny D, Kiswele KP, Vedder A, Freeman K (1998b) A survey of gorillas (Gorilla

gorilla graueri) and chimpanzees (Pan troglodytes schweinfurthi) in the Kahuzi-Biega National Park lowland sector and adjacent forest in eastern Democratic Republic of Congo. Int J Primatol 19:207–235

Hall JS, White LJT, Williamson EA, Inogwabin BI, Omari I (2003) Distribution, abundance and biomass estimates for primates within the Kahuzi-Biega lowland and adjacent forest in eastern DRC. Afr Primates 6(1–2):35–42

Inogwabini BI (2006a) A preliminary checklist of mammals and plants: conservation status of some species in the Salonga National Park, Democratic Republic of Congo. Endanger Species Updat 23(3):104–117

Inogwabini BI (2006b) Fire as management tool in cattle raising concession: a preliminary assessment to help develop a system that sustains biological diversity in the southern in the Lac Télé - Lac Tumba Swampy forest, DRC Segment. Field report submitted to WWF and CARPE-USAID

Inogwabini BI (2008) Etudes des possibilités de transmission des maladies entre les animaux domestiques, les animaux sauvages et les populations humaines – protocole de collecte des données pour le programme Santé Animale. Typescripted document submitted to WWF US and USAID

Inogwabini BI, Matungila B (2009) Bonobo food items, food availability and bonobo distribution in the Lake Tumba Swampy forests, Democratic Republic of Congo. Open Conserv Biol J 3:1–10

Inogwabini BI, Thompson JAM (2013) The Golden-bellied Mangabey (Cercocebus chrysogaster): distribution and conservation status. J Threat Taxa 5(7):4069–4075

Inogwabini BI, Matungila B, Mbende L, Abokome M, Tshimanga WT (2007a) Great apes in the Lake Tumba landscape, Democratic Republic of Congo: newly described populations. Oryx 41(4):532–538

Inogwabini BI, Bewa M, Longwango M, Abokome M, Vuvu M (2007b) The bonobos of the Lake Tumba–Lake Maindombe hinterland: threats and opportunities for population conservation. In: Furuichi T, Thompson J (eds) The bonobos behavior, ecology, and conservation. Springer, pp 273–290

Kingdon J (1997) The kingdon field guide to african mammals. Academic Press

Konstant WR, Mittermeier RA, Rylands AB, Butynski TM, Eudey AA, Ganzhorn J, Kormos R (2002) The world's top 25 most endangered primates. Neotropical Primates 10(3):128–131

Magnuson L (2005) Conservation of African monkeys. Int J Primatol 26(3):511–513

McGraw S (1994) census, habitat preference, and polyspecific associations of six monkeys in the Lomako Forest, Zaire. Am J Primatol 34:296–307

Mittermeier RA, Rylands AB, Konstant WR (2004) IUCN/SSC Primate Specialist Group Report 2001–2004. International Union for Conservation of Nature (IUCN)

Mittermeier RA, Valladares-Pádua C, Rylands AB, Eudey AA, Butynski TM, Ganzhorn JU, Kormos R, Aguiar JM, Walker S (2006) Primates in Peril: the world's 25 most endangered primates, 2004–2006. Primate Conserv 20:1–28

Oates JF (1986) Action plan for African primates conservation: 1986–1990. The International Union for the Conservation of Nature, Switzerland

Oates JF (1999) Myth and reality in the rain forest: how conservation strategies are failing in West Africa. University of California Press

Reinartz G, Inogwabini BI, Mafuta N, Lisalama WW (2006) Effects of forest type and human presence on bonobo (Pan paniscus) density in the Salonga National Park. Int J Primatol 27(2):603–634

Thomas S (1991) Population densities and patterns of habitat use among anthropoid primates of the Ituri Forest, Zaire. Biotropica 23:68–83

Tutin CEG, White LJT (1999) The recent evolutionary past of primate communities: likely environmental impacts during the past three millennia. In: Fleagle JG, Janson C, Reed KE (eds) Primate communities. Cambridge University Press, pp 220–236

Tutin CEG, Ham RM, White LJT, Harrison MJS (1997) The primate community of the Lopé Reserve, Gabon: diets, responses to fruit scarcity, and effects on biomass. Am J Primatol 42:1–24

Uehara S, Ihobe H (1998) Distribution and abundance of diurnal mammals, especially monkeys, at Kasoje, Mahale Mountains, Tanzania. Anthropol Sci 106:349–369

Van Krunkelsven E, Inogwabini BI, Draulans D (2000) A survey of bonobos and other large mammals in the Salonga National Park, Democratic Republic of Congo. Oryx 34(3):180–187
Walker EP, Warnick F, Lange KI, Uible HE, Hamlet SE, Davis MA, Wright PF (1964) Mammals of the world, Vol II. The Johns Hopkins Press
White LJT (1994a) Biomass of rain forest mammals in the Lopé Reserve, Gabon. J Anim Ecol 63:499–512
White LJT (1994b) Sacoglottis gabonensis fruiting and the seasonal movements of elephants in the Lope Reserve, Gabon. J Trop Ecol 10:121–125
White LJT, Edwards A (2000) Conservation research in the Central African rain forests: a handbook. Wildlife Conservation Society
Whitesides GH, Oates JF, Green SM, Kluberdanz RP (1988) Estimating primate densities from transects in a West African rain forest: a comparison of techniques. J Anim Ecol 57:345–367

# Chapter 14
# Elephants in Lake Tumba Landscape: Malebo, Ngiri, and Bolombo-Losombo

**Abstract** The chapter describes a survey of large mammals in the Lake Tumba landscape that focused on forest elephant (*Loxodonta cyclotis*). The survey used both line transect sampling and forest reconnaissance to collect data on elephant dung piles in three major forest blocks of the Lake Tumba Landscape (Malebo region, Ngiri Triangle and Bolombo—Losombo). The Malebo region, as it is referred to throughout this book, is the south of the Lake Tumba down to the Kwa-Kasai River and bordered in the east by the Lake Maindombe. The Ngiri Triangle, comprising the Ngiri River, is located between the Congo River and the Ubangi River from where the two rivers join together. Bolombo–Losombo is located between the Congo River and the Lulonga River. After the completion of the survey, data were not sufficient to provide densities of elephant populations in these forest blocs. However, the calculated relative abundance was $\tau = 0.33$ dung piles/km at Lukolela, $\tau = 0.03$ dung piles/km at Ngombe–Lake Tumba and $\tau = 0.04$ dung piles/km at Mbanzi–Malebo. Ngombe–Lake Tumba and Mbanzi–Malebo are located in the wider area of Malebo region. For both Ngiri and Bolombo–Losombo, respective values of $\tau$ were rather low but elephants were present.

**Keywords** Forest elephant · Dung pile · Encounter rate · Distance sampling · Conservation status

## 14.1 Introduction

As it was the case for all the wildlife species of the Lake Tumba Landscape, there was no data for anything, even the absence and presence of species were essentially guessed, albeit some of these speculations being informed ones. So, the very first activity that was initiated was to conduct a general preliminary survey to be conducted throughout the Lake Tumba Landscape to document the presence of any species. That general preliminary presence–absence survey discovered two discrete populations of elephants. The first was located in the northern section of the Lake Tumba Landscape, located between the Congo River and the Ubangi River, the area that is currently known as the Ngiri Triangle. The second population is that of the Malebo region, in

 
B.-I. Inogwabini, *Reconciling Human Needs and Conserving Biodiversity: Large Landscapes as a New Conservation Paradigm*, Environmental History 12,
https://doi.org/10.1007/978-3-030-38728-0_14

the southern part of the Lake Tumba Landscape. Based on these findings, long-term work for conserving these populations as conservation targets were initiated.

However, being located in isolation means that the two populations of elephants have ecologically adapted to distinct environments. Indeed, as it is said in Chap. 5, Ngiri Triangle and Malebo are in completely different biomes. So, there was a need to document the abundance of these two populations separately, in accordance with the distinctiveness of habitats and the potential importance of these populations. The knowledge of the distribution and precise population estimate would have to help to formulate an appropriate conservation action targeting the species that has to take place in these two zones. Therefore, the objectives of the study on the elephants in the Lake Tumba Landscape were to undertake a population's study of elephants in the three blocs of the forest of the Lake Tumba Landscape, to understand elephant's movements both in the southern and northern Lake Tumba Landscape and to study the dynamics of the human–elephant conflict in the Malebo, southern Lake Tumba Landscape.

## 14.2 Elephant Survey in the Forests of Lake Tumba Landscape

There are three forest blocks of the Lake Tumba Landscape. The first one of these areas is the forest area south of the Lake Tumba down to the Kwa-Kasai River and bordered in the east by the Lake Maindombe, which region is referred throughout this book as Malebo region. The second forest stretch is the Ngiri Triangle, which is located between the Congo River and the Ubangi River from where the two rivers join up to a latitudinal line drawn from Mobeka (above Makanza) to the Ubungi. This bloc of the forest is crossed by the Ngiri River, which is why it is called the Ngiri Triangle. The third forest bloc is called Bolombo–Losombo and it is located between the Congo River and the Lulonga River. These forest blocs were proposed, with the exception of the Ngiri Triangle which was proposed and became a natural reserve, to become community-managed areas and was identified to harbor remnant populations of elephants.

As indicated above, the elephant populations in these three forest blocs were isolated and confined in forests where they could hardly be connected with other elephant populations in the region of Central Africa: for the elephants of the Ngiri Triangle being located between the Congo and Ubangi River meant that they were forced to remain in this area because of the barrier made of the Congo River and the Ubangi. Equally, those of the Bolombo–Losombo are encapsulated by the Congo and Lulonga where movements were limited by these rivers. Finally, Lakes Tumba and Maindombe, the rivers (Congo River and Kwa-Kasai) constituted major physical and ecological barriers for the elephants of Malebo. Hence the survey had to be conducted in these different zones and conclusions drawn for the entirety of the landscape rather than sampling just one forest bloc and generalizing the conclusions.

## 14.3 Methods to Document Elephants in the Lake Tumba Landscape

The survey design followed the general wildlife survey methods used in the context of Central African forests. These standard methods are made of a sampling design that is known as Distance Sampling Methods (White and Edwards 2000; Walsh and White 1999; Hall et al. 1997; White 1994; Buckland et al. 1993), which was already described in Chap. 6. However, in the case of elephants, signs to be counted and whose perpendicular to distances to the transect line were to be measured were the dung piles. Sample sizes for each forest bloc were generated by the use of DISTANCE after a pilot study has determined elephant dung encounter rates.

The widespread acceptance of the use of the Line-transects methods is thought to be based on the assumption that, contrary to many survey methods, they provide an unbiased estimate of mammalian abundance, allow simple spatial modeling of objects in relation to ecological and human influence covariates, while removing spatial and habitat bias thus allowing for valid abundance estimates by habitat type. Apart from the dung piles, which were the key data recorded along the transect line with accurate measurement of perpendicular distance from the transect center line other signs of elephant presence were also recorded even though they were not to be included. The other evident signs of elephants included fresh footprints, signs of rubbing on trees, food remains, direct physical encounters, and elephant calls. But before getting to the proper study design, it was important to work on the key parameters that help calculate elephant abundance, which is the dung deposition and the dung decay rates

## 14.4 Dung Deposition and Decay Rate

To convert numbers dungs into elephant numbers, two factors are key: (1) daily dung deposition rate and (2) dung decay rate or the time it takes for dung deposited to disappear (Barnes 1993). The dung deposition rate is mainly a function of food composition (Tchamba 1992) whereas dung decay rate is a function of multiple factors such as food content, season including humidity, exposure to the sun, rainfall regime, and insect and other animal's actions on dung piles (Nchanji and Plumptre 2001; White 1995; Barnes et al. 1997; Barnes and Barnes 1992). These factors are site-specific (Blake 2002), although some of them might remain constant over a large area within a landscape. Reviewing issues related to different measurements of dung decay rates, the MIKE dung decay task force suggested the use of the retrospective model, which requires only a binary response (i.e. whether a dung pile is present or absent) and producing an age-dependent survival function and meantime for decay (Laing et al. 2003), with a requirement that marked fresh dung piles be visited 5–6 times prior to surveys to the survey midpoint, and the status of marked piles (present or absent) is determined. For the Lake Tumba Landscape, a dung decay study was conducted at the Malebo region, at the southern limit of the landscape where the habitat is predominantly a forest-savannah mosaic (see Chap. 10). Throughout the dung decay study, fresh elephant dung piles were located by following elephant trail systems. Fresh dung bolus sighted, were be marked with numbered tags. At the first sighting of a tagged fresh dung pile, the following data were collected: (1) presence of stale oil, (2) presence of insects or fungi, (3) presence of fruits in the bolus' composition, (4) exposure of the bolus to the sun, (5) the morphological shape, (6) boluses dispersal radius, and (7) the habitat wherein the dung bolus is located. As suggested by Buckland et al. (2001), a minimum of 50 dung piles was identified across the study area. At first sight, the heights of each dung pile were measured using a 1 mm-error tape measure. Then the team visited the dung piles on a weekly basis and measured subsequent heights. Apart from heights, the rainfall data were collected from the field base situated approximately 15 km, straight-line distance south of the trail. The mean dung decay ($\mu$) was calculated as the number of days it took for all the dung samples to completely decay. Then individual measures of dung heights were plotted and fitted to an exponential model $f(x)$. As suggest by Barnes et al. (1997), dung heights were also plotted again the rainfall regime in the region in order to see whether patterns of rainfall influenced the decay rate. With these functions properly calibrated, the team then moved on to the proper survey design, which is shortly described below.

## 14.5 Survey Area and Sample Effort

This section of the chapter is short because the rationale behind the elephant surveys is the same as the reasoning guiding surveys of great apes described both in the case of bonobos (Chap. 6) and as will be described below in the case of chimpanzees (Chap. 11). In terms of the geographic coverage, the area of Malebo comprises the entirety of the southern Lake Tumba Landscape and is large of the 40,000 km$^2$. The totality of the Ngiri Triangle is large of 7000 km$^2$ and the zone of Bolombo–Losombo equals approximately 4000 km$^2$ proposed to become a community-managed area. These zones were stratified based on human influence rather than ecological or geographical boundaries because, as it is indicated in the introduction to this part of the book, it is human actions that override other ecological determinants in defining the distribution and abundance not only of forest elephants (Blake 2002; Barnes et al. 1991) but also of the other biodiversity components and, furthermore, it is usually human influence that defines conservation management actions (Barnes et al. 1991). In this respect, a survey with two strata was designed based on the distances traveled by humans to complete their daily tasks. Indeed, the survey was helped by the socioeconomic study of the Lake Tumba Landscape which came up with an average distance five kilometers around the villages as being the maximum walk distance for livelihood activities. Hence, each zone that was falling within this 5 km distance from the nearest village or other settlement was considered as a stratum adjacent to humans, which was distinguished from anything that was beyond this distance, which was considered to be the core forest. For the Malebo region, a total effort of 80 km of sampling effort was obtained and translated into 80 transects using a systematic trackline survey. For the Ngiri and Bolombo–Losombo areas, the same methods generated 75 transects to be laid throughout the two areas, with 40 transects in Ngiri and 35 in the Bolombo–Losombo.

To repeat what is already described in Chap. 10, using the methods of Buckland et al. (1993), these above numbers of sampling efforts were generated using the formula $cv(Đ) = \sqrt{\left[\frac{b}{L(n_o/l_o)}\right]}$, wherein cv(Đ) is the coefficient of variation, which was taken to have the value of 25%. In the formula above Đ is projected estimate of density; b is of the range 1.5–3; and as suggested by Buckland et al. (1993), the best value for planning reasons was set to be $b = 3$. L predicts the total length of transects, and $n_o/l_o$ is the encounter rate from the pilot study. The design (transect deployment) has been generated by the use of Distance 4.0 software that produces survey designs for given sign encounter rates by stratum (Laake et al. 1994; Buckland et al. 1993). Transects for this survey will be systematically segmented Track-lines, which allow an even distribution of transects across each stratum, ensuring equal coverage probability. The allocation of sampling effort by stratum will be a function of the expected abundance of signs per each stratum.

The analytical framework of the survey data was felt to be essentially through the use of DISTANCE, which is based on transcribing perpendicular distances between lines-transects and elephants dung piles in a project data browser. The project data spreadsheet defines the study area and each stratum ($km^2$), and effort per transect and observations. In normal conditions, Distance would have provided the fits of the distribution curve into four mathematical models (half-normal with a cosine adjustment, hazard with a polynomial adjustment, uniform with a cosine adjustment, and half-normal with a hermit adjustment). Before model fitting, data that are far away from the line transect are truncated because they are difficult to model and provide very little additional information. After the truncation, which makes researchers lose about 5% of the data, the densities would have been generated. Unfortunately, as reported by Inogwabini et al. (2011), the data collected from all the 3 areas, were not sufficient enough to run DISTANCE. So, the data were analyzed as simple encounter rates, which are defined in Chap. 13 as being the relative index of abundance and consists of ratios of sighting of the mammalian sign over the sampling effort (White and Edwards 2000).

The second objective of the study was to understand the patterns of movements of elephants in the Lake Tumba Landscape. Indeed, the movements of elephants determine the species interactions between humans and elephants. They determine the elephant population's home range and critical areas needed to preserve the species. Hence, the study did not have only to determine the geographical extent of forest elephant movements in the Lake Tumba Landscape but also to study the human influence on forest elephant movements. As methods for this part of the study, systematic monitoring of elephant trails at Malebo was implemented. This consisted of counting cumulated presence and absence of elephant dung piles on different permanent trails in order to follow the elephant movements in the region. A bonus of using this method has been the possibilities it offered to cheaply monitor the elephant relative abundance in the region, becoming, therefore, a simplified monitoring tool.

The third objective of the elephant study was to concentrate on the human–elephant interactions. As described by Inogwabini et al. (2014), data collected consisted of (1) interviews with local populations about the occurrence of field raiding by elephants in the vicinities of their villages and (2) visits to fields to collect evidence of elephant intrusion. Interviews consisted of a simple and clear structured interview questionnaire about what species of crops are grown in their villages, how often elephants happen to come to their fields, and on which species the elephants raid most frequently. Villages were selected randomly and those that reported higher intensities of the Human–Elephant Conflict were subsequently visited throughout one year to confirm events they declared. Fields in selected villages were visited at random; any sign of elephant encountered near the field was recorded. Another set of data consisted of the distribution of key fruit plant species in an area of 50 m around the targeted fields. Elephant key fruit plant species were defined as plant species whose fruits were identified in higher frequencies in elephant dung bolies. These plant species were monitored throughout the study in order to see whether their fruiting seasons are related to the crop-raiding events. Another aspect of the study that the survey was to assess was about the perception of the human–elephant conflict by communities. While administering the questionnaire, the respondents were asked to give their age, sex, and level of education. These data were recorded against the specific questions on the degree of the felt human intrusion

The analytical framework of this part of the study consisted of four major steps of which the first step was the analysis of the Human–Elephant Conflict, which involved mapping human settlements along major roads while the second step comprised overlaying maps of human settlements and fields with the elephant ranging patterns. The third analytical process was the calculation of the relative frequencies of crop-raiding events. The fourth analytical procedure was to calculate the phonological frequencies of fruiting seasons for key plant species in the region. Finally, frequencies were compared using non-parametric statistics ($\chi^2$—test for association) to examine relationships between different variables between fruit seasons and events of crop-raiding by elephants (Ennos 2000). The same procedure, using $\chi^2$—test for association, maintained as an independent variable, was applied to check if sex, age groups, education, and professions (Wasilwa 2003) influenced the perception of human–elephant conflict.

Finally, the elephant study in the Lake Tumba Landscape aimed at seeking a sound mitigation scheme either through compensation or other methods to solve the elephant–Human conflict. A preliminary questionnaire indicated that for communities in the southern Lake Tumba Landscape the best approach to mitigate the elephant–human conflict was to provide some compensation. However, the compensation needed to be fair and justifiable. Hence, there was a need to collect data on losses incurred by farmers. These data were necessary and consisted of not only the market value of crops raided by elephants but also to establish a list of preferred staples that may need special areas of cultivation in order to avoid being exposed directly to elephant movement routes. Collected data consisted of destroyed field area in hectares, quantities of raided crops both in terms of numbers and weight. Then an evaluation of raided item prices on the local and regional market was collected. These were collected through a survey, using a standard questionnaire that was triangulated by focus groups, targeting income generated by local communities from the items proven to be raided by elephants. The sampling unit of the crop-raiding study was the households and provide quantified losses incurred were calculated by households rather than by individuals. Sessions of negotiations followed the study to share the results with the concerned local communities and discuss the best ways to mitigate the situation.

## 14.6 The Distribution and Abundance of the Elephant in the Lake Tumba Landscape

As has been reported by Inogwabini et al. (2012), the overall mean dung decay rate $\mu = 109$ days, within the range [42–182] days. In the dry season, the mean decay rate $\mu = 127.6$ days within the range [70–182] days and in the rainy season, $\mu = 87.33$ days, within the range [42–126] days. There was no sufficient data to provide densities of these two populations of the elephant. However, the calculated relative abundance indicated the highest relative abundance was $\tau = 0.33$ dung piles/km at Lukolela while it was lower in the areas of Ngombe–Lake Tumba and Mbanzi–Malebo with respective encounter rates of dung piles being, respectively, $\tau = 0.03$ dung piles/km and $\tau = 0.04$ dung piles/km. The monitoring of the permanent elephant trails indicated that elephants used their trails in the area of Mbanzi with a frequency of 0.52 passages/day within the margins of [0.3–0.5]. Four-year records from this monitoring scheme indicate that all elephant trails in the Mbanzi–Malebo area lead to the north. We hypothesized that the direction may indicate a long-ranging movement between the three blocks.

## 14.7 The Distribution and Abundance of the Elephant in the Lake Tumba Landscape in Perspectives

The difference between the decay rates was significant between the rainy and dry seasons, indicating that not only rainfall but also humidity influenced the decay rate. This finding was in perfect agreement with the results from other study sites in Central African. All these studies (Nchanji and Plumptre 2001; Barnes et al. 1997; White 1995; Barnes 1993, 1987; Barnes and Barnes 1992) have identified rainfall and humidity as being among the most important factors in degrading dung piles deposited by the elephants across all the microclimates of Africa. Neither exposition to the sun nor by the slope of the soil influenced significantly the differences in times taken by dung piles before complete disappearance. In fact, both were relatively similar across the study site of Malebo. Indeed, as described in Chap. 5 of this volume, the Malebo area is better described as a forest-savannah ecotone; most of its wooded savannah is lightly covered at the canopy level and the sun rays hit the ground at almost the same intensities in all points. Also, the study area is located on the plateau where the relative fluctuations in slopes are relatively a lot less steepy, which equates to having the very similar physical disposition of dung piles across the samples. The study did not collect sufficient data on the potential effects of insects and the composition of the elephant diets in the region; hence no comparisons can be made on these two factors. However, the geographical and weather conditions at Malebo are different from those of the Ngiri Triangle and Bolombo–Losombo forest bloc, where swampy forests are predominant. The implications are that the data and results presented in the case of Malebo should not be sensibly applied to Ngiri and the Bolombo–Losombo areas. Therefore, there was a need to conduct other local dung decay studies, which was not possible at that time given the financial constraints.

As stated above, the highest relative abundance of dung piles/km (was $\tau = 0.33$) was calculated for the area of Lukolela. This figure is a significant one for the conservation of the elephants across the country because it was comparable, in magnitudes, to the encounter of 0.39 dung piles per km from the Salonga, several kilometers away east of the study site. Salonga National Park had been felt to be a stronghold for elephants in Central Africa for so long but the results of the recent surveys (Maisels et al. 2013; Blake et al. 2007) have demonstrated that the populations of this species in that region are declining in alarming speeds. Hence, even though that encounter rates of elephants in the Lake Tumba Landscape remain lower compared to other sites where elephants were surveyed in Central Africa such as Nouabale–Ndoki (Republic of Congo), Zangha National Park (Central African Republic) and Minkebe (Gabon), the finding that the Lake Tumba Landscape holds populations of elephants that are similar to those found in Salonga is a critical one since it means that the Lake Tumba Landscape is an important site for the national plans for the survival of the species.

Generally and theoretically, however, lower relative abundance of elephants in the Lake Tumba Landscape, and particularly in the area of Malebo, should be interpreted against the background of ecological determinants of the distribution of elephants in Central Africa. Such ecological determinants include the distance from major settlements and roads (Blake et al. 2008; Barnes et al. 1991). Against the human route paradigm, the situation in the southern Lake Tumba Landscape is of particular interest because all the three locations (Ngombe–Lake Tumba, Lukolela, and Mbanzi–Malebo) are close to Congo River, the main route for commerce. Apart from being crossed by this major human route, the southern Lake Tumba Landscape is located near Kinshasa, a town of about 11,000,000 people whose pressure on natural resources is felt long distances far away. Indeed, a study on the exploitation of fuelwoods to complement the needs for energy for the communities of Kinshasa has indicated that the influence of Kinshasa is felt up to 600 km away. With these realities, it would be expected to see elephants locally extinct because Malebo is, in fact, located within this width strip of the massive influence of Kinshasa. Hence, the presence of elephants in this zone calls for an explanation that is different from the human route paradigm. This paradigm has been put forward by Barnes et al.

(1991) in a famous paper entitled 'man determines elephant distribution' and goes that elephants would be found in large numbers only in very remote locations. Hence, nearer roads and nearer human settlements, elephants should occur in very light densities. Of course, elephants in Malebo do not occur with massive densities but they do happen in relatively comparable relative abundance with remote areas such as the Salonga National Park, as described above. Taken from this view, the finding of elephants in the Malebo area suggests several options to explain the distribution of the species. One plausible way is to say that the explanation is not the distance but the easiness of accessibility to one location. Indeed, the Ngombe–Lake Tumba and Lukolela, where elephants were found in sizable relative abundance, are swampy zones. Swamps are a difficult area to hunt in, and this may have naturally buffered the zone from intensive poaching.

Explaining the finding of elephants' presence in the southern Lake Tumba Landscape by the difficulty to access the area outside of main routes and the uneaseness to hunt would also provide interpretational ground to the other findings of the same study. Firstly, the fact that elephant trails were documented to indicate a north–south migration route; data collected on the presence of elephants in the Malebo area indicated that elephants were present in the area throughout the year but their presence was minimal (0.3 passages/day per trail) during the dry seasons. Indeed, during the dry seasons, elephants of the southern Lake Tumba migrated to the north, which is in the zone around Lukolela and near Lake Tumba. This migration can be attributable to the classic factors such as the search for water and other food items. But there is no reason these factors should exclude the fact that elephants might also be moving to safer areas during the dry seasons. Secondly, elephants are also present in the Ngiri and Bolombo–Losombo areas. These areas are major swampy zones of the Lake Tumba Landscape; large areas of the Ngiri Triangle are inundated even during dry seasons and a large portion of the Bolombo–Losombo area becomes a huge muddy area in the dry seasons too. These conditions may explain why even closer to the Congo River, which is a major human route, elephants can still be found because it would be difficult for poachers to easily act in such difficult conditions.

## References

Barnes RFW (1993) Indirect methods for counting elephants in forest. Pachyderm. 16:24–30

Barnes RFW, Barnes KL (1992) Estimating decay rates of elephant dung-piles in forest. Afr J Ecol 30:316–321

Barnes RFW, Barnes KL, Alers MPT, Blom A (1991) Man determines the distribution of elephants in the rain forests of northern Gabon'. Afr J Ecol 29:54–63

Barnes RFW, Asamoah-Boateng B, Naada-Majam J, Agyeio-Hemng J (1997) Rainfall and population dynamics of elephant dung-piles in the forests of southern Ghana. Afr J Ecol 35:39–52

Blake S (2002) The ecology of forest elephant distribution and its implications for Conservation. Ph.D. Thesis submitted to the University of Edinburgh. Edinburgh, United Kingdom

Blake S, Strindberg S, Boudjan P, Makombo C, Inogwabini BI, Omari I, Grossmann F, Bene-Bene L, de Semboli B, Mbenzo V, S'hwa D, Bayogo R, Williamson EA, Fay M, Maisels F (2007) Forest elephant crisis in the Congo Basin. PloS Biology 5(4):1–9

Blake S, Deem SL, Strindberg S, Maisels F, Momont L, Inogwabini BI, Douglas-Hamilton I, Karesh WB, Kock MD (2008) Roadless wilderness area determines forest elephant movements in the Congo Basin. PLOS ONE 10:e3546:1–9. www.plosone.org
Buckland ST, Andersen DR, Burnham KP, Laake JL, Borchers DL (2001) Introduction to Distance Sampling: Estimating abundance of biological populations. Oxford. University Press
Buckland ST, Anderson DR, Burnham KP, Laake JL (1993) Distance sampling: estimating abundance of biological populations. Chapman and Hall
Ennos R (2000) Statistical and data handling skills in Biology. Pearson & Prentice Hall
Hall JS, Inogwabini BI, Williamson EA, Omari I, Sikubwabo C, White LJT (1997) A survey of elephants (Loxodonta africana) in the Kahuzi-Biega National Park lowland sector in eastern Zaire. Afr J Ecol 35:213–223
Inogwabini BI, Mbende L, Abokome M (2011) The relic population of forest elephants near Lac Tumba, Democratic Republic of Congo: abundance, dung decay, food items and movements. Pachyderm 49:40–47
Inogwabini BI, Abokome M, Kamenge T, Mbende L, Mboka L (2012) Preliminary bonobo and chimpanzee nesting by habitat type in the northern Lac Tumba Landscape, Democratic Republic of Congo. Afr J Ecol. https://doi.org/10.1111/j.1365-2028.2012.01323.x
Inogwabini BI, Mbende L, Bakanza A, Bokika JC (2014) Crop damage done by elephants in Malebo Region, Democratic Republic of Congo. Pachyderm 54:59–65
Laake JL, Buckland ST, Anderson DR, Burnham KP (1994) DISTANCE - User's guide V2.1. Colorado Cooperative Fish & Wildlife Unit. Colorado State University.
Laing SE, Buckland ST, Burns RW, Lambie D, Amphlett A (2003) Dung and nest surveys: estimating decay rates. J Appl Ecol 40:1102–1111
Maisels F, Strindberg S, Blake S, Wittemyer G, Hart J, Williamson EA, Aba'a R, Abitsi G, Ambahe RD, Amsini F, Bakabana PC, Hicks TC, Bayogo RE, Bechem M, Beyers RL, Bezangoye AN, Boundja P, Bout N, Ako ME, Bene LB, Fosso B, Greengrass E, Grossmann F, Ikamba-Nkulu C, Ilambu O, Inogwabini BI, Iyenguet F, Kiminou F, Kokangoye M, Kujirakwinja D, Latour S, Liengola I, Mackaya Q, Madidi J, Madzoke B, Makoumbou C, Malanda G-A, Malonga R, Mbani O, Mbendzo VA, Ambassa E, Ekinde A, Mihindou Y, Morgan BJ, Motsaba P, Moukala G, Mounguengui A, Mowawa BS, Ndzai C, Nixon S, Nkumu P, Nzolani F, Pintea L, Plumptre A, Rainey H, De Semboli BB, Serckx A, Stokes E, Turkalo A, Vanleeuwe H, Vosper A, Warren Y (2013) Devastating decline of forest elephants in Central Africa. PLoS ONE 8(3):e59469:1–13
Nchanji AC, Plumptre AJ (2001) Seasonality in elephant dung decay and implications for censusing and population monitoring in southwestern Cameroon. Afr J Ecol 391:24
Tchamba M (1992) Defaecation by the African forest elephant Loxodonta africana cyclotis in the Santchou Reserve, Cameroun. Mammalia 56:155–158
Walsh PD, White LJT (1999) What will it take to monitor forest elephant populations? Conserv Biol 13:1194–1202
Wasilwa NS (2003) Human-elephant conflict in the Masai Mara dispersal areas of Trasmara District. Ph.D. Thesis, Durrell Institute for Conservation and Ecology, University of Kent, United Kingdom
White LJT (1994) Sacoglottis gabonensis fruiting and the seasonal movements of elephants in the Lope Reserve, Gabon. J Trop Ecol 10:121–125
White LJT (1995) Factors affecting the duration of elephant dung piles in rain forest in the Lope Reserve, Gabon. Afr J Ecol 33:142–150
White LJT, Edwards A (2000) Conservation research in the Central African rain forests: a handbook. Wildlife Conservation Society

# Chapter 15
# Developing a Threat Index for Documented Large Mammal Species

**Abstract** The information on biodiversity and its distribution is of limited importance if levels of threats to which each biota and each species are exposed remained poorly quantified. However, determining the species conservation status remains difficult for conservation practitioners because of high technicalities of how to generate threat indexes for species. This study endeavored to solve this problem by developing a sensibly simple method using the ordinary vector calculus to set levels of threats to which species in the Lake Tumba Landscape were exposed to. The method yielded results that were similar to the conventional listings though they had also shown that the relationships between abundance indexes were not straightforwardly parallel with the levels of threats, calling therefore for more attention to be devoted to each species rather than generalizing from lay wisdom that lower signs equate to higher treats. The message being conveyed by this finding is that ecological metrics are more important in setting threat indexes than just bulky presence indices.

**Keywords** Vector analysis · Large mammals · Abundance · Threats · Threatened species · Biodiversity

## 15.1 Introduction

With prospects to delineate different land use units with the segment, which is the prime objective of the landscape approach, the second most important data needed for biological conservation purposes were on the biological diversity and its distribution. However, the information on biodiversity and its distribution would be of limited importance if levels of threats to which each biota and each species are exposed remained unknown. This is because assessing threats under which species live is a critical process in conservation biology; levels of threats determine the species status and, therefore levels of efforts (political will, financial, and human resources) devoted to the preservation of any given species and its habitat. Levels of threats have been used in several classification exercise, including the now world widely used the red list of the World Union for Conservation of Nature (IUCN) and other regional and national listing systems.



B.-I. Inogwabini, *Reconciling Human Needs and Conserving Biodiversity: Large Landscapes as a New Conservation Paradigm*, Environmental History 12,
https://doi.org/10.1007/978-3-030-38728-0_15

However, stating that quantifying threats remains weak is a mild statement. This is so because of the dependence of many conservation biologists on methods that is often dictated numbers of mammals, habitat availability, felt or quantified human pressures, etc. There are few instances where those types of information are pulled together to define what is meant by being threatened. In many cases, information is hardly comparable because it is, significantly and at its best, by informed guesses. Therefore, throughout the work undertaken in Lake Tumba, I concentrated on developing a simple yet robust way to quantify species threat index for different documented species by use of vectors field and its laws to forge is a new attempt of defining a simple yet coherent and comparable method to determine a threat index for large mammals.

## 15.2 Data on Large Mammals

Large mammals were identified by direct sighting using Kingdon's 1997 guide for mammals. Unseen monkey species were identified by calls, using the audio CD-ROM of the Central African primate call repertoire recorded by Gauthier-Hion et al. 1999). The study also relied on indirect unquestionable evidence such as dung piles or pellets clusters fresh spurs (Parnell 2000). Data were collected either along with line-transect methods (Buckland et al. 1993), reconnaissance routes, and opportunistically. Examples include the presence of species skins (e.g. *Felis serval*, *Civettictis civetta vivera*, skins sighted at different villages), and dead specimens (e.g. *Uromanis tetradactyla* seen at roadsides, being sold as bushmeat). Local trackers were also used to identify plant species in local languages, and then converted into the scientific nomenclature using Hulstaert (1992). Hulstaert (1992) incorporates three variants of *Lomongo* spoken in the region, and used museum collections to identify species (Inogwabini 2005), and therefore was appropriate. The following human signs were recorded to document the extent of human activities and threat level on each species: permanent campsites, recent machete cuts, snares (nylon and cable), used shotgun cartridges, human encountered in the forest and their possessions (spears, bows, accompanying dogs, hunting nets, machine guns, machetes, etc.) and open permanent human footpath. The survey documented also the extent of logging extractive activities. Teams estimated cleared area, extracted species, and numbers of marked trees in concessions. They also estimated human populations in logging camps and number of logging machines (vehicles and others). To get an idea of perceived level of threats for each species, informal interviews were organized with hunters to gather this information: what species they specifically targeted, what species were commonly hunted by use of which means. Hunters in logging concessions also responded to the questionnaire.

Dung piles, pellet clusters, sightings, and calls were used to estimate abundance indexes. Spurs (even the fresh ones) were not accounted for to avoid difficulties related to their conversion into abundance. Perpendicular distances to transect lines were measured to the center of each dung piles or pellet clusters, following previous similar studies (e.g. Blom et al. 2004; Inogwabini et al. 2000; Hart and Hall 1996).

## 15.3 Assumption and Methods on Threat Index

The method proposed here assumes that there are forces, internal and external, that drive species conservation status. Positive forces are forces such as specific reproduction rates, survival rates, longevity, etc. Negative forces are those forces driving species to extinction such as habitat fragmentation, hunting, epidemics, etc. Considered as forces, all these parameters can be added using laws governing additions of vectors to produce a modulus, which represents a level of the threat under which the species happens to live over time. The second assumption considers the species abundance as a point on which all these forces act (Fig. 15.1). To simplify the case, remarkable angles were used in such a way that all forces pulling maintaining the species alive over time (direct forces) drag the species upward in an angle of $\theta = 90°$ whiles those acting indirectly to keep the species alive drive the species upward but by the angle $\theta = 45°$ (Fig. 15.1). Similarly, forces directly pulling the species toward abundance diminution act in an angle of $\theta = 270°$ while those act indirectly are posed to be at the angle of $\theta = 315°$ (Fig. 15.1).

Based on those assumptions and as shown in Fig. 15.1, it can be written as follows:

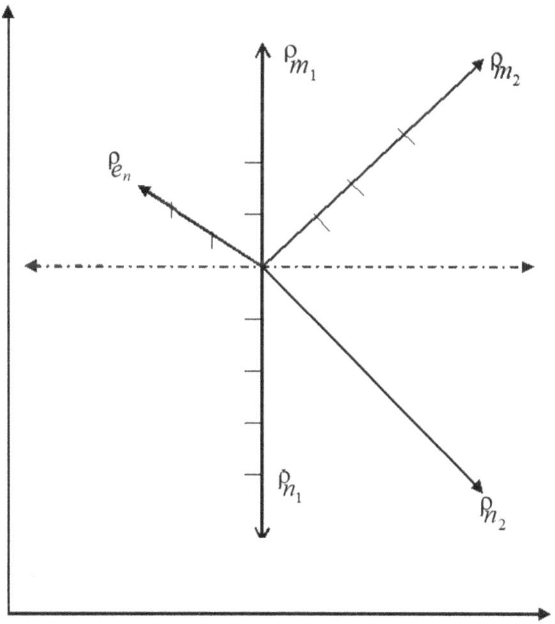


**Fig. 15.1** Different forces acting on species abundance, the species is thought immersed in a field of forces symbolized by vectors, which are acting on individuals in conflicting directions. Persistence of the species over time is obtained by calculating the modulus of the total forces acting on the species

$$\vec{\Gamma}_x = \sum (\vec{n}_1 + \vec{n}_2) + (\vec{m}_1 + \vec{m}_2) - (\vec{e}_1 + \vec{e}_2) \quad (15.1)$$

$$= (\sum \vec{n}_i + \sum \vec{m}_i) - (\sum \vec{e}_n) \quad (15.2)$$

Taking the strength as a modulus, and generalizing for x parameters

$$\vec{\Gamma}_x = \left\| \sum \vec{n}_i + \sum \vec{m}_i) - (\sum \vec{e}_n) \right\| \quad (15.3)$$

## 15.4 Parameterization and Definitions of Threat Levels

To solve the Eq. (15.3) ranks categories of threats under which the species have to survive. To solve this Eq. (15.3), values of $\vec{n}_i$, $\vec{m}_i$, and $\vec{e}_n$ can be assigned to each species. These values ranged from 1 to 4. Definitions of $\vec{n}_i$ were constructed from known threats such as hunting, increases in human populations, and habitat fragmentation. Definitions of $\vec{m}_i$ taking account of known conditions that enable threatened populations to become sustainable in long-term perspectives such as direct management and high prospects for reintroducing individuals from captivity to the wild, augmented by the existence of a normative legal framework to sustain the conservation of the species both nationally and internationally. Definitions of $\vec{e}_n$ account for intrinsic potential of the population to bounce back from severe decreases should conditions that brought them to decrease improve. In the case of this exercise, $\vec{n}_1$ were threats acting directly on the species, i.e. a technique of hunting targets a given species or outbreaks of zoonoses that target specific species. As for $\vec{n}_2$, it was meant to represent threats that acted indirectly on a given species, i.e. habitat loss or fragmentation. Similarly, $\vec{m}_1$ represents direct opportunity for the species to survive while $\vec{m}_2$ is designed to account for indirect opportunities, i.e. existence of laws and their implementation. The general definitions and values of $\vec{n}_i$, $\vec{m}_i$, and $\vec{e}_n$ are as follows.

## 15.5 Negatively Contributing Factors

$\vec{n}_1$: defined the hunting pressure on any given species and has four values; $\vec{n}_1 = (1, 2, 3, \text{and } 4)$, which equate respectively to (1) only subsistence hunting with traditional tools, (2) Subsistence hunting with some barter; involves arms, (3) hunting with modern tools; primarily for trade but with a large stake remaining for local consumption and (4) hunting entirely for trade and substantively relying on automatic weaponries and possible destructive armaments. As vectors, all $\vec{n}_1$ are directed at 180° and, hence are negative because effects of $\vec{n}_1$ directly act against the species maintenance. $\vec{n}_2$:

defined the state of the habitats and the presence of human settlements in the region for any given species and has four values; $\vec{n}_2 = (1, 2, 3, \text{and } 4)$, which equal respectively (1) human density in the area is $\geq 1$ individual/km$^2$, (2) habitat is fragmented by roads, rails, or any other major infrastructure, (3) habitat is being actively logged or actively mined and (4) a major natural catastrophe, landslide or inundation affecting more than 50% of the species niche or human made catastrophe such a destruction of habitats by bombs such as the napalm. As vectors, all $\vec{n}_2$ are directed at an angle of 135° and, hence are all negative because effects of $\vec{n}_2$ act against the species survival; the angle denotes the fact that this action is tangential. $\vec{n}_3$: defined the possibility of the species to suffer from an epidemic disease and has four values; $\vec{n}_3 = (1, 2, 3, \text{and}$ 4) depending on whether (1) the species happens in area where the some diseases had already been reported on wildlife species, (2) the species belonged to a wildlife guild that has been already reported to suffer from known diseases, (3) the species has been already reported to suffer from some epidemics but with no massive losses resulting from these epidemics outbreaks, and (4) species has been already reported to suffer from some epidemics with massive losses resulting from these epidemics outbreaks. As vectors, all $\vec{n}_3$ were laid to pull the species they acted on toward an angle of 225°.

## 15.6 Positively Contributing Factors

$\vec{m}_1$ defined the management of any given species; $\vec{m}_1 = (0,1, \text{and } 2)$, one or zero respectively referring to whether any subset of the species population was directly managed within the landscape or not while 2 referred to the potential of (highly) successful species reintroduction program. All the vectors denoted by $\vec{m}_1$ were directed 0°, directly opposed to the vectors of $\vec{n}_1$ because direct management was felt to provide support to the sustainability of the species. $\vec{m}_2$ defined everything falling under the legal framework; $\vec{m}_2 = (1, 2, 3, \text{and } 4)$. These values correspond to (1) just a specific national law protects the species has been officially voted and published within either the country or the provinces where the landscape is located, (2) the species is protected by an international law, (3) the species occurs within a protected area and (4) laws (national and international) and the protected areas to protect the species are fully operational. As these are supposed to keep the species up only tangentially, $\vec{m}_2$ vectors are oriented at 45°. $\vec{e}_1$ defined the intern potential for the species to regenerate its population. Given the fact that many other factors implicated in the ability of the species to regenerate and most of them being difficult to measure, it was deemed necessary to factor only the potential for the species to procreate under natural conditions. The crude measure used to measure this potential was the inter-birth period. Hence, $\vec{e}_1 = (1, 2, 3, \text{and } 4)$, respectively for the (1) species reproduce very slowly (inter-birth period greater than or equal to 5 years), (2) the species reproduce slowly (inter-birth period greater than 3 years), (3) species whose inter-birth period falls within less than 2–3 years and (4) the species whose inter-birth period is within 0–1 years. All $\vec{e}_1$ vectors were oriented 315°.

## 15.7 Factoring the Unknown in Threat Index: Penalty for the Unmeasured Effects

Two major threats affected all the species in the Lake Tumba Landscape. The first one, with particular geographic focus on the southern part of the Lake Tumba Landscape was habitat loss. The intensive extraction of timber, and particularly the extraction of the Wenge (*Millettia laurentii*), the black wood species that is localized in this part of the country is a cause for most of the extensive logging in the landscape. The logging industry has, as it would be expected, opened up large tracks of forests and installed large camps far from market points. Hence, logging crews within the Lake Tumba Landscape were directly involved in hunting and contributed to bushmeat trade in the region. Logging roads also had paved the way for massive movements of professional bushmeat traders in the region; instances of this situation were documented and hunting, which in its majority is a combination of cable-snaring and armed poaching was confirmed in roads that ultimately were leading to logging camps. Hence, in establishing the threat index for each species, the modulus of the vectors affecting species in relationship with habitat loss was set to be equal for all species regardless of their ecological responses to the stimulus imposed by habitat loss .

The second major threat that affected species in the Lake Tumba Landscape was intensive hunting. However, armed hunting was treated separately with other forms of hunting. Armed hunting is not the only threat for species such as elephants and

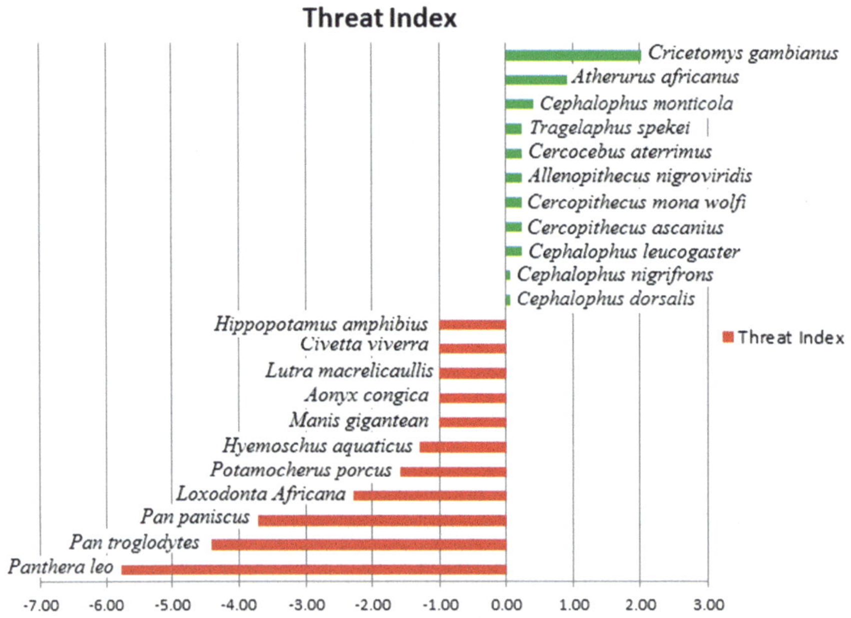


**Fig. 15.2** Threat index for species in the Lake Tumba Landscape

hippopotami; there were instances of traditional techniques targeting these species, including poisoning that is used for very rare cases. That being said, however, armed hunting was widespread in the region. All stakeholders were in agreement that the wars that spread through the country left quantities of automatic weapons that became easily accessible to most people. The same stakeholders also concluded that species that were previously protected and sold in hidden environments have become easily available and sometimes in open environments. This agrees with what has been documented by De Merode and Cowlishaw (2006) in the areas around the Garamba National Park where they found out that during wartime, the sales of protected species in the urban markets increased fivefold because the military officers fled, leaving behind an open-access system that led to a massive increase in the exploitation of protected species. Hence, it can sensibly be concluded that the easiness to access guns and ammunition placed a high toll on other species, particularly on primates. In establishing the threat index for each species, the modulus of the vectors affecting species in relationship with armed hunting was differentiated with regard to whether species hunted by guns were first targets such as is the case for large-bodied species such as elephants, buffaloes, red river hogs, etc. or not. The modulus of vectors applied to each large-bodied species was 50% larger than the ones applied for small-bodied species. The modulus of the vector for elephants, buffaloes, and red river hogs was 0.75, at the scale of 0–1, that of species like African brushed-tail porcupine was 0.25, as a contribution to the total penalty added to Eq. (15.3).

Snaring was also brought into the calculation of specific threat index for different species. There are, generally, three types of snaring and these depend on the material used: the cable-snaring, nylon-snaring, and natural liana-snaring. Although often described as petty hunting by local communities, cable-snaring can become very dangerous; it is unselective of its victims and often leaves non-targetted mammals in severe hindrance. Indeed, most species of mammals (with probably the exception of elephants) can fall prey to cable-snaring. The cable-snaring often seriously wounds any caught species; bonobos have been seen with amputated hands that were cut by cable snares. Also, snaring does also differ depending on the parts of the body of the victims it is meant to capture. In general, snares are made to capture either the legs or the necks of their victims. Without minizing the effects of snares in general, being captured at the neck is much more dangerous than being captured by the leg. Because of the intricacies of dissociating these effects, snaring has been factored in the calculations of the threat index in equal magnitude regardless of the material a snare was built on or the targeted bodypart. To dissociate effects on each species, the penalty modulus of the vector of snaring was set at the highest value of 0.25 for ungulates whereas the vector's modulus for other species, including great apes was set 50% below that threshold. In the case of elephants, the modulus was set to zero.

Both large-scale cable-snaring and armed poaching are due to diverse factors, which would need a full-fledged study to disentangle. But suffice it to be said here that they depend on the availability of tools being used, the demand in wild meat, accessibility to the hunting sites and effectiveness of law enforcement. Tools used in large-scale cable-snaring and armed poaching are metallic cables, automatic weapons, and ammunition. Metallic cables are widely available throughout the region of the Lake

Tumba Landscape; they are often brought in by boats to keep their barges together and to bundle their loads, particularly the timber logs. On their way to their official duties, these boats sell the metallic cables to hunters in exchange of bushmeat that they carry back to major towns in the south of the Congo River (Kinshasa and Brazzaville). Studies on effects of logging on bushmeat are often confined within the premises of habitat losses it directly causes and the collateral effects of bringing in poachers via roads logging companies open in forests but ignore the fact that boats owned by logging companies do indeed cause much more problems as they spread cables throughout the region. They also contribute to transporting bushmeat to the markets, fuelling, therefore, the demand. Whereas cables are provisioned by boats, arms, and ammunition have been a legacy of the multiple cycles of the wars through which the Democratic Republic of Congo has gone over the last 2 decades. Indeed, the wars have made arms and ammunition easily and cheaply available, with dire consequences on wildlife species (Draulans and Van Krunkelsven 2002). In the particular case of the Lake Tumba Landscape, the presence of a military training center at Irebu, located at the geographic center of the landscape has contributed significantly to the widespread availability of automatic weapons and ammunition throughout the landscape. Indeed, poorly paid, lightly trained, and wildly undisciplined soldiers that occupied the center between 1996 and 2010 were selling or lending arms and ammunition for trophies of wildlife species. This contributed significantly to disseminating arms and ammunition in the region. The Lake Tumba Landscape is located near the Congo River, which is a traditional highway for commerce, which has been in use for centuries. Hence, the location is easy access for poachers of all walks and products of poaching also easily access the markets in larger human settlements located within (Mbandaka, Bikoro, Lukolela, Bolobo, Inongo, etc.) and outside of the landscape (Kinshasa, Brazzaville, etc.). Indeed, with increasing human densities in these major human settlements and the collapse of both the rule of the law and the global economy in the Democratic Republic of Congo over the last two decades contributed to the increase in demands of bushmeat.

Other forms of hunting include digging holes, collective net-hunting, dog-hunting and, in very limited cases, poising water or food items for large mammals. Hole-trapping consisted of the ground, placing sharpened sticks on the lower end of the hole and covering these holes with light dead leaves. Obviously, hole-trapping is unselective and is deadly for any species of large mammal, including elephants for which it is usually constructed. However, hole-trapping is not commonly practiced, maybe because of the threat it poses for humans themselves. Because of its rarity, it has been factor as a limited threat with its vector's modulus being 0.01 for all the species, as additional penalty to the Eq. (15.3) above. Collective net-hunting is a traditional hunting technique and it is disappearing in most of the Lake Tumba Landscape, being practiced only in the southern part of the landscape particularly among the Bateke. It is often combined with the dog-hunting whereby dogs are used throughout the forest to chase out hiding preys, which are then either captured by dogs or slaughtered by owners of dogs. Because the impacts of both collective hunting and dog-hunting or their combination are rather very limited, the modulus of their impact on large mammals of the Lake Tumba Landscape has been set to be very minimal, at 0.25, as a penalty to be added to Eq. (15.3) above. Poisoning has been reported only in several instances where elephants and other wild species were seen as nuisances because they were felt or found to destroy fields of human staple cultures. Despite the occurrence of this practice being very limited in areas where human–wildlife conflicts were prevalent, the impact of such a practice on wildlife species can be but disparaging and critical because it is likely to affect many other species as well. Because of that the modulus of the vector for poisoning has been set very high and was 0.2 for all species, as a penalty to be added to Eq. (15.3).

A type of threat was particular to primates only and consisted of the trade in monkeys as animals of company (pet). This specific threat would not come as a surprise to wildlife specialists who happen to work in the Democratic Republic of Congo and the wider Congo Basin. Indeed, as one walks along the streets of major human settlements in the Democratic Republic of Congo is very likely to encounter people holding individuals of species of monkeys that they try to sell as pet (Inogwabini and Thompson 2013). Of these individuals are captured from the field; and the survey in the Lake Tumba Landscape indicated that species of monkey were captured unselectively in the landscape after shooting and killing some individuals from diurnal monkey groups. Hence, not only collecting younger individuals poses a serious threat to the species but also the fact that large numbers of individuals are killed before young individuals are captured is very damaging for the survival of any given species. Indeed, species observed on roads include species of endangered species and in some cases even the bonobos or chimpanzees. Some buyers do buy individuals from wildlife species in compassion while others do buy them to express their wealth. While those who believe in helping the species to survive may have genuine concerns for species survival, they ignore that buying these individuals fuels the markets located, mostly in long-distance major towns such as Kinshasa, Mbandaka, and Brazzaville. Being hunted for pet has been factored in the threat analysis by a vector of modulus valued at 0.5, as a penalty added to the Eq. (15.3).

Finally, a broader yet difficult to measure threat has been the potential effects of diseases on wildlife species. This single threat can, indeed, wipe out significant populations of wildlife in a short span of time; a potential effect that has been documented in the recent history of biodiversity conservation in Central Africa when the epidemy of Ebola knocked down nearly 50% of the wild populations of the western lowland gorillas (Walsh et al. 2003, 2007; Bermejo et al. 2006 ). Hence, it would not make any sense to talk about threat to wildlife species without emphasizing the importance of zoonoses among major threats. Indeed, an outbreak of epidemic zoonoses like Ebola as a much higher potency of destroying the totality of species where they occur and in a short time than many other threats, which act silently and slowly such as poaching. This is particularly the case if one would consider species that are located within one political entity as bonobos that occur only the Democratic Republic of Congo.

## 15.8 Putting All of It Together

After setting different penalties needed to factor the unknown and correcting for specific types of threats and their potential effects on different species, as described in the section above, the Eq. (15.3), which can be rendered as followed:

$$\vec{\Gamma}_x = \left\| \sum \vec{n}_i + \sum \vec{m}_i) - (\sum \vec{e}_n) \right\| - \prod \tag{15.4}$$

$\prod$ in the above equation is the penalty, which was measured as described above. For this formula to be operationally used, Eq. (15.4) was decomposed in its components as follows:

$$\begin{aligned} \vec{\Gamma}_x &= \vec{n}_1\mathrm{Cos}(180^o) + \vec{n}_2\mathrm{Cos}(135^o) + \vec{n}_3\mathrm{Cos}(225^o) + \vec{m}_1\mathrm{Cos}(0^o) \\ &\quad + \vec{m}_2\mathrm{Cos}(45^o) + \vec{e}_1\mathrm{Cos}(45^o) - \Pi \end{aligned} \tag{15.4a}$$

$\prod$ is a momentum function and can be obtained by multiplying all the factors that are difficult to measure and different types of penalties. Hence, $\prod$ can be calculated as

$$\prod = \frac{\Sigma \Pi_i}{n_i^i}$$

In the particular case of the Lake Tumba Landscape, as the values of each penalty were already set to correspond with being hunted by arms ($\prod_1$), by snaring ($\prod_2$), by trapping holes ($\prod_3$), by collective hunting ($\prod_4$), by poisoning ($\prod_5$) and being hunted for pet ($\prod_6$), the total penalty added to the threat index for each species was obtained as

$$\prod = \frac{\Pi_1}{1^1} + \frac{\Pi_2}{2^2} + \frac{\Pi_3}{3^3} + \frac{\Pi_4}{4^4} + \frac{\Pi_5}{5^5} + \frac{\Pi_6}{6^6}$$

## 15.9 Mammalian Diversity

Fifty-four species of mammals were identified in the Lake Tumba Landscape. These include 7 species of diurnal primates: (1) black mangabey (*Lophocebus aterrimus*), (2) blue monkey (*Cercopithecus ascanius*), (3) Allen's swamp monkey (*Allenopithecus nigroviridis*), (4) Wolf's monkey (*Cercopithecus mona wolfi*), (5) Angolan pied colobus (*Colobus angolensis),* (6) black mangabey (*Lophocebus aterrimus)* and (7) Tshuapa red colobus (*Piliocolobus tholloni)*; these species are an addition to the bonobo and chimpanzees. Other mammals of conservation concern present in the Lake Tumba Landscape were: (1) forest elephant (*Loxodonta africana cyclotis*),

(2) giant pangolin (*Smutsia gigantea*), (3) African forest buffalo (*Syncerus caffer nanus*), (4) sitatunga (*Tragelaphus spekei*), (5) blue duiker (*Cephalophus monticola*), (6) bay duiker (*Cephalophus dorsalis*), and (7) water chevrotain (*Hyemoschus aquaticus*). The leopard (*Panthera pardus*) has been reported by local hunters but was not documented independently.

## 15.10 Species Abundance

A total of 120 antelope pellets, when clustered as a guild yielded an estimated pellet density = 0.58 pellets $km^{-2}$, with the confidence intervals CI = [0.39428–0.84444] pellets $km^{-2}$ (effort = 25.5 km). Elephant dung piles were absent near the shores of both lakes but were present in relatively remote forests though in the southern part of the Lake Tumba Landscape they were reported near new human settlement (see chapter 14). An overall total encounter rate of 0.03 signs $km^{-1}$ was calculated and it is comparable to that of the Salonga National park where similar encounter rate was converted into 357.28 dung piles $km^{-2}$ or a low mean elephant density of 0.25 elephant's $km^{-2}$. As described by Inogwabini et al. (2007a); see also Chapts. 6 and 10 in this volume), a total of 7 discrete communities of bonobos was identified in the southern portion of the Lake Tumba Landscape and was described in the forest ranging from the Irebu Channel, the forests between the Lake Tumba and the Lake Maindombe and bordered in the west by the Congo River and in the south by the Kwa-Kasï River. These 7 discrete communities were: (1) Botuali-Bosango, (2) Iswelo, (3) Mponga, (4) Ngelo, (5) Ngembo-Bakumu, (6) Bobala, and (7) Malebo. Mean distances between communities were 37 km [SD $\pm$ 12]. Because of insufficient data replicates, distance could not be used to estimate the population for all communities. However, adjusting encounter rates with published densities to account for patchiness and patch width (Inogwabini and Omari 2005), the bonobo population in the Lake Tumba Landscape would total a minimum of about 700 bonobos, within the margins [450–1000] (Inogwabini et al. 2007a) and see also Chaps. 6 and 11 in this volume), compelling a complete re-reading of both methods used to estimate bonobo populations in the wild and a resetting of conservation action priorities. Linear regression showed that there was a negative relationship between mammalian signs and human activity signs ($r = -0.66$, slope $m = -0.30$) though this relationship was not significant ($p = 0.33 > 0.05$, df = 2). These results provide insights for a new species-targeted approach for bonobos conservation in general.

## 15.11 Threat Index by Species

Of the fifty-four species of mammals were identified in the Lake Tumba Landscape, the threat index has been produced only for 23 species (Fig. 15.2). Using just the total vector, only 11 species had positive index, which equates to having better conservation perspectives. These eleven species included not only Sitatunga ($\vec{V} = 0.24$), blue duiker ($\vec{V} = 0.41$), bay duiker ($\vec{V} = 0.06$), black-fronted duiker ($\vec{V} = 0.06$), white-bellied, duiker ($\vec{V} = 0.24$), African brush-tailed porcupine ($\vec{V} = 0.91$), and Gambian pouched rat ($\vec{V} = 2.02$) but also species a cohort of 4 diurnal primates, including Redtail monkey, Wolf's mona monkey, Allen's swamp monkey, and Cercocebus aterrimus who all had the modulus of $\vec{V} = 0.24$. All other species had negative vectors. Particular cases were those of the landscape species, which are the bonobos (Chap. 6), the chimpanzees (Chap. 11), the elephants (Chap. 14), and the lions (Chap. 12). The bonobos had $\vec{V} = -3.71$ and chimpanzess had $\vec{V} = -4.41$. Elephants and lions, respectively, had $\vec{V} = -2.29$ and $\vec{V} = -5.77$.

But when penalties are added into the equation, the picture completely changed, with only the African brush-tailed porcupines and the Gambian pouched rat having positive values, respectively $\vec{\Gamma}_x = 0.39$ and $\vec{\Gamma}_x = 1.50$. Two other species have near zero total threat index but still negative; these are the blue duiker and sitatunga, which respectively $\vec{\Gamma}_x = -0.05634$ and $\vec{\Gamma}_x = -0.23$. Unsurprisingly, all the landscape species (bonobos, chimpanzees, elephants, and lions) had gloomiest total indexes, which were respectively, $\vec{\Gamma}_x = -4.98$, $\vec{\Gamma}_x = -5.68$, $\vec{\Gamma}_x = -3.56$, and $\vec{\Gamma}_x = -9.04$ (Fig. 15.2).

## 15.12 Species Threat Index in the Biodiversity Conservation Perspectives

Based on this analysis, the species most threatened in the Lake Tumba Landscape where those with the highest modulus of $\vec{\Gamma}_x$. In this respect, and unsurprisingly, the lions were the most threatened species followed by the chimpanzees. Bonobos were the third most threatened species while elephants were the fourth in decreasing order. However, it should be equally said that near all the species assessed through this process all negatively impacted by human activities and can be said to be threatened. Major exceptions of this were the cases of the African brush-tailed porcupines and the Gambian pouched rat that both had positive values of $\vec{\Gamma}_x$. This is a rather gloomier picture, which should appeal for significant conservation efforts to be invested not only in charismatic landscape species but also in other species that are generally felt as being very abundant throughout the country. Stating so also points to a more general problem, which is that often species are categorized in degrees of threats they face on natural ranges without necessarily factoring local situations. Indeed, while all the landscape species are categorized as being endangered throughout their respective natural ranges, some other species are felt to be less threatened in that context. Good examples of such instances include species such as blue duikers and red river hogs which are considered to be in great numbers and are felt to be very abundant in areas like the Salonga National Park (Inogwabini 2006) would be described here, as it is the case for most of the species, as being threatened.

Threat indexes account for many variables, including biological responses of a given species to habitat alterations. Other components of the index include vulnerability to given hunting techniques, market demands and specific cultural culinary values, which all instilled by human actions toward any given species. The index provided in this context was essentially made of susceptibility to habitat loss, hunting degrees, and the capacities for a species to reproduce under natural conditions. The advantage of this approach is that it is simple and can be done even by people without necessary high level of mathematics and a deep understanding of the biological processes. The other advantage of the method is that it can be easily used to drack back the contribution of each factor to the overall threat on a given species. A second advantage of this method is that it provides two levels of looking at the specific situation of each species. The notion of penalty introduced in this method helps factor unforeseen impacts and can be expanded to include other factors such as ones described at the end of the next section of this chapter. The rationale behind the calculation of the $\prod$ is that factors with the highest impact are those that come first in the order while those that would have but meek effects would come later in the series.

## 15.13 Effects of Human Activities in the Southern Lake Tumba on Wildlife Species

Two main human activities have been documented to be seriously threatening the biological diversity of the Lake Tumba Landscape. These included habitat loss by logging activities and hunting for bushmeat trade. Logging is practiced by large-scale logging companies and both local communities. The large-scale logging in the Democratic Republic of Congo is mostly developed in the western part of the country; which fact is driven by the logistical needs. In fact, traditionally, there have always been two general routes for the exportation of goods from the Democratic Republic of Congo: via East Africa and the Congo River, the exports via the Republic of Congo and other Central African countries being somewhat more difficult and limited in its scope. With decades of conflicts, the outlet via Easter Africa greatly decreases in the case of timber though there were reports of some products (including timber) being illegally transferred out via that pathway during the conflict. Wars meant that the thin road infrastructure eroded to near nonexistence; hence most of the logging activities were reshaped to rely on short land transportation combined with long legs of water transportation. Hence, the development of logging industry in the Lake Tumba Landscape is attributable to the fact that the landscape is crisscrossed by an extensive network of navigable rivers that allow easy transport of the logs from the area of their extraction to the point of their export. Indeed, the logistics of the timber industry in the Democratic Republic of Congo are generally designed in such a way that the timber logs are carried by boat to Kinshasa, where they are then transported by road to Matadi and uploaded to boats for exportation mostly (80%) to European countries (CTB 2010).

The very same logic of the logistics of logging activities is also placed at the heart of the artisanal logging. Communities located near major rivers are the most active in this activity, which has been expanding more and more over the last few decades. This is particularly, the case for the Lake Tumba Landscape where most of the communities live near major water bodies and have been invested in artisanal logging as means of obtaining financial gains over the last 2 decades. Indeed, most of the timber that is used for domestic consumption originates from artisanal logging. But that is not the entire story; a great number of cubic meters of the timber exported by industrial logging also come from artisanal logging, which sells a great portion of its products to industrial logging companies. In fact, to walk around the regulations and for reasons of cheap labor, a relatively large number of industrial logging companies do rely on artisanal logging to get the wood located outside of their concessions and avoid paying taxes (Lescuyer et al. 2014). According to Lescuyer et al. (2014), artisanal logging is a very lucrative activity and brings a total net financial income of 111 million American dollars annually to four major categories of stakeholders; of that financial manna about 47% is left with local communities, 42% goes to the industrial logging companies as benefit made from resale, and 8% is paid in different forms of taxes to the public administration. In a country whose total budgets are always below 10 billions American dollars annually, this much money does represent a major stake and explains why it has been difficult for the biodiversity conservation organizations have difficulties to curb the expansion of the artisanal logging and explains why regulations on artisanal timber extraction had been very difficult to be implemented. Industrial logging companies in the Lake Tumba Landscape have been operating a long time before landscape as a biodiversity conservation concept emerged and logging of the hard black timber during the colonial era. Most of the Wenge (*Millettia laurentii*) is, indeed, extracted from the Southern part of the Lake Tumba Landscape and it has been extracted throughout the political history of the country. There has not been any clear-cutting logging practice in the Lake Tumba Landscape and the readings of the forest cover in the region are relatively closer to the national annual rates, which is around 0.3% of the last decades. But effects of logging on biodiversity should not be looked at simply from the perspective of forest cover; logging can insidiously affect ecological functions of many habitats in many ways. For example, an insidious effect of logging on ecological functions in the Lake Tumba Landscape have been documented to be essentially around increased water turbidity for small rivers in the southern Lake Tumba Landscape where there were current logging activities by industrial logging companies as compared to those areas where there was no logging . In fact, Inogwabini and Mputu (2008) designed a study that allowed the collection of data from three rivers of which one was felt to have very little impact of human activities, the second was subjugated under cattle rising as the principal land use and the third one was within logging concessions. Data collected included physical characterizations of the rivers (habitat types around the river and river substrates), physic-chemical data (pH, turbidity, Dissolved Oxygen, Alkalinity, NH3, Dissolved $CO_2$, Salinity, etc.) and biotic elements (fish species and other forms of life that were resident in these water bodies. After collecting a 1-year data, they looked at the dry season data set and discovered that the water turbidity in the river where there was lower human activities and the one where near an intensive cattle husbandry had lower turbidity value (10 JTU) while the value tripled in the river located within the logging concessions (30JTU). High levels of turbidity have been said to impact some ecological interactions, particularly between prey and predator fish species (Bonner and Wilde 2002); they have been found to reduce the productivity, to reduce species richness and impact the composition of communities (Kemp et al. 2011). Hence, an impact of logging on biodiversity has been occurring in the Democratic Republic of Congo without being properly documented and factored in the debate. Everything has been on either forest cover or the bushmeat crisis that generally follows unproperly enforced regulations about what can and cannot be done within logging concessions.

The above discussion about the impact of logging on the quality of freshwater does mean only one essential thing: measures of threats are rather complex and what we often use do not necessarily take into account multiple effects one single human activity can have on the wider biodiversity. Indeed, in some cases and when there are no appropriate studies, conservation biologists are left to summing up physically visible impact and think of their potential effects on biological assemblages can be. What often lacks on these circumstances is the information on what links exist between different elements of the system and the quantification of how the whole system would react. This method of quantifying threats is not safe for that weakness because it does not have any means to factor the effect of logging on freshwater to account for the general system. That is why to include unseen effects of physical destructions of habitats or species it was felt that the best option was to present the index as being a minimal level of threats.

Of course, the most studied and publicized effects of logging on biodiversity are those related to the bushmeat trade, which has been often described in terms of collateral effects of opening roads that would stimulate bushmeat markets. As will be discussed in Chap. 18, bushmeat trade has to be requalified and its two major components clearly distinguished. Firstly, bushmeat trade by local communities as part of their local total economy might have an impact on wildlife species but can be controllable if appropriate measures are put in place. This contributes to the local economy by bringing in cash to pay for services that communities need most such as medical care, school fees, and the purchase of other necessary and important commodities that communities genuinely need. In the Lake Tumba Landscape, hunting for sustainable community livelihood is practiced using different techniques, including such traditional techniques as dog, nets, and spears and bow. Some people use guns either locally made or imported; in most cases these are gauge-12. Apart from the locally made gauge-12 gauge guns, the socioeconomic studies in the Lake Tumba Landscape found out that close to 90% of gauge-12 guns possessed by local communities were imported from France and Russia; 100% of those who reported possessing imported

French-made and Russia-made hunting guns acknowledged that they acquired their guns via the Republic of Congo. It was also found out that a high proportion (45–60%) of guns were a communal property of whole families. This poses the question of how to measure the impact of one gun if it can be used alternatively by more than 5–10 members of one community. Indeed, in view of such a finding, factoring numbers of guns in the region into the threat index does not provide real picture of the potential effect a gun in terms of hunting.

Yet, there is another type of bushmeat trade that warrants concerns not only from a biodiversity conservation perspective but also the angle of the national security as it involves automatic weaponries and organized criminality (Nellemann et al. 2016). This type of bushmeat trade can easily be qualified of mafia-like type of bushmeat trade and it is difficult to tackle. This is the form of bushmeat trade that is widely decried and is principally constituted of large gangs and involves very sophisticated circuitries, which often involve higher military hierarchies and political complexities. In some remote areas of the Democratic Republic of Congo, it is brought by traders that focus on natural resources such as fish and bushmeat and who invest significant resources to arrive in remote areas where they set camps wherein they would dwell for months (Inogwabini 2013). In some circumstances, the gangs are brought in the forests when logging companies open up large tracks of forest to permit the transport of the timber they cut toward their export points. That is why often bushmeat trade has been described as a collateral effect of timber logging though in some cases it can be arguably a direct effect as logging crews in camps can be involved too. The patterns and effects of the mafia-like bushmeat trade have been studied in several forest areas in Central Africa. In their condensed versions, these patterns can be said to be the following: first of all, large and organized bands of people, often involving retired or former military or militia personnel and, sometimes even officials in some countries gather together with the aim to exploit bushmeat as commercial means to make money. They then invest monies and other resources in the acquisition and the use of significant quantities of automatic weapons and ammunition. They also invest in independent means of transportation as hunting often occurs in remote areas where public transport is hardly available and the logistics to sustain large numbers of hunters in the forest for a long time. Then the recruited hunting gangs are deployed in the field and proceeded with massive decimation of large-bodied mammals (elephants, buffaloes, great apes, crocodiles, primates, diverse other ungulates, etc.) in remote forests, which represent higher market values (De Merode et al. 2004). According to Colom et al. (2006), the species most financially valued and the most sold in the markets in the southern part of the Lake Tumba Landscape include *Cephalophus dorsalis castaneus* (the bay duiker), *Cephalophus nigrifrons* (the black-fronted duiker) *Tragelaphus spekeii* (the marshbuck or sitatunga), *Tragelaphus scriptus* (bushbuck), *Potamochoerus porcus* (the red river hog), Hyemoschus aquaticus (the water chevrotain), and (5) *Cephalophus monticola* (blue duiker). On these species should be added diverse species of primates among which the most prevalent in the markets throughout the Lake Tumba Landscape were De Brazza's monkey, red tail monkey, Mona monkey, and Allen's swamp monkey. Other species of diurnal primates, particularly the Angolan pied colobus, the black mangabey, and the Tshuapa red colobus were present in markets of large human settlements such as Mbandaka, located at the heart of the Lake Tumba Landscape, but bushmeat traders openly declared that close to 75–80% of these species were brought from very remote areas such as the Salonga National Park. However, local communities acknowledged that these species were, indeed, residents of the Lake Tumba Landscape but their numbers have been significantly reduced over the last 2–3 decades. A more interesting case in this line was that of the golden-bellied mangabey *Cercocebus chrysogaster*, which was talked of in historical terms as having been present but totally disappeared completely in the area (Inogwabini and Thompson 2013).

Bushmeat trade in the Lake Tumba Landscape was not selective enough. Apparently value-less species, in terms of financial income to generate, such as African brush-tailed porcupine (*Atherurus africanus*), civets (*Civetta vivera*), Clawless otter (*Aonyx congica*), Congo squirrel, Demidoff's galago, the Gambian pouched rat (*Cricetomys gambianus*), the Nile monitor (*Varanus niloticus*), Potto, Thomas' galago, Thomas's rope squirrel, and a cohort of other species such as bats are also present in the markets and are sold throughout the year. A combination of biological knowledge, principally the reproduction rates in combination with external threat analysis made using vector in combination with mathematical series led to seeing that even species often thought not being threatened in general were found to be threatened in the Lake Tumba Landscape.

The conclusion above shows that reliance on biodiversity charismatic species would overlook the threats that other species live under. Indeed, often principally large mammals such as apes, elephants, and other big species are discussed in conservation fora while other species are often not even mentioned though their fate is always lurking in the minds of any true conservation practitioner. The message being sent here is that a simple tool or method as the one described in this chapter would ease the understanding of the concept of threat for biodiversity because managers, wardens, and eco-guards can easily plug numbers in the formula and monitor the conservation status of each species very easily.

## References

Bermejo M, Rodriguez-Teijeiro JD, Illera G, Barroso A, Vila C, Walsh PD (2006) Ebola outbreak killed 5000 gorillas. Science 314:1564

Blom A, Van Zalinge R, Mbea E, Heitkönig IMA, Prins HHT (2004) Human impact on the wildlife population within a protected Central African forest. Afr J Ecol 42:23–31

Bonner TH, Wilde GR (2002) Effects of turbidity on prey consumption by prairie stream fishes. Trans Am Fish Soc 131:1203–1208

Buckland ST, Anderson DR, Burnham KP, Laake JL (1993) Distance sampling: estimating abundance of biological populations. Chapman and Hall

Colom A, Bakanza A, Mundeka J, Hamza T, Ntumbandzondo B (2006) The socio-economic dimensions of the management of biological resources, in the Lac Télé—Lac Tumba Landscape, DRC Segment: a segment-wide baseline Socio-Economic study's Report. Submitted to the World Wide Fund for Nature

Coopération Technique Belge (CTB) (2010). La forêt congolaise, la gouvernance et le commerce du bois: FLEGT! https://www.btcctb.org/fr/casestudy/t-congolaise-gouvernance-commerce-du-bois-flegt. Accessed 10 June 2016
De Merode E, Homewood K, Cowlishaw G (2004) The value of bushmeat and other wild foods to rural households living in extreme poverty in Democratic Republic of Congo. Biol Cons 118(5):573–581
De Merode E, Cowlishaw G (2006) Species protection, the changing informal economy, and the politics of access to the bushmeat trade in the Democratic Republic of Congo. Conserv Biol 20(4):1262–1271
Draulans D, Van Krunkelsven E (2002) The impact of war on forest areas in the Democratic Republic of Congo. Oryx 36(1):35–40
Gautier-Hion A, Colyn M, Gauthier JP (1999) Histoire naturelle des primates d'Afrique Centrale. Multipress, Ecosystèmes d'Afrique Centrale. Union Douanière et Economique d'Afrique Centrale
Hart JA, Hall JS (1996) Status of Eastern Zaire's forest parks and reserves. Conserv Biol 2:316–327
Hulstaert G (1992) Onomastique Mongo. Annales Aequatoria 13:161–275
Inogwabini BI (2005) Common fishes of the Salonga national park, Democratic Republic of Congo: preliminary survey and conservation issues. Oryx 39(1):78–81
Inogwabini BI (2006) A preliminary checklist of mammals and plants: conservation status of some species in Salonga National Park. Endangered Species Update 23:104–117
Inogwabini BI (2013) Bushmeat, over-fishing and covariates explaining fish abundance declines in the Central Congo Basin. Environ Biol Fishes 97(7):787–796
Inogwabini BI (2006a) A preliminary checklist of mammals and plants: conservation status of some species in the Salonga National Park, Democratic Republic of Congo. Endangered Species Update 23(3):104–117
Inogwabini BI (2006b) Fire as management tool in cattle raising concession: a preliminary assessment to help develop a system that sustains biological diversity in the southern in the Lac Télé–Lac Tumba Swampy forest, DRC Segment. Field report submitted to WWF and CARPE-USAID
Inogwabini B.I. and Mputu D.A (2008). Evaluation de la qualité de l'eau des rivières de basse altitude dans la zone de Malebo: case de Bambu, Lebomo et Bongo, Sud du paysage Lac Tumba. République Démocratique du Congo. Rapport Soumis au WWF US, Washington DC et USAID-CARPE, Kinshasa, République Démocratique du Congo
Inogwabini BI, Omari I (2005) A landscape-wide distribution of Pan-paniscus in the Salonga National Park, Democratic Republic of Congo. Endangered Species Update. 22:116–123
Inogwabini BI, Hall JS, Vedder A, Curran B, Yamagiwa J, Basabose K (2000) Conservation status of large mammals in the mountain sector of Kahuzi-Biega National Park, Democratic Republic of Congo in 1996. Afr J Ecol 38:269–276
Inogwabini BI, Matungila B, Mbende L, Abokome M, Tshimanga WT (2007a) Great apes in the Lake Tumba landscape, Democratic Republic of Congo: newly described populations. Oryx 41(4):532–538
Inogwabini BI, Thompson JAM (2013) The Golden-bellied Mangabey (Cercocebus chrysogaster): distribution and conservation status. J Threat Taxa 5(7):4069–4075
Kemp P, Sear D, Collins A, Naden P, Jones I (2011) The impacts of fine sediment on riverine fish. Hydrol Process 25:1800–1821
Lescuyer G, Cerutti PO, Tshimpanga P, Biloko F, Adebu-Abdala B, Tsanga R, Yembe-Yembe RI, Essiane-Mendoula E (2014) Le marché domestique du sciage artisanal en République démocratique du Congo: État des lieux, opportunités, défis. Document Occasionel Centre de Recherche Forestière Internationale (CIFOR), p 110
Nellemann C, Henriksen R, Kreilhuber A, Stewart D, Kotsovou M, Raxter P, Mrema E, Barrat S (eds) (2016) The rise of environmental crime: a growing threat to natural resources, peace, development and security. A UNEP-INTERPOL Rapid Response Assessment. United Nations Environment Programme and RHIPTO Rapid Response–Norwegian Center for Global Analyses, www.rhipto.org

Parnell RJ (2000) Information from animal tracks and trail. In: White LJT, Edwards A (eds) Conservation research in the African rain forests: a technical handbook. Wildlife Conservation Society, pp 153–184

Walsh PD, Abernethy KA, Bermejo M, Beyers R, De Wachter P, Akou ME, Huijbregts B, Mambounga DI, Toham AK, Kilbourn AM, Lahm SA, Latour S, Maisels F, Mbina C, Mihindou Y, Obiang SN, Effa EN, Starkey MP, Telfer P, Thibault M, Tutin CE, White LJ, Wilkie DS (2003) Catastrophic ape decline in western Equatorial Africa. Nature 422(6932):611–614

Walsh PD, Breuer T, Sanz C, Morgan D, Doran-Sheehy D (2007) Potential for ebola transmission between gorilla and chimpanzee social groups. Am Nat 169:684–689

# Chapter 16
# Synopsis of Freshwaters, Species Diversity, and Conservation Issues

**Abstract** Freshwaters of Democratic Republic of Congo are poorly studied because of different factors, including the geographic spread of the country, lack of expertise, and financial resources to invest in research. As a review of the current conservation status of the freshwaters of the Lake Tumba Landscape, the chapter did not provide details of species, water habitats, etc. Drawing mostly on peer-reviewed published materials and unpublished reports, the chapter established that common to the waters of the region of the Lake Tumba Landscape are dark-brown, hot and acidic, characteristics attributed to the immensity of organic materials carried by waters in their course down streams. Most of the existing studies expressed concerns over fish stocks, changes in fish community structures and fish species. Declines in fish stocks were reported and widely felt communities. However, this feeling should be demonstrated by strong scientific evidence. Morphometric measurements collected from the culinary most preferred fish species at all sites had significantly smaller mean maximal lengths when compared to standard maximum growth measurements. The chapter concludes that increase in human populations throughout the Lake Tumba Landscape affected freshwater environment and fish diversity.

**Keywords** Freshwater · Lakes · Rivers · Fish species · Over-fishing · Freshwater dependent species

## 16.1 Introduction

Through its course to the Atlantic Ocean, the Congo River is separated into three natural divides whose boundaries are mostly constituted of rapids and chutes that are major barriers for species dispersal (Chapman 2001). These three main portions are: (1) the upper Congo, (2) the mid Congo, and the (3) lower Congo (Bailey 1986; Groombridge 1992). These three segments combined have more fish species than any other African river (Groombridge 1992). Communities throughout the Democratic Republic of Congo, particularly those that live along the Congo River dependent on fish resources both for substance and commerce (Banister 1986); fish contribute a large portion of protein to these people who consume about 700,000 tons/year of fishes in the entire country (Leonard 1987; CEFDHAC 2001).



B.-I. Inogwabini, *Reconciling Human Needs and Conserving Biodiversity: Large Landscapes as a New Conservation Paradigm*, Environmental History 12,
https://doi.org/10.1007/978-3-030-38728-0_16

However, a major current characteristic of freshwaters of Democratic Republic of Congo is that fish species, habitats, and other biological traits have not been properly documented yet (Bailey 1986), owing to different factors that include the geographic spread of the country, lack of expertise and financial resources to invest in research. The objective of this chapter is not to provide details of species, water habitats, etc. but rather to review the current conservation status of the freshwaters of the Lake Tumba Landscape after several studies were conducted. The chapter will draw on peer-reviewed published materials as well as unpublished reports; it will describe current felt and documented threats to biodiversity and the water quality.

## 16.2 Rivers and Lakes of the Lake Tumba Landscape

In relationship with the first point of this introduction, Lake Tumba Landscape is in the second part of the three major divides of the Congo River, which is known as the Mid-Congo River. This is the 1700 km segment of the Congo River flowing from Kisangani to Kinshasa; it is sometimes known as the Central Basin or the Cuvette Central (Banister 1986; Chapman 2001). Its habitats are characterized by the presence of several tributaries that flow into the Congo River throughout the Cuvette Centrale on both of its sides (Northern and Southern respectively). The major of these tributaries are Ubangi in on the northern side and four others in the southern side (Lomami, Lulonga, Ruki-Tshuapa, and the Kasai-Lukenie-Kwango). This is vast a flat zone (altitude: 300 m–700 m a.s.l). In its course, through this area, the Congo crosses major swamps, including Ngiri swamp as well as those located at the junctions of the Congo and its tributaries, and expansive wetlands ($\leq$5–10 km width on both sides) constituted of areas of seasonally flooded and permanently inundated areas (Evrard 1968; Bailey 1986; Gauthier-Hion et al. 1999; Inogwabini et al. 2012). Characteristics forest types have open understory, with the most important plant communities including *Raphia sese*, *Pandanus*, *Guibourtia demeuse*, *Uapaca guineensis*, *Uapaca heudelotii*, etc. (Evrard 1968). Other freshwater-dependent species who reside in this zone include Hippopotamus (*Hippopotamus amphibus*), African slender-snouted crocodile (*Crocodylus cataphractus*), Nile crocodile (*Crocodylus niloticus*), giant otter shrew (*Potamogale velox*), Congo clawless otter (*Aonyx congica*), sitatunga (*Tragelaphus spekei*), water chevrotain (*Hyemoschus aquaticus*), and the endangered Hartlaub's duck (*Pteronetta hartlaubii*) (Groombridge 1992; Toham et al. 2006).

Apart from the Congo Rivers, its tributaries and the swampy forests, the freshwater habitats of the Lake Tumba Landscape also have particular habitat types, which are offered by the lakes in the region. There are several lakes, the most important being, of course, Lake Tumba and Lake Maindombe. Lake Tumba is located within the administrative territory of Bikoro and is 765 km$^2$ wide; it is a shallow lake with average depths ranging between 4 and 7 m. The water of Lake Tumba is dark; the water transparency, as measured by the use of the Secchi Disc is about 0.75 m near the shores and varies between 0.90 and 1.10 m when one moves away from the edges (Dubois 1959). Dubois (1959) measured the temperatures of the Lake

Tumba according to the depths and found that while surface temperature varied between 26.70 and 27.55 °C, the variations of temperatures between the surface and the bottom of the Lake Tumba were minimal and ranged between 26.50 and 27.05 °C, leaving very slim probabilities of the Lake being ecologically stratified by temperature of different depths. The waters of the Lake Tumba are acidic, with a water pH that varies depending on the period of year. The reported values depend on authors but overall all figures confirm the variations as a function of periods of the year. Dubois (1959) reported the range of 4.5–4.9, whereas Corsi (1984a) gave a range of water pH between and 4 and 4.5. As Bailey (1986) and Inogwabini et al. (2009) reported, Lake Tumba is chemically impoverished while its planktonic availability is sparse. However, the water of Lake Tumba contains a lot of vegetable debris (Bailey 1986), which were estimated by Dubois (1959) to represent 0.87% of dried materials and waș carried into the Lake by several tributaries that flow from adjacent forests. The catchment basin of Lake Tumba has been estimated to be 7,380 $km^2$ (Marlier 1958), which is a little over 10 times the current size of the lake itself. Forests adjacent to the Lake Tumba, whether located in this catchment basin or outside of it, have been logged for timber over the last 45–50 years; which period had seen an alarming rapidity in human population growth (see Chap. 18). According to Béné et al. (2007), there are thirty-one villages located on the shores or in the proximities of the Lake Tumba, with the village population ranging from 500 to 11,000 people (Inogwabini and Lingopa 2006), totaling about 60,000 people. This rapidly growing population relies heavily on natural resources and particularly on fishing, which had led to increasing fishing rates, which are indeed very high and fuel markets both in Mbandaka and Kinshasa (Corsi 1984b).

The habitat around the Lake Maindombe is a mosaic of dunes of bare sand, herbaceous savannas, and degraded rain forest, with large expanses of *Raphia* swamps. Maindombe means black water; and as its name indicates, the water is colored black (Corsi 1984b), which is due, according to Hardman and Stiassny (2008), to the fact that the majority of its carbon input comes from allochthonous sources, principally from terrestrial plant litter and associated invertebrates, which decompose to stain the water dark-brown. That coloration indicates low levels of mineral content. The town of Inongo is located on the central peninsula of the lake while Kutu is at the southernmost tip of the lake. Large towns are known to discharge their wastes into lakes, threatening thus the biological diversity of the lake water. High levels of fishing are also reported from the lake to feed both the local and the national market.

Apart from the two main lakes of Maindombe and Tumba, the landscape does also counts other smaller lakes within it. Unfortunately, despite their respective sizes being larger than the Lake Tele in the Republic of Congo, these smaller lakes have been attracting very little attention; this is partly because some of them are often seen as part of the general Maindombe ecosystem. These smaller lakes are the Lomako (in the northern part of the Lake Tumba Landscape, within the Ngiri swamps) and five other lakes in the southern part of Lake Tumba Landscape (Ikenge, Besako, Ilungu, Nkoloetulu, and Ndeke-Ngelo). Inogwabini and Lingopa (2007) described the smaller lakes of the Southern Lake Tumba Landscape in a report submitted to the World Wide Fund for Nature in generic terms as follows. Ikenge (geographic coordinates: [−15057 and 104240]) is a 27 $km^2$ wide and 3 m deep (shallow) lake, with dark-brown, humic, and acidic (mean pH ≈ 4.46) waters. Ikenge is separated from the Lokoro River by a 500-m-long channel through which the lake's waters are carried out while its external water intake comes essentially from the Bokoka River that crosses a densily swampy forest area. With the bottom of the lake varying from being sandy (in the north) and muddy in the south, the average surface water temperature had been measured at 26 °C in the dry season (May–August), which is the coolest period of the year. This means that the temperature of the lakes in the region can become very high during the months of rainy seasons.

The second smaller lake of the Maindombe ecosystem is the 35 $km^2$ and the 2.6 m deep Lake Besako (geographic coordinates: [−13136 and 183349]), which is located at a 4.5 km distance from Lake Maindombe on the Lutoyi River. Contrary to Lake Ikenge which has only one major and permanent human settlement on its shores, Lake Besako has two middle-size villages built on a specific geologic pebbly formation, which formation allows the growth of deep green algae. By middle-sized villages it is meant 30–40 households with average household size being itself around 8 people. Once more, the waters of Lake Besako are acidic (mean pH ≈ 3.5) and their temperature very high (27.3 °C) on a sandy bottom. The third small lake is the 10 $km^2$-wide and 2.2 m-deep Ilungu (geographic coordinates [−13142 and 183913]) on the Lutoyi River; it is located within the same forest type as the above lakes but has permanently small islands within it. Once more, Ilungu is acidic (pH ≈ 3.4) and its mean surface water temperature is very high (27.6 °C) and the substrate of the bottom of the lake is thickly covered with mud. The closest human settlement to Ilungu is at 2.7 km distance. The fourth small lake in the Lake Maindombe ecosystem is Lake Nkoloetulu, a shallow (mean depth = 2.47 m) dark-brown (water transparency = 0.60–0.70 m), hot (mean daily temperature 26.3 °C), and acidic (pH ≈ 3.5) water body of 25 $km^2$ located at the geographic coordinates −13259 and 184335 at about 1.9 km distance from the Lutoyi River and it is immersed within the same dense tropical inundated forests where three semipermanent small villages (Nkoloetulu, Ndobo, and Botangela) can be identified. The substrate of the bottom of Lake Nkoloetulu is mostly sandy though the center of the Lake is muddy. The fifth lake in the larger Maindombe ecosystem is the 8 $km^2$-wide Ndeke-Ngelo (geographic coordinates [−12735 and 184045]) and distanced by Lutoyi of a 1.25 km width. The waters of this shallow lake (mean depth = 1.9 m) are, once again, acidic (pH ≈ de 3.4), dark-brown (water transparency = 0.50–0.60 m), and very hot during the daytime (highest water temperature 30 °C). The surrounding environment is that of permanently inundated forests in the north-western edges while the remaining part is on terra firma terrain where 2 permanent human settlements (Ngelo and Nkile respectively located at the distances of about 0.5 km and 5.86 km from the lake shores). The substrates of the bottom are a mixture of sands and mud though sands can be noticed to predominate. Lake Ndeke-Ngelo does receive very limited inflow water as there is no significant river flowing through it.

As indicated above, there is the Lomako Lake, which is in the northern part of the Lake Tumba Landscape. Very little is known about this lake; most of what is said comes from inferences made of the general status of the swamps of the Ngiri Triangle, a bloc of forest laying between the Congo River and Ubangui. Lake Lomako is located on the Mpoka River, which flows into the junction of the Congo River and Ubangui at the point where both rivers meet amid many islands and where there are a dozen of semipermanent human settlements. According to Inogwabini and Lingopa (2006), the area where Lake Lomako is located is characterized by very sparsely dispersed human populations with a density of 4 inhabitants per $km^2$ (PNUD/UNOPS 1998) and having dark-brown water with limited transparency and holding significant quantities of biological debris as it is the case for most of waters of the tropical forests in this zone. The habitats of the swamps around Lake Lomako are essentially composed of water trees and other vegetation types, which are, as described above and according to Evrard (1968) and Gautier-Hion et al. (1999), composed of *Raphia sese*, *Guibourtia demeuse*, *Uapaca guineensis*, *Uapaca heudelotii*, etc.

Being dark-brown, hot, and acidic is a major characteristic common to the waters of the region of the Lake Tumba Landscape. The dark-brown coloring of the waters in this region has been often attributed to the immensity of organic materials (e.g. leaves from trees, tree branches, etc.) that are carried by waters in their course downstream. That the waters of these lakes are essentially hot can be explained by a combination of several factors that play together: their geographic situation, their depth and their color. Indeed, these lakes are near the equator line and receive significant daily radiation from the sun. These quantities of radiations are then absorbed by the water that is darker because, as it is generally known, darker surfaces exposed to the sun rays absorb more radiation than lighter colored ones. The heat, therefore absorbed, is entirely transmitted downward because, these lakes are rather shallow ones, and the radiation heat the waters. The acidity of the waters in these lakes is, as argued by Butler (2012), caused by the very origins of these waters on the ancient soils, which have no minerals to increase water hardness, which once added to the tannins released from decaying leaves, augment the acidity.

## 16.3 Fish Species Diversity in the Rivers and Lakes of the Lake Tumba Landscape

Characteristic species of the area of the Mid-Congo River include *Protopterus dolloi*, *Hydrocyon vittiger*, *Hydrocyon goliath*, *Hydrocyon vittatus*, *Phractolaemus ansorgii*, *Ctenopoma fasciolata*, *Channa obscurus*, *Lates niloticus*, *Claria buthupogon*, *Synodontis nigriventris*, *Auchenogianus punctatus*, *Schilbe marmoratus*, etc. (Banister 1986; Chapman 2001). The fish diversity in the Lake Tumba, as it the case for many water bodies in the Democratic Republic of Congo, is still subject to variations. Past research documentations have been quoting different figures, varying from 67 species (Marlier 1958) of 86 fish species (e.g. Aveling et al. 2003) present in the Lake Tumba Landscape but that has been reduced down to 30 species (Inogwabini and Lingopa 2006). Reasons for these reductions are difficult to ascertain for the time being because of some changes in the nomenclatures of species, which have been changing over years. Also, despite some species previously not being currently observed, it would be difficult to conclude their local extinction since proof of extinction would take dozens of years to be confirmed. Indeed, some species were collapsed into families and species from which they were previously distinguished. This reality is not singular to Lake Tumba, however; it does emphasize the need for more updated research in the region as well as in the country. Despite these changes, however, most common species of the Lake Tumba, which remain constant over different reporting periods, are: *Auchenoglianis occidentalis*, *Clarotes laticeps*, *Gephyloglianis congicus*, *Claria buthopogon*, *Distichodus*, *Channa obscurus* etc. (Corsi 1984b).

The figure of 41 fish species has been quoted by Corsi (1984b) for Lake Maindombe, of which three endemics but, as it is the case for many areas in this zone that is characterized by being poorly known (Schliewen and Stiassny 2006), Kutshukina and Micha (2013) reported more than fifty fish species in a market study from the same lake. Apart from the widely spread species of Clariidae (*Clarias angolensis, Clarias* gabonensis, etc.), other frequently encountered species that were listed by Kutshukina and Micha (2013) include *Auchenoglanis occidentalis, Gephyroglanis congicus, Synodontis ornatipinnis, Distichodus fasciolatus,* and *Mormyrops nigricans*. As it was the case for Lake Tumba and as it discussed in the section of threats to freshwater in this chapter, reports are that fish stocks have decreased over the last decades. Decreasing fish stocks can be captured from surrogate methods; in the case of Lake Maindombe, the situation can be depicted by the fact that nets with smaller mesh sizes are wildly and widely used. Indeed, Kutshukina and Micha (2013) found that nets of greater than or equal to 2 mm and 2–2.5 mm mesh sizes had been used in higher frequencies (respectively 32% and 38%). Nets of small mesh sizes, particularly of the range 1–2.0 mm, are often used when catches per unit effort are low and cause depletion in fish stocks as they would literally comb any species regardless of the age and the size of the fish.

## 16.4 Wildlife Species Depending on the Rivers and Lakes of the Lake Tumba Landscape

Other freshwater-dependent species of the Lake Tumba Landscape include crocodile species (*Crocodylus cataphractus* and *Crocodylus niloticus*), which are commonly observed in the Lakes and rivers of the Lake Tumba Landscape. Hippopotamuses (*Hippopotamus amphibious*) are present in Mid-Congo River and were observed to occur in somewhat sizable numbers between the town of Mbandaka and Makanza; they were present in Lake Tumba but are currently almost absent but they were spotted in much-reduced frequencies in Maindombe. Water Chevrotain (*Hyemoschus aquaticus*) and Sitatunga, also known as Marshbuck (*Tragelaphus spekii*), are widely distributed, though Water Chevrotain seems to appear in lower frequencies when compared to other sites as the Salonga National Park (Inogwabini Inogwabini 2006a, b). Giant Otter Shrew (*Potamogale velox*) and the Congo clawless otter (Aonyx congica) are present but their conservation statuses are difficult to assess, as there is very little knowledge on their ecology in the region.

## 16.5 Freshwaters of the Lake Tumba Landscape: Threats and Current Status

Few studies exist from the Lake Tumba Landscape on the threats the landscape faces as a global conservation priority area. However, the majority of these existing studies have been expressing concerns over fish stocks, changes in fish community structures, and fish species being felt to veer toward local extinction. Claims of declining fish stocks have been expressed by studies as those of Milton (2015), Colom et al. (2006), and Inogwabini and Lingopa (2006). Despite the widespread observation by communities and other observers that the catch per unit effort has declined over the last few years. Béné et al. (2007) argued that declining fish stocks needed to be proven and, even if it would be proven true, a need to link that decline with over-fishing should be demonstrated in a rigorous way. While there is need to conduct further studies, as it is appealed for in Chap. 24, it would be sensible to argue back and indicate that the widespread acknowledgment by communities that fish resources have been declining is rather genuine; the single fact of having some many more new fishermen due to both the increase in human populations and lack of alternative employment has increased the pressure on the fish stocks. Obviously, as Béné et al. (2007) argue, using knowledge acquired from other sites throughout the world, the variations in catches per unit efforts can also result from variations in environmental conditions. But it is also true that the changes in environmental conditions are to be traced back to the increased human densities and the pauperization of the communities. Indeed, pointing at the environmental changes does not exclude the possibilities for these changes being jointly operating along with over-fishing. In fact, a study conducted by Inogwabini (2013) throughout the Cuvette Central indicated high fishing pressures and correlated 88% of observed differences in fish relative abundance with the mean number of fishing nets while 60% of these differences were attributable to distance from the nearest market. While these results absolve effects of human density as being only marginally associated with fish abundance, the other facts are clearly about fishing. The introduction of new fishing techniques such as the destruction of reproduction sites is also a testimony to the decrease of fish stocks in the Lake Tumba Landscape. Indeed, if fish stocks were abundant there would be no need for fishermen and fisherwomen to conduct such hardworking to extract minimal outputs for large efforts invested.

Clearly, at a broader scale, increase in human populations throughout the Lake Tumba Landscape had affected the freshwater environment. For example, Inogwabini and Lingopa (2013) have studied the impacts of people residing on the Congo islands as a consequence of both increased populations and of the civilian turmoil that the country went through for nearly 2 decades. That preliminary assessment had as objective do document island habitat types, fish species richness and effects of human activities on fish distribution. Inogwabini and Lingopa (2013) found that humans permanently settled 76% of the islands and 70% were cultivated. Permanent human settlements and agriculture were related to erosion on ~50% of islands. The most significant results of that study were that when taken separately, numbers of fish morphotypes captured significantly near islands decreased with increasing sizes of fields on islands and that only 3 species were observed when field sizes were larger than 2.5 ha. Also, when taken separately, fish morphotype numbers significantly increased with decreasing water depths, which was farther away from the shores. These results were a clear indication that human-induced habitat destruction and modifications cause if not losses but at least perturbations of aquatic biodiversity.

Studies on fish communities' structures are rather absent though human communities claimed that nowadays fishes they captured were smaller and smaller as compared to previous catches (Colom et al. 2006). The study of structure is time-consuming and requires repetitious measurement of species in the same areas for a long span of time. That is what Inogwabini et al. (unpublished data) have done from twelve different sites where they morphometric measurements. The sites of the *Cuvette Centrale* were: (1) Etate, (2) Lokofa, (3) Lake Maindombe, (4) Lake Tumba, (5) Lulonga, (6) Ngiri, (7) Ubangi, (8) Congo River—between and the mouth of Lulonga, (9) Bambou River, (10) Lebomo, (11) Bongo, and (12) swamps adjacent to the Lake Tumba. Together, these twelve sites represent a wide variety of habitats of this ecosystem, comprising lakes, major rivers, swamps, and small streams. Data from Salonga (Etate and Luilaka) were collected from November 2001—May 2003 (Inogwabini 2005) while those from other sites have been collected repetitively between June 2005 and July 2009. Each site was revisited at least twice and in two different seasons, and fish specimens were measured each time they were caught.

Results of these morphometric measurements were that all of the most preferred species of fishes (for culinary reasons, mostly) at all sites had significantly smaller mean maximal lengths when compared to standard maximum growth measurements: Etate ($t = 0.0031$, $df = 9$), Lokofa ($t = 0.017$, $df = 5$), Lake Tumba ($t = 0.000$, $df = 9$), Lulonga ($t = 0.000$, $df = 9$), Ngiri ($t = 0.000$, $df = 8$), Maindombe ($t = 0.000$, $df = 9$), Congo River ($t = 0.000$, $df = 9$), Lake Tumba adjacent swamps ($t = 0.0081$, $df = 9$), Ubangi ($t = 0.000$, $df = 9$), Bambou ($t = 0.000$, $df = 9$), Lebomo ($t = 0.000$, $df = 9$), and Bongo ($t = 0.000$, $df = 9$). Seventy-five percents of fishes measured showed a decreased growth of the range 30–69%. Other than that, the following species lost more than 70% of their expected maximum growths in different sites: *Clarias buthupogon* (70% at Lokofa), *Chrysichthys punctatus* (76%, Lake Tumba), *Distichodus sexfasciatus,* and *Distichodus sexactus* (respectively 74% and 73% at Lulonga and 75% for each in the Congo River), *Hydrocyon goliath* (78%, 80%, 88%, at Congo River, Lulonga, Maindombe, and Lulonga respectively), *Mormyrus delicious* (78% at Maindombe), *Clarias longipinis* (77% in the Congo River), *Clarias angolensis* (80% in the Lake Tumba and adjacent swamps), *Malapterurus electricus* (86% in the Lake Tumba adjacent forest swamps), *Rheoglanus dendrophorus* (79%

and 84% at Bambou and Bongo respectively), *Schilbes grenfelli* (at Bongo) and *Auchenoglianis punctatus* (75% at Bongo).

The trends in fish community structures being described for the Lake Tumba Landscape and other areas of the Cuvette Central are similar to those from the Benoué River Basin (Corsi 1984a) and those from commercial fishing in the oceans (Welcome 2005). There are few studies on the contribution of fishes in the bushmeat crisis. Yet, data from socioeconomic studies indicate that in most of the areas in the Central Congo Basin, fishes may contribute more than it is actually acknowledged by studies that biased for large mammals. General changes in fish community structures may be an alarming signal for the fate that fishes are facing as part of the general bushmeat.

There are multiple causes that impact fish community structures including habitat destruction, changes in water chemistry, and over-fishing (Welcome 2005). A study of breeding sites (Inogwabini et al. 2009) in Lake Tumba indicated that these were targeted by fishing activities that were destructive to large expanses of habitats. Another study in Lake Tumba described phenomena such as decreasing water depths due to changes in the rainfall regimes. Also in the last 39 years, the Ubangi basin water depths decreased by 4.5%, which entered a continuous drought phase (IRD 2002 ). These factors were attributed to global climate change effects. All these factors, which may depict the general trends in the region of the Central Congo Basin, may have affected fish reproduction cycles and, therefore may have influenced fish community structures. Nevertheless, their causation links and magnitudes are yet to be described and proscribe thus any firm conclusion. What remains certain, however, is that results being reported in this paper show that fish species allover the Central Congo Basin have seen their community structures changed over the last years and the only factor that has been thoroughly documented is that all species described here are the most abundant (Inogwabini et al. 2009), the most marketed and the most widespread species (Chapman 2001) in Central Africa. Preferred fishes in the region were hypothesized to be commercially over-fished given their abundance in markets (Inogwabini et al. 2009). Data being presented here lend support to this hypothesis and indicate severe structural changes, which are known to be caused by, among other factors, over-fishing. The over-fishing is a consequence of insufficient paid jobs in the DRC, which began since 1990s subsequent to the political and socioeconomic crisis that had paralyzed the normal economic system (Newbury 1986; Lutala et al. 1993) pushing 80% of the workforce into parallel productive activities (Newbury 1986). This situation dragged people into fishing activities that have become of the most important income-generating activities in forests of the Central Cong Basin in particular (Toham et al. 2003; Inogwabini 2005; Inogwabini et al. 2009; Colom et al. 2006). The lucrative motivations had led to an increase in both numbers of people leaving only of fishing activities from 25% reported in the late 1980s (Kambay 1995) to 65–70% respondents in 2006 (Colom et al. 2006) from the Central Basin. Increase in dependence on fishing had been accompanied by the introduction of new fishing gears and methods whose consequences are decreasing catch per unit efforts in both lakes and some rivers where time series data were available.

Based on the above argumentation, destructive fishing techniques and global climate change effects on water abundance and chemistry may decrease fish stocks in the region. Given the proportions taken by fishing and fishes in the economy of the region, trends describes here will very likely accelerate movements of people around the region to seek good fishing conditions in water basins that will still harbor sufficient fish stocks. Those movements will dismantle the fragile equilibrium of land and water allocation essentially based on traditional rights. Conflicts over fish resources would occur thereafter, requiring a state intervention. A strong conservation message conveyed by this study is, therefore, that sound planning and conservation measures to regulate levels of off-takes and fishing techniques are urgently needed to curve the actual trends in fish community structures to avoid acting in emergency. There is no third layer of the Congo River that would harbor fishes dying of old ages as layman rumors present the situation of the fishes in the Central Congo Basin, fish stocks are declining and the issue should be taken seriously.

Finally, some fish species were reported by communities to become rater and rarer (Colom et al. 2006). Does this mean species are veering toward local extinction? This is a difficult question to respond to. However, there are some principles in conservation biology that should guide anyone trying to provide any sensible answer to this feeling. Indeed, getting rarer is not equivalent to going extinct; there is a period over which an unseen species can be said extinct. While the feeling by fishermen and fisherwomen in the region should be taken as genuine because they are the ones that know what species they have been catching over decades, this is an area where thorough scientific investigation is needed. As described above, some numbers of species may be collapsed just because of synonymy while others may be moved to other areas because of changes in environmental conditions, as discussed above and by Milton (2015).

## 16.6 Freshwaters of the Lake Tumba Landscape: Prospects for the Future

Despite questions raised by Béné et al. (2007), at best, the situation in the Lake Tumba Landscape remains globally uncertain for the aquatic biodiversity. Hence, the precautionary principle would call for practical biodiversity conservation measures. But these measures needed to be grounded on scientific knowledge. That is why Inogwabini et al. (2009) conducted a study to guide the delineation of aquatic protected zones in Lake Tumba. The study used *Tilapia congica* as an umbrella species to help identifying breeding zones. They found that 60% of nest sites were clumped within the range of 51–150 m, which was an indication that there was a relationship between nesting sites and the distance from edges. They also found that the largest group spread (width) group was 300 m and that one site was 10 km long. When multiplying the width and the length they identified a zonal spatial spread of 300 ha, with a core reproduction zone of 100 ha. Hence they had proposed that zone and other smaller ones to become aquatic protected areas with the aim to ensure that fish of different species would come and breed in these zones. This reasoning was taken because *T. congica* shared a great chunk of the breeding territory (at least 41.81%) of its nesting sites with other fish species, leading to the conclusion that protecting the species habitats would provide the umbrella for the conservation of other species.

Apart from the above traditional conservation approach, a more popular approach that has been being pushed for the Lake Tumba Landscape and its massive water bodies has been focused on the organization of local communities in associations and local fishing committees (Béné et al. 2007). Béné et al. (2007) suggested that these non-traditional conservation methods have had very limited sustainability; they often failed to continue to operate once projects implemented by the international conservation organizations closed and any other external support came to an end. While the lack of sustainability in initiatives of community fisheries was true in many ways, the rationale behind the thinking remains plausible as an option for better management of aquatic resources of the Lake Tumba Landscape. Of course, flaws of the communities associations were substantively the same as those that can be pointed out in the case of the general development paradigm, including the fact that the actions are defined remotely without proper consultation with local communities, the one-size-fits-all approaches that superbly ignore the contextualization of any approach and, more importantly, the disjunction between the expert's knowledge and the local communities' knowledge. Solutions proposed by Béné et al. (2007) to fix the failures of the community fisheries were to (1) focus on the commercialization of fish, (2) increase accessibility to financial resources through microfinancial scheme, and (3) sustainable community-based monitoring of fish stocks. The value of the third point of this proposed solution is unquestionable because, apart from the traditional development paradigm indicated above, one of the major reasons why community fishery has been failing in the area of the Lake Tumba Landscape is partly because of limitation of knowledge of the available resources. Hence, encouraging communities to become more involved in searching and distributing knowledge will create a sense of ownership. But for that to work, the sense of ownership needs to be backed by a sense of legality as well; hence communities law enforcement should become an essential part of the monitoring in order to ensure that all members of the communities are playing by the same rules. Only a genuine community law enforcement mechanism would ensure that there are no free riders and that the tragedy of the commons does not become a true reality. This view is in alignment with the views expressed elsewhere and particularly focusing, as suggested by (De Merode 2005), on devolving responsibility to local institutions.

The first solutions proposed by Béné et al. (2007) to fix the failures of the community fisheries, namely (1) the focus on the commercialization of fish and (2) the increase accessibility to financial resources through microfinancial scheme can call to some arguments. Focus on commercialization without any control and with no clear ownership rights for communities would lead to the depletion of the aquatic resources. This is true for obvious reasons that increased commercial value of fishes would bring in more fishermen and fishermen and would create a system whereby large commercial enterprises take over or exploit the smaller ones. This situation is perfectly illustrated by commercial bushmeat, which profits more to larger traffickers that make the most out of the bushmeat business than local communities that involved in the trade chain (Cowlishaw et al. 2004; De Merode et al. 2004) and increased financial returns drive the demands for bushmeat (De Merode et al. 2004; Rowcliffe et al. 2004); hence increased financial incomes from fish would very likely quadruple the current fishing intensities and drive to further reduced stocks. Stating this obvious reality does not mean to plead for locking out fishing zones throughout the Lake Tumba Landscape but it rather emphasizes the fact that the whole issue of community-related usages of diverse natural resources is not solvable by simple straight-line approaches; the use of natural resources to benefit local communities and communities at large is rather a system issue and should be dealt with as such.

A more important issue for local communities has been the rights on fishing basins, which they claimed to have owned for centuries as felt that current basins were properties of their ancestors for that long. As will be discussed in Chap. 19, the issue of ownership of fishing basins, as it is the case of the land, has several implications for communities and the way they would work toward conserving the biological diversity they shelter. Indeed, one of the reasons why fishing has intensified has been that fishermen and fisherwomen travel long distances from main towns and dwell for long periods where they find better catches per unit efforts (Inogwabini 2013). Because of the ownership regime in the country, local communities do not have any right to prevent these migrant fishing communities to exercise their activities wherever they please. Hence, they contribute to the depleting resources wherever it is available and migrate to other locations. This is, in fact, one of the facets of the reality on fish stocks that have been marginally addressed by Béné et al. (2007) and which warrants further documentation and increased attention.

## References

Aveling C, Botaka B, Hall JS, Hart JA, Hart TB, Inogwabini BI, Plumptre A, Wilkie D (2003) Democratic Republic of Congo Environmental Analysis—Final Report. Type-scripted report submitted to the United States Agency of International Development, Washington DC

Bailey RG (1986) The Zaïre River system. In: Davies BR, Walker KF (eds) The ecology of river systems. Junk Publishers, pp 201–214

Banister KE (1986) Fish of the Zaïre system. In: Davies BR, Walker KF (eds) The ecology of river systems. Junk Publishers, pp 215–224

Béné C, Brummett R, Bungubetshi G, Gordon A (2007) Preliminary assessment of Lake Ntomba fisheries. WorldFish Center: typescript report commissioned by WWF DRC

Butler R (2012) The forest River: river types. http://rainforests.mongabay.com/0602.htm. Accessed 21 Sept 2016

Chapman LJ (2001) Fishes of African rain forests: diverse adaptations to environmental challenge. In: Weber W, White LJT, Vedder A, Naughton-Treves L (eds) African rain forest—ecology and conservation: an interdisciplinary perspective. Yale University Press, pp 263–290

Colom A, Bakanza A, Mundeka J, Hamza T, Ntumbandzondo B (2006) The socio-economic dimensions of the management of biological resources, in the Lac Télé—Lac Tumba Landscape, DRC

Segment: a segment-wide baseline Socio-Economic study's Report. Submitted to the World Wide Fund for Nature

Conference on the Central African Moist Forest Ecosystems (CEFDHAC) (2001) Regional Strategic Action Plan and Biodiversity Resources of the Congo Basin Ecosystems. UNDP/IUCN/GEF, Yaoundé, Cameroon

Corsi F (1984a) Développement des pêches dans le fleuve Zaïre. Rapport préparé pour pour le projet d'Assistance à l'Office National des Pêches et Amélioration de la Pêche Artisanale. Food Aid Organization, Rome

Corsi F (1984b) Zaïre: Développement et aménagement des pêches du lac Mai-Ndombe et ses affluents. Technical report submitted to the Food Aid Organization, Rome

Cowlishaw G, Mendelson S, Rowcliffe JM (2004) The Bushmeat Commodity Chain: patterns of trade and sustainability in a mature urban market in West Africa. ODI Wildlife Policy Briefing 4:1–4

De Merode E (2005) Protected areas and decentralisation in the Democratic Republic of Congo: a case for devolving responsibility to local institutions. In: Homewood K (ed) Rural resources and local livelihoods in Africa. James Currey Publishers, pp 36–58

De Merode E, Homewood K, Cowlishaw G (2004) The value of bushmeat and other wild foods to rural households living in extreme poverty in Democratic Republic of Congo. Biol Cons 118(5):573–581

Dubois JT (1959) Note sur la chimie des eaux du lac Tumba. Note présentée par M.J. Kurfferath). Academie Royale des Sciences d'Outre-Mer. Bulletin des Séances 6:1321–1334

Evrard C (1968) Recherches écologiques sur le peuplement forestier des sols hydromorphes de la Cuvette Centrale congolaise. Série scientifique 110. Office National de la recherche Scientifique et du Développement du Ministère Belge de l'Education Nationale et de la Culture

Gautier-Hion A, Colyn M, Gauthier JP (1999) Histoire naturelle des primates d'Afrique Centrale. Multipress, Ecosystèmes d'Afrique Centrale. Union Douanière et Economique d'Afrique Centrale

Groombridge B (1992) Global biodiversity—Status of the Earth's living resources. The International Union for Conservation of Nature

Hardman M, Stiassny MLJ (2008) A sexually dimorphic species of Chrysichthys (Siluriformes: Claroteidae) from Lac Mai-Ndombe, Democratic Republic of the Congo. Ichthyological Exploration of Freshwaters 19(2):175–184

Inogwabini BI (2005) Common fishes of the Salonga National Park, Democratic Republic of Congo: preliminary survey and conservation issues. Oryx 39(1):78–81

Inogwabini BI (2013) Bushmeat, over fishing and covariates explaining fish abundance declines in the Central Congo Basin. Environ Biol Fishes 97(7):787–796

Inogwabini BI, Dianda M, Lingopa Z (2009) The use breeding sites of Tilapia congica (Thys & van Audenaerde 1960) to delineate conservation sites in the Lake Tumba, Democratic Republic of Congo: toward the conservation of the lake ecosystem. Afr J Ecol 48(3):800–8006

Inogwabini BI (2006a) A preliminary checklist of Mammals and Plants: conservation status of some species in the Salonga National Park, Democratic Republic of Congo. Endangered Species Update 23(3):104–117

Inogwabini BI (2006b) Fire as management tool in cattle raising concession: a preliminary assessment to help develop a system that sustains biological diversity in the southern in the Lac Télé—Lac Tumba Swampy forest, DRC Segment. Field report submitted to WWF and CARPE-USAID

Inogwabini BI, Lingopa Z (2006) Les inventaires des poissons dans le Lac Tumba, Congo et Ngiri, Paysage Lac Télé—Lac Tumba. Segment RDC, Rapport produit dans le cadre du Programme CARPE/USAID Kinshasa

Inogwabini BI, Lingopa Z (2013) Fish species occurrence, estimates and human activities on the islands of the Congo River, Central Africa. Environ Biol Fishes. https://doi.org/10.1007/s10641-0136-4

Inogwabini BI, Lingopa Z (2007) Evaluation de la biodiversité de poissons au lac Maindombe, Nord-Est/Août 2007. Paysage Lac Télé—Lac Tumba, Segment RDC. Rapport produit dans le cadre du Programme CARPE

Inogwabini BI, Abokome M, Kamenge T, Mbende L, Mboka L (2012) Preliminary bonobo and chimpanzee nesting by habitat type in the northern LacTumba Landscape, Democratic Republic of Congo. Afr J Ecol 50(3):285–298

IRD (Institut de Recherche pour le Développement) (2002) Congo River Basin: geology and soil types influence drought. In: European Water Management News, January 16, 2002. http://www.ewaonline.de Accessed 30 Nov 2005

Kambay B (1995) Demain le Zaïre: Choix idéologiques et barrières mentales. Centre de Recherche Eugemonia, Kinshasa

Kamdem-Toham A, D'amico J, Olson D, Blom A, Trowbridge L, Burgess N, Thieme M, Abell R, Carroll RW, Gartlan S, Langrand O (2006) A vision for biodiversity conservation in Central Africa: biological priorities for conservation in the Guinean: Congolian forest and freshwater region. World Wide Fund

Kutshukina FL, Micha JC (2013) Analyse des modes d'exploitation des ressources halieutiques du Lac Mai-Ndombe en République Démocratique du Congo. Geo-Eco-Trop. 37(2):273–284

Leonard C (1987) République du Zaire: Conjoncture Economique (26). Ministère de l'économie et de l'industrie, Imprimerie Pirotte

Lutala M, Kintambu M, Mvudi M (1993) Zaïre. In: Adepoju A (ed) The impact of structural adjustment on the population of Africa. UNFPA, Heinemann and James Currey, pp 130–143

Marlier G (1958) Recherches hydrobiologiques au lac Tumba (Congo Belge, Province de l'Equateur). Hydrobiologia 10(1):352–385

Milton AD (2015) Forest resilience for livelihoods and ecosystem services. PhD Thesis submitted to George Mason University, United States of America

Newbury C (1986) Survival strategies in rural Zaire: realities of coping with crisis. In: Ntalaja N (ed) The crisis in Zaire: myths and realities. Africa World Press, pp 99–112

Programme des Nations Unies pour le Développement (PNUD) and United Nations Office of Project Services (UNOPS) (1998) Programme de national de relance du secteur agricole rural (PNSAR) 1997–2001: Monographie de la Province de l'Equateur. Ministères de l'Agriculture et de l'Elevage, du Plan, de l'Education Nationale et de l'Environnement, Conservation de la Nature, Forêts et Pêche. République Démocratique du Congo

Rowcliffe JM, de Merode E, Cowlishaw G (2004) Do wildlife laws work? Species protection and the application of a prey choice model to poaching decisions. Proc Biol Sci 271:2631–2636

Schliewen U, Stiassny MLJ (2006) A new species of *Nanochromis* (Teleostei: Cichlidae) from Lake Mai-Ndombe, Central Congo Basin, Democratic Republic of Congo. Zootaxa 1169:33–46

Toham KA, Olson D, Blom A, Burgess N, D'Amico J, Abell R, Carroll RW, Gartlan S, Langrand O, Mussavu RM, O'Hara D, Strand H, Thieme M, Trowbridge L (2003) Biological priorities for conservation in the Guinean-Congolian forest and freshwater region. Science 299:346

Welcome R (2005) Impacts of fishing on inland fish populations in Africa. In: Thieme ML, Abell R, Stiassny MLJ, Skelton P, Lehner B, Teugels GG, Dinerstein E, Toham AK, Burgess N, Olson D (eds) Freshwater ecoregions of Africa and Madagascar: a conservation assessment. Island Press

# Chapter 17
# An Abridgement of the Birds Throughout the Diversity of Habitats

**Abstract** Direct observations were used and checked with species described in several ornithological field guides, albeit with only one specifically focused on the Democratic Republic of Congo. The field observation documented 97 bird species belonging to 31 families in North Lake Tumba Landscape (Lake Tumba, the Ubangi River and Ngiri River) and 54 species in the South Lake Tumba. Both the crested guineafowl (*Guttera pucherani*) and the helmeted guineafowl (*Numida meleagris*) were reported in both parts. In the north, the numerically most important order was that of Passeriformes, which counted 11 species in six families (*Corvidae*, *Laniidae*, *Muscicapidae*, *Paridae*, *Ploceidae*, and *Turdidae*). The second most important order was that of Coraciiformes, which comprised 3 families and 9 species and was followed by the order of Ciconiiformes, with its two families (*Ardeidae* and *Scopidae*) and two species (*Ardeola ibis* and *Scopus umbretta*, belonging to the two families, respectively). The chapter attributes differences in species between the northern part and the southern part to the differentiation of habitat types in the south and in the north.

**Keywords** Bird species · Passeriformes · *Melaniparus leucomelas* · *Cercotrichas podobe* · *Psittacus erithacus* · CITES

## 17.1 Introduction

The landscape conservation paradigm needs as many data sets as possible. Put in simple terms, this need is linked to the fact that ideally, the planning process is to include the necessities of each component of biological diversity and each human community. Unfortunately, however, there will be no way to collect data on each individual species and all the wildlife guilds. That is why the idea of umbrella species has been put forward with the idea that protecting the umbrella species would have positive effects on all other species that reside in the same ecosystem as the charismatic species being protected. That is also why throughout the data collecting process, each data set that is specific to any guild can only be warmly hoped for and should be sought for as the more data available the better and the more robust the planning will be. Hence, responsible landscape management should seek to collect as many data



B.-I. Inogwabini, *Reconciling Human Needs and Conserving Biodiversity: Large Landscapes as a New Conservation Paradigm*, Environmental History 12,
https://doi.org/10.1007/978-3-030-38728-0_17

sets as possible. That is what was tried in the context of the Lake Tumba Landscape, with the caveat that nothing was tried in such critical areas as data on amphibians and reptiles. This chapter is brief and it is about the meager data collected on birds of the Lake Tumba Landscape.

## 17.2 How Insights on Birds of the Region Gathered?

The methods used to collect data on birds in the Lake Tumba Landscape were those developed by Inogwabini et al. (2007) for the Bombo–Lumene Game Reserve, which was a modified and contextualized version of the methods used by Dowsett-Lemaire (1995) and, for the specific case of freshwater habitats, by Dodman (1997). It consisted of calmly walking along preestablished trails between 6:00–11:00 and 15:00–17:00 while using Bushnell binoculars to observe birds and systematically taking notes of characteristics of the species. Essential characteristics recorded were body color (or rather assemblage of colors, as it is often the case), beak, legs, morphology, and an estimate of the size. In some instances, vocal calls were also recorded. Of the utmost importance in identifying birds have been the characteristics of the beaks and legs because these parts can provide essential indications on the feeding ecology of the species and the entire species' lifestyle.

Direct observations were then matched with species described in Borrow and Demey (2002), Olivier (1998), Lippens and Wille (1976), and Serle and Morel (1970). While the field guide of Lippens and Wille (1976) was the only document specially focused on the Democratic Republic of Congo, the other field guides were felt to be important complements because they were judged to provide additional information that was not included in Lippens and Wille (1976). Furthermore, given the reality that ecological knowledge has increased since the publication by Lippens and Wille (1976) of their pioneering work, it was only normal to have updated information factored in the work being done in the Lake Tumba Landscape. Hence, field guides that were chosen were so because provides color illustrations and species distribution maps. Color illustrations were felt to be critical because they would help match the species directly spotted by observers and the image they represented. Species distribution maps were important because they can help as a selection criterium where doubts would emerge in matching species with what their illustrations.

Particularly, bird's identifications plates in Borrow and Demey (2002) are followed by descriptions covering voice, habits, similar species, and status; common names in English and French. The field guide by Borrow and Demey (2002) covers most of West Africa (Senegal, Gambia, Guinea-Bissau, Guinea, Sierra Leone, Liberia, Ivory Coast, Burkina Faso, Ghana, Togo, Benin Mauritania, Mali, Niger, and Nigeria) while also providing details on extensive parts of Central Africa where its extent runs from Cape Verde Islands, Bioko and Sao Tome and Principe up to Chad and then down to Cameroon, Central African Republic, Equatorial Guinea, Gabon, and Congo.

Because of the diversity of habitats in the Lake Tumba Landscape, all the preliminary surveys of birds were designed to be linked with habitat types; particularly the northern and the southern parts of the Lake Tumba Landscape. Birds were also dissociated according to the general divide between freshwater habitats and terra firma habitats.

## 17.3 Bird Species in the Region

For the northern part of the Lake Tumba Landscape, which is the region covering Lake Tumba, the Ubangui River and Ngiri River, Inogwabini and Mbenzo (2007) reported 97 bird species belonging to 31 families. The most numerous order was that of Passeriformes, which counted 11 species in six families (*Corvidae*, *Laniidae*, *Muscicapidae*, *Paridae*, *Ploceidae*, and *Turdidae*). The second most important order was that of Coraciiformes, which comprised 3 families and 9 species and was followed by the order of Ciconiiformes, with its two families (*Ardeidae* and *Scopidae*) and two species (*Ardeola ibis* and *Scopus umbretta,* belonging to the two families, respectively). All other orders had one family and either two or one species. Orders with two species were the Accipitriformes (*Gypohierax angolensis* and *Polyboroides radiatus*), the Caprimulgiformes (*Caprimulgus climacurus* and *Caprimulgus inornatus*), the Cuculiformes (*Bubo lacteus* and *Cuculus clamosus*), the Galliformes (*Guttera edouardi* and *Numida meleagris*), and the Piciformes (*Gymnobucco calvus* and *Pogoniulus atro-flavus*). Orders with one species were Apodiformes (*Apus aequatorialis*), Coliiformes (*Colius striatus*), Columbiformes (*Treron australis*), Lybiiformes (*Prodotiscus insignis*), Psittaciformes (*Psittacus erithacus*), and the Strigiformes (*Asio capensis*).

As it has been reported elsewhere (e.g. Brosset and Erard 1986), these bird species were encountered in four major types of habitats and/or specialized niches. The first habitat was that of herbaceous species of Graminae (particularly species of *Panicum maximum* and *Saccharum officinalis*) that were often associated with swamped meadows and other clearings along rivers (Chap. 5) and water pounds. Birds were also found along the rivers and streams and open lakes. This was particularly the case of the African fishing eagle (*Haliaeetus vocifer*) that was frequently observed in both Lake Tumba and Lake Maindombe. In terra firma forests, they occupied different niches such as high treetops and forest canopies and middle storey trees and forest ground floor's shrubs. For the southern part of the Lake Tumba Landscape, particularly around the Malebo region, Colson (2008) identified 54 species using Sinclair and Ryan (2003), in the three main habitats of that region (Chap. 5), which are the mosaic of open woodland savannah, thickets, and forest galleries. Both the crested guineafowl (*Guttera pucherani*) and the helmeted guineafowl (*Numida meleagris*) were reported in both parts of the Lake Tumba Landscape but in substantively more numbers in the south.

## 17.4 Bird Species Described in the Region in the Biodiversity Conservation's Perspectives

Birds are the best-known group of vertebrates (Bennun and Howell 2002). However, bird inventories in the Democratic Republic of Congo, in general, and, particularly, in the Lake Tumba Landscape are very limited in their geographic scopes. Hence, the possibilities of identifying new species remain widely open. This is what happens recently when Kylin et al. (2010) documented the presence of the Nominate Lesser Black-backed Gulls in the Lake Tumba Landscape after a very short stay.

That the most numerous family was that of Passeriformes was hardly surprising. Indeed, previous studies (Lippens and Wille 1976) have given the figure of about 11000 bird species for the Democratic Republic of Congo, with 128 species being migrants. Among these species, *Passeriformes* had a total of 602 species. Hence, the massive presence of this family in the Lake Tumba Landscape is a reflection of the country-wide situation. But that distribution could, itself, be explained by a cohort of different ecological explanations of which the adaptability of species of this family to different ecological niches and conditions would count among the most plausible. Indeed, species of this family had been identified across all habitat types in the landscape, which is a clear indication that the species of this family can survive in different habitats.

That some bird species were observed more in one of the parts of the Lake Tumba Landscape than in the other maybe, obviously, due to the differentiations of habitat types in the south and in the north. Indeed, as indicated in the previous section, both the crested and the helmeted guineafowls were present in both parts of the Lake Tumba Landscape; but they were more frequently encountered in the south than in the north. These species happen throughout the sub-saharan Africa but prefer open habitats where significant quantities of sunrays reach the ground; these conditions are perfectly matched with conditions in the southern part of the Lake Tumba Landscape where the habitat has been described as a combination of forest galleries and the mosaic of open woodland savannah. Indeed, thickets found in some of the areas in this part of the Lake Tumba Landscape resemble with characteristic habitats of the southern Africa savannahs where the two species proliferate. The resemblances of the habitats in the Southern Lake Tumba and the larger southern Africa savannahs can also explain why a species such as the white-winged black tit (*Melaniparus leucomelas*) whose distribution is essentially in the Southern and Eastern Africa savannahs has been documented to be present in this location. The same reasoning applies for the black scrub robin (*Cercotrichas podobe*), which does occur principally in dry areas of the Sahel and other countries where the habitats are dry savannahs.

Of importance among the species described in the Lake Tumba Landscape was the presence of the African grey parrot (*Psittacus erithacus*), a species that has been being proposed to be moved from Category II to Category I of the IUCN Convention on International Trades in Endangered species of wildlife and plant species by the World Parrot Trust (WPT) and WCS (2016). WPT and WCS (2016) argue that this demand has been supported by the marked declines in grey parrot populations observed over the last decades; which declines are driven by patterns of exploitation and decreases in habitat quality and area. WPT and WCS (2016) argued that in 2012 it was uplisted to Vulnerable on the IUCN Red List of Threatened Species on the basis that 'the extent of the annual harvest for international trade, in combination with the rate of ongoing habitat loss, means it is now suspected to be undergoing rapid declines over three generations whereas recent accounts indicate that the species population declined by more than 50% within the same period in many range states. A generation of parrots is about 15 years.

The plight of the grey parrot is well known in the Democratic Republic of Congo though there is very limited scientific documentation of their conservation status in the field. The limited recent evidence from the field is, however, that the species has been being exploited throughout the country in order to feed the national and international markets of parrots (Hart et al. 2015). That quandary has several explanations, first among them being the market of the species which, as it is the case for other species of the wildlife in the Democratic Republic of Congo, is very porous and law enforcement mechanisms very weak. The second cause of the dim situation of grey parrot in the country is that the species' habitat is currently under diverse forms of assaults, from people leaving their villages rendered insecure by years of conflicts to establish new villages in remote and safer areas to logging industry chopping down extensive areas that were previously more or less pristine. These factors are combined with the reality of the Congolese rapidly expanding populations, which are in need of more space and more lucrative activities to sustain their livelihoods. In essence, causes for the current declining of populations of the grey parrots in the Democratic Republic of Congo are the very same as those that apply to other species; if there is a need to sort these out, they should be done globally.

## References


Bennun LA, Howell K (2002) Bird sampling techniques. In: Davies EG (ed) African forest biodiversity: a field survey manual for vertebrates. Earthwatch Institute (Europe), Cambridge, pp 131–161

Borrow N, Demey R (2002) A guide to the birds of Western Africa. Princeton University Press

Brosset A, Erard C (1986) Les oiseaux des régions forestières du nord-est du Gabon. Volume 1: Ecologie et comportement des espèces. Société Nationale de Protection de la Nature

Colson F. (2008) Bonobo Observation Research Stay 20.05 – 24.06.2008—Final Report Submitted to WF-Belgium and WWF-DRC as part of the cooperative volunteering work.

Dodman T (1997) A Preliminary Waterbird Monitoring Strategy for Africa. Wetlands International Publication 43, Wetlands International Wageningen, The Netherlands

Dowsett-Lemaire F (1995) Inventaire Ornithologique du Parc National d'Odzala. Ecosystème Forestier d'Afrique Centrale (ECOFAC). République du Congo. Groupement Agreco/CTFT

Hart J, Barnard A, Abani R, Salumu L, Martin RO (2015) Harvesting and trade of Grey Parrots in DR Congo: is sustainable exploitation possible? In: Presentation to symposium 'Towards a sustainable and legal wildlife trade', 18–19 June 2015. Durrell Institute for Conservation and Ecology. School of Anthropology and Conservation, University of Kent at Canterbury, United Kingdom

Inogwabini BI and Mbenzo A. (2007). Le rapport technique sur les observations des oiseaux dans le Landscape Lac Tumba. Rapport soumis a WWF – US, Washington DC et USAID CARPE, Kinshasa, Democratic Republic of Congo.
Inogwabini BI, Amube J, Bagurubumwe N, Bakanseka JM, Bakanza A, Basinyire P, Bikanza M, Bikumbu H, Bongwele E, Chibembe A, Iyomi BI, Karamba J, Kasongo N, Kayeye H, Lola P, Madidi J, Mafuta N, Mangoni G, Masasu F, Mayifilua J, Mbende ML, Mbenzo PA, Ngomba MN, Nlemvo B, Ntumba B, Ohole AN, Pasi MS, Yata F, Zanga NL (2007) Etat d'habitats, distribution et abondance de la faune et état de la conservation du Domaine de chasse et Réserve de Bombo Lumene, République Démocratique du Congo. Rapport scientifique de la formation Dans le cadre du Programme des systèmes de gestion d'information sur les aires (SYGIAP). World Wide Fund for Nature, Democratic Republic of Congo
Kylin H, Louette M, Herroelen P, Bouwman H (2010) Nominate Lesser Black-backed Gulls (Larus fuscus fuscus) winter in the Congo basin. Ornis Fennica 87:106–113
Lippens L, Wille H (1976) Les oiseaux du Zaïre. Belgique, Linod Tielt, Bruxelles
Olivier G (1998) Echassiers, Canards, et Limicoles de l'ouest Africain. Castel, Paris, France
Serle W, Morel GJ (1977) A field guide to the birds of West Africa. Collins
Sinclair I, Ryan P (2003) Birds of Africa South of the Sahara. Princeton University Press
World Parrot Trust (WPT) and Wildlife Conservation Society (WCS) (2016) Proposal to transfer African Grey Parrots (Psittacus erithacus) from CITES Appendix IIto Appendix I. FAQ.

# Part III
# Not Only Biodiversity but Also Human Communities

Conservation biology is about both biodiversity and people though the earliest versions of biodiversity conservation endeavors have tempted to propagate the opposition between these sides of the same coin. Indeed, this view of conservation is being left being despite the existence of tendencies that continue to see biodiversity conservation as essentially a matter of preserving species even at direst costs for human communities. Landscape approaches, as detailed in Chaps. 1 and 2, are to help biodiversity conservation community to embrace the new attitudes, to ensure that human communities are becoming seen more and more as part of the solution rather than being the sole conservation problem. That is what this part of the book intends to do: show how human local communities can become part of the design making processes and how they can benefit from biodiversity conservation activities. The part starts with Chap. 18, contrary to the traditional approaches, by presenting a global reading of the political economy of the landscape. In the traditional approaches, this part of the book would be presented as a socioeconomic monograph of the Lake Tumba Landscape whereby threats posed by communities to the biological diversity of the area would be given and solutions to each of the identified threats elaborated. The monograph would then proceed to give the list of the tabulated needs of communities and potential to ensure that these needs are taken care of or substitutes provided. The idea of starting with a somewhat full-fledged Political Economy of the Lake Tumba Landscape deviates from that tradition for several good reasons. The first obvious reason is that Landscapes, as they are envisioned in the context of Central Africa, are larger than protected areas in their geographic extent and call for different and enlarged conceptual frameworks. Viewing things in terms of threats versus threat mitigation actions is but very basic to embrace issues at the scales of landscapes. Secondly, landscapes are not only larger than protected areas but are also territorial administrative management units, which have more than just one cultural, economic, and social objective. Because of this reality, the relevance of limited analytical framework is, logically, nearly obsolete and questionable because limiting the human–biodiversity interactions under the threat versus threat mitigation mindsets ignores all other relevant issues that are equally morally justifiable such as the needs for local communities to sustainably live on the natural resources located where they happen to occur, as discussed in Chap. 3. Thirdly, as was discussed in Chap. 15, threats often described for species are very limited and do not account for all the penalties that seemingly non-harmful activities pose on entire systems. This too calls for a broadened approach. The needed enlargement of the thinking around the biodiversity conservation that is imposed by the scope and magnitude of biodiversity conservation in a Landscape needs also a tool that can embrace larger cultural, economic, and social issues; the enlargement of the thinking demands more than just tabulated human needs. What is needed is a framework whereby the inter-activities between different components of the landscape and different stakeholders are brought to light by a thorough analysis, which the global political economy of the landscape seems to offer. Indeed, the political economy framework, as enlarged to incorporate the needs for protecting biological diversity, is ample enough to include issues that are traditionally avoided by the simplifications imposed by tabulations and limited focus on species. In fact, the value of using the political economy analysis in this context has been revealed when applying methods such as reflected image whereby absent stakeholders are factored in the analysis and their roles in conserving biodiversity or the impacts of their activities on the global livelihoods of the Landscape depicted with some degrees of clarity that tabulating threats would not arrive at.

The second chapter of this part (Chap. 19) comes straight because of the global political economy analysis which precedes it. Indeed, for communities throughout the Lake Tumba Landscape, the most important gain they felt like being made from the entire Landscape approach was around the issue of legal access to and legal ownership of natural resources. Chief among the natural resources of the Lake Tumba Landscape was land, which also includes water. Chapter 19 deals with this vexing issue at both the theoretical and the practical points of view. In so doing, it does show the limits of theoretical approaches that are not grounded on true field assessments. In fact, a major lesson to be learnt from confronting the two is that theoretical approaches to the issue of land should be used only as a basic level from which to start discussions with all parties. This chapter is a great example of why the one-size-fits-all models would not work in many circumstances dealing with natural resources even in the same landscape. Of course, the Lake Tumba Landscape crosses many different cultural nexuses and different biological profiles, which may be the reason why what is planned in the northern part is not necessarily acceptable in the southern part of the landscape. However, the reality is also that major parts of communities in the Lake Tumba Landscape do also share extensive cultural commonalities. What differentiates them in terms of needs and plans for livelihoods would be reduced to differences in the geography of the locations where they reside and the types of economic activities they perform in these areas. A good example of this situation is that while communities in the southern part of the Lake Tumba Landscape allocate greater proportions of their lands for cattle husbandry, those in the north have no such land function. Conversely, while the communities in the northern part of the Lake Tumba Landscape have allocated a greater chunk of their land as fishing basins, the communities in the south do not have such functional units. This leads to the conclusion that that geography or rather environment does influence human activities.

# Chapter 18
# The Political Economy of the Landscape

**Abstract** For landscapes to conserve and take care of needs of local communities, understanding the total local economy of the landscape is imperative. Hence, to know who is where and who has what stake in the landscape is so important. Equally important are the knowledge of types of natural assets present in the landscape and what asset contributes to fulfill what need. Weighting values of that stakeholder's place on assets, finally, is also important to understand the dynamics of the landscape and how that would, ultimately, affect biodiversity. Assessing who is where and who has what weight in the Lake Tumba Landscape, the chapter describes an ensemble of 150 organizations located within the Lake Tumba Landscape with 69% made of national and local non-governmental structures and 13% of churches. The study identified 9% being international structures, including the UN Agencies, whereas the state was among the minority of structures. Power-wise, churches, particularly the Catholic Church, were the most important force; and lack of political leadership has also meant diminishing sense of purpose as a community at both the national and local levels. Traditional authority and the churches filled the vacuum left by the state in providing the sense of purpose for whole communities. Churches and ambulant merchants were found to possess most of the economic power: The message conveyed by the chapter is that to conserve biodiversity; these non-traditional biodiversity conservation actors have to become an essential part of the biodiversity conservation equation.

**Keywords** Political economy · Landscape total economy · Stakeholders' analysis · Power analysis · Local needs

## 18.1 Introduction

To make a sensible use of the landscape approach as conservation paradigm that ensures that biodiversity is conserved while local needs are taken care of, it is imperative to understand the total economy of the landscape itself and how local communities view their own livelihood. In this perspective, the exercise should be done in three major parts: the first one is on who is where and has what stake in the landscape, the second is about what asset contributes to fulfill what need and the third part is



B.-I. Inogwabini, *Reconciling Human Needs and Conserving Biodiversity: Large Landscapes as a New Conservation Paradigm*, Environmental History 12,
https://doi.org/10.1007/978-3-030-38728-0_18

putting everything of the two first parts together to gather the sense of where the weighted values lay.

Obviously, the statement above is what is meant to be done by the Political Economy Analysis. In its formal definition, the practical Political Economy, which is what is needed in the context of the landscapes, is the analysis of who gains what after losing what (Chambers 1983) and what would be the total gains. Hence, this chapter would first discuss issues of who is where in the Lake Tumba Landscape before moving on to the who has what stake where and what each of the stakeholders would gain and lose and what their respective total gains would look like when they embark in the process of the landscape. In order to get a better understanding of the total weighted strengths of each stakeholders, the first part of this analysis needs to go beyond the numbers of existing organizations and structures in the landscape; it has to look deeply into the structures of the powers and how different partial powers come together to determine the directions of affairs in the Lake Tumba Landscape. There is no better place, in the context of the Democratic Republic of Congo, to start the second part of the exercise but to look at the use of non-timber forest products and its contribution to the general economy (including both the informal and the formal economies). The final part of the chapter will pull together the weighted powers and the economy of non-timber forest products to provide a general perspective of who gains what and who loses what throughout the process of implementing the landscape as a conservation paradigm. Indeed, there has been a tendency to promote landscape approach as a win-win kind of trad between biodiversity conservation and human needs for the communities in Central Africa. The reality is that this is not so for everything, there are things that biodiversity conservation has to give up in order to win on other issues, which is also the case for local communities. This chapter will ultimately come up with these in order to lay bare the constraints being placed on each stakeholder. This is but a normal outcome, those who have had the chance to attend the training on the Microeconomy of Biodiversity Conservation delivered by the Woods Institute of the Stanford University must have learned from that course that the first law of economy is known as 'the no free cup of coffee' rule and this basically says that whatever you end up gaining comes with a price, as taught by Professor David Johnson from Harvard University. Professor Johnson also adds that free riders are, at the end of all, either forced out of the system by being rejected from the system for good or are doomed to pay if they want to continue to be part of the society.

## 18.2 Defining the Concept of Stakeholder for the Context of Biodiversity Conservation

Before going any further with the political economy analysis, it is important to define the concept of stakeholder broadly and as it is understood in the landscape approach. Reformulating the definition given by Freeman (1984) to the context of biodiversity landscapes, a stakeholder is any group or individual who can affect or is affected by the achievement of biodiversity conservation's objectives. However, as Mitchell et al. (1997) say, a review of the theory of stakeholders, offers a maddening variety of methods to identify stakeholders, which leads to the need to contextualize the concept. Hence, to identify stakeholders in the context of a biodiversity context, the most appropriate place from which to start is in defining what and who affect and what and who are affected by the achievement of biodiversity conservation's objectives. Who and what affect the achievement of biodiversity conservation's objectives fall within the area technically known as threats to biodiversity. Hence, the approach for the biodiversity conservation landscapes has focused primarily on defining threats to biodiversity as a way to identifying who or what really counts.

In the context of the Congo Basin Forests Partnership threats to biodiversity are primarily anthropogenic threats. There are philosophical problems with this way of looking at things because it makes humans look like being outside of the ecosystems. Indeed, while having something more than being just another species, humans are also part of the ecosystems in which they happen to occur and where they have their activities, which is nearly everywhere on earth. Extracting them out of the whole system is partly the reason why biodiversity conservation efforts have always been felt as opposing the need for communities. This does not have to be so and that is why the landscape approach is welcomed.

That requalification being made, in the conceptual mindset of the Congo Basin Forests Partnership threats to biodiversity, as defined above, is categorized as direct and indirect. Direct threats identified by the Congo Basin Forests Partnership include poaching and its consequent trade in bushmeat, agriculture, unplanned and unmanaged logging, and over-fishing. Indirect threats, in the same conceptual mindset, include human conflict and subsequent displacements of populations, growths in human populations, corruption and prevalent bad good governance, lack of awareness of the scale of problems, and lack of capacity of national and local institutions, Non-Governmental Organizations (NGOs) and Community-Based Organizations (CBOs) to manage national natural resources. A threat to biodiversity can be or may become a threat to local populations' well-being. In other cases, a search for subsistence opportunities and the absence or disappearance of non-extractive economic alternatives may become or may pose already a threat to the sustainability of local natural resources.

With the above precisions made, the stakeholders in the Lake Tumba Landscape were communities, groups, and individuals that could influence one of the defining parameters of threat to biodiversity in one direction or the other. Hence, the landscape's socioeconomic study aimed at understanding how each of communities, groups, and individuals understand their environments (local, national, and international), the mechanisms they use to extract natural resources and the ways in which they manage these resources. Stakes were defined in such a way that everything that would sustain the well-being of a community, an interest-group or an individual would be addressed as an important reason for that specific community, interest-group, or individual to fight for the access to that resource should remain open. But what the landscape approach has brought in the discussion was to ensure not only access but also the availability of the given resource for a much longer period of time. Hence, discussions were first centered on what communities, interest-groups, and individuals see as their present conditions and what future threats to their livelihoods they foresaw in different time horizons. Understanding the varied contexts in which every stakeholder operated was an essential part of the exercise. Then, discussions moved on to changes and adaptations communities, interest-groups, and individuals would have to operate in their own conducts to ensure that natural resources their livelihoods and other economic activities dependent on to be maintained over a long time as possible. But doing so also preordained to look at existing levels of organization at the village, tribal and territorial level, and the type of dynamics between internal and external drivers of change, local, and non-local stakeholders. Understanding how local communities, interest-groups, and individuals perceive and use their natural environment was important because studies (e.g. Almquist et al. 1993) have shown that views held by communities correlated with expectations on economic development they held. So, these views and expectations would have to be included in sustainable management plans to help adapt conservation and to the landscape's social reality.

## 18.3 Methodological Approach for the Political Economy Analysis

As methodological approach, the Political Economy Analysis for the Lake Tumba Landscape started with the study of the spatial distribution of livelihood activities, which was mapped with the inclusion of government, national researchers, and Non-Governmental Organizations. Employing appropriate Rapid Rural Appraisal techniques (Pretty et al. 1995), communities whose members depend on goods and services originating from selected water and forest areas were identified. Having identified communities associated with natural resources, a process of Participatory Community Appraisal (PCA) was undertaken in at least 30% of landscape segment villages thought to be representative communities, disaggregated by wealth, gender, and ethnicity status, in each site to assess the role of natural resources derived goods and services in livelihoods of local people; the vulnerability context of each community and an assessment of assets was established. This included a Poverty and Wealth Analysis (PWA) (Grandin 1988) to assure local meanings of well-being and poverty were taken into account in assessment of assets and benefits from the fisheries. A Gender Framework Analysis will be used to address specific gender issues. Institutional analysis was undertaken to assess relationships, motivations, and legal frameworks governing natural resources exploitation, including water and forest systems exploitation and management. To identify opportunities for capacity building in current policy formulation and linkages were investigated and interactions between local and centralized decision-making were understood; stakeholder analysis, an assessment of Community-Based Organizations, Non-Governmental Organizations and institutions, and overview of market networks was used to investigate the role of power relations, market forces and processes related to ecosystem functioning, livelihood strategies, institutional arrangements, and the representativeness of community institutions to all local user groups. Outcomes were presented to local, national, and regional stakeholders (including the USAID-funded Central African Regional Program on Environment and relevant ministries etc.) in workshops at each study site and feedback used to check and verify findings. Understanding of existing data collection programs and, where applicable, recording by local communities was useful in developing the Pilot Conservation Program and the inventory and indicators for assessments of the successes of the landscape as biodiversity conservation paradigm. This assessment helped craft the following views on the total political economy of the landscape and ensured that the strategy to approach all stakeholders operating within the Lake Tumba Landscape was smoother and acceptable by all.

Using the framework provided by Mitchell et al. (1997), the analysis of stakeholders was conducted using the three major categories, which are constituted of power—legitimacy-urgency. As will be detailed below, the stakeholders with power are institutions or individuals with different attributes of power (legal authority, financial power and positional (social capital) and individuals, or groups with knowledge. Building on the listing of institutions, the role of key actors and stakeholders in regulating and managing the natural resources was being assessed, as was relevant policy and legislation both at the regional and national levels (see Chap. 4). Hence, with regard to the power, the analysis of stakeholders, run in parallel with the investigation of the national policies, laws and decision-making processes of access and use of natural resources. The depth of investigation enabled identification of tensions between local, national and international policy and legislation and existing natural resources management and current realities and practices that concrete stakeholders and communities, in particular, witness on ground. The aim of doing so was to identify areas of potential opportunities to reconcile identified problems before embarking on the process of convening stakeholders.

Identified areas of potential opportunities to reconcile identified problems included the formulation of agreed-upon action plans, identification of appropriate methods, and approaches through the association of different stakeholders, including civil society and local community and government representatives. Indeed, stakeholders identified as possible strategies to achieving the action plans to include workshops, focus groups and in-depth and topic-specific discussions. Considering the need to reconcile multiple uses, an assessment was made, incorporating outcomes of the assessment of current uses and users of natural resources in the Lake Tumba Landscape. Secondary data and stakeholder enquiry, based on Participatory Rural Appraisal techniques, were used to assess how access and use patterns have changed over time, what events or factors moderated this change, and what changes to access and uses are foreseen for the future. To provide a basis for negotiation and to ascertain the perceptions of different people concerning the different uses of natural resources, work has been undertaken with users, local communities, government representatives, the military and logging operators to assess the value ascribed to functions, goods and services (including non-use values) derived from natural resources, and the multiplicity of ecosystems. The perceived impact of factors such as environmental degradation, changing demand for goods and services, and modified management practices on these values were also assessed in some significant details.

## 18.4 Who Is Where and Who Has What Weight in the Lake Tumba Landscape?

The first thing that was done to identify different stakeholders present in the Lake Tumba Landscape was a simple listing of villages, centers of administrative sectors and administrative territories, and other major towns. Then the listing continued to collect names of state institutions, non-governmental organizations, industrial companies, traders, and any other form of activities. The analysis of power was then run through by use of different techniques. Power was analyzed taking into account its different forms: the positional power, the financial power, the internal power, and the knowledge power. Indeed, whoever has had a practical experience interacting with different forms of power would recognize these four essential forms; all other forms can be one way or the other reduced to these essential forms.

The assessment of who is where has led to identify an ensemble of 150 organizations located within the Lake Tumba Landscape. The majority (104, representing 69%) of this large number of organizations were made of national and local non-governmental structures. The second most important group was composed of the 13% that composed different structures belonging to churches. The third largest group was made of the 9% of the international structures present in the landscape, including the UN Agencies (FAO, UNICEF, UNOPS, and the World Food Program). The remaining 8% were essentially the administrative and state structures but also include industrial actors such as logging companies. This dispatch, however, is dependent on the number only and does not convey the weighted balance of power within the Lake Tumba Landscape. Indeed, the fact that the majority of stakeholders is made of local and national non-governmental organizations should not be surprising at all; it rather testifies of a new mentality born from the decade-long crisis; lack of employment and the influxes of international agencies and aid organizations have pushed communities to create many parallel structures to capture whatever would be brought in by mechanisms of aid. Of course, as discussed below, lack of community structuration has been identified truly as one of the major problems in delivering the aid and planning for development. And there is a need for communities to be structured in such a way that their voices are being channeled in a manner that would be as coherent as possible and the issues of common interests treated as appropriately as possible. Despite that need, however, the long economic crisis and the deficiency of opportunities that it created has resulted in a mentality that could be lethal for the health of a truly participative management approach; that mentality is the fact that communities are organized around structures that are dusted everywhere in the country, and particularly in the landscape, with the expectations that the international structures would bring in the aid to support. There is, of course, no need to criticize the aid that has been provided for the less abled people to cope with the crisis by the international structures; but the adverse effects this sustained aid created have led to a situation whereby incentives to work hard or to cohere in smaller numbers of structures would need to be very strong to move whole communities out of the current attitude. This is even so when one considers the weaknesses of the State of the Democratic Republic of Congo to deliver even the very basic services to its communities.

The presence of the scattered local and national non-governmental structures across the landscape does not necessarily imply they have that much of the power. Financial, the power lies not even with the state structures. Indeed, the administrative organization of the country is such that in each village, in each administrative sector and territory as well as in each province, there are political and administrative authorities who have law enforcement forces and other security services in their command. *De jure*, these authorities have the effectiveness of the legal power. Indeed, they often use that positional power whenever they want to use it and often negatively to stop something from happening or to coercively demand whatever they believe is their right. In essence, the positional power in the Lake Tumba Landscape could be said being shared between the state institutions, the churches, and the UN Agencies that are present in the area. The true center being around the state institutions, particularly different security services, inclusive of the civil intelligence. But where military and civil intelligence services are in cohabitation, the military intelligence service has an upper hand on everything and can become a much more difficult entity to deal with. In theory, however, the positional power can be said to include civilian administrative and political authorities at its core though in terms of exercising the imperatum, it would be right to state that the true positional force centers around intelligence services then at the second circle comes the administrative and political authority. The UN Agencies, as indicated above, can also be said to have a positional power which is linked to the prestige of that their mission confers to them and the fact that they can actually meet with different intelligence services, the administrative and political authorities to discuss issues they want. Apart from these three concentric circles, there is also a niche in the fourth concentric center that is filled by the international structures present in the landscape. This positional power is effective because of the linkages these structures have with the UN Agencies within the landscape and the fact that they are often seen as representing the interests of their countries within the landscape. Indeed, numerous projects funded within the Lake Tumba Landscape between 2002 and 2010 were funded by the US Agency for the International Development, the Belgian Cooperation and other countries. The projects were part of the portfolios funded by either the bilateral or multilateral aid mechanisms. It was just awesome, for example, to see how many flags of the USA were displayed in the form of advertising signs across the landscape as part of the communication and visibility strategies of each of the implementing organizations. These advertising signs, as displayed, convey a strong message of how closer the international agency implementing projects are with the donors. In a sense, they conveyed a message that the power that the donors hold by their stronger position was in some ways transferred to those who were doing the fieldwork. In that power scenery, very little was left as real power to communities. De facto, however, a much larger part of the positional power in the Lake Tumba is with the traditional authority, which kept its historical prerogatives despite years and efforts to dismantle it by the colonial and the current modern powers. In fact, a combination of factors ensures that traditional leadership is still highly regarded: one of these is the sentiment of vacuum that has been created by year of neglect by the central government of the Democratic Republic of Congo. Years of lack of clear political leadership and the weak presence of legal institutions in most of the country have left communities fare for themselves and, because nature abhors a vacuum, the natural authority to fill in the gap left open was the recourse to the traditional authority. The traditional authority provides a sense of security in combination with that of unity and purpose; within the war circumstances where lack of protection cornered people back to seeking the sense of their identity in whatever would provide it, such a prominent institution was at the easiest reach of people and provided the feeling of being part of community.

Lack of political leadership has also meant diminishing sense of purpose as a community at both the national and local levels. While the traditional authority has been filling the vacuum left in the security and the sense of belonging, the churches have taken the niche that provide the sense of purpose for whole communities. Indeed, in the Lake Tumba Landscape, there is nowhere where the Catholic Church does have either a physical presence or that of its influence. In fact, as will be discussed below, the Catholic Church has taken the true place of the state and has become so influential that there is nearly nothing that can be achieved within the landscape without it being part of the negotiations. Indeed, one would argue that this does only reflect the general reality of the country, which is that the Catholic Church has become an unavoidable stakeholder even if sometimes its princes seem to be following the same path as the laymen or wormy politicians. Also, it should be mentioned that the very logic that brought back the traditional authority to management of the communities acted also to bring churches and faith-based organizations at the front line of the social power, which is designated here as the positional power. Indeed, over the last 3 decades, evangelical churches and Muslim faith organizations have emerged and rapidly spread in the Democratic Republic of Congo and have been occupying the social niches that were previously occupied singly by the Catholic Church and multiple denominations of the Conventional Protestant Churches. The scramble for converting new believers has been accompanied by a message of immediate salvation, which rings loudly in the ears of Congolese citizens. Positionally, speaking, these churches and other faith-based organizations are not far behind the Catholic Church or its counterparts in the Conventional Protestant Churches. Taken together, churches and other faith-based organizations such as Muslim Organizations and Black-Nationalistic oriented believers are a major positional (social) power; in some areas, they have way stronger impacts on communities than would have entitled political leaders.

Stating that international non-governmental structures that are present in the Lake Tumba Landscape had more power than communities they are meant to work with is also grounded to the fact that positional power alone is not sufficient in itself to make things happen. One has to remember the motto that 'mission equals means'. Without means, even the positional power becomes at best very diminished. This is what happens in the Lake Tumba Landscape when it comes to executing true sustainable development activities. In this instance, the positional power held by local administrative and political authorities as well as the intelligence services do no longer match what other stakeholders have as power. There is a total shift of power when it comes to financial power, which is essentially held by international actors. Looking from this angle, the most powerful stakeholders in the Lake Tumba Landscape have been different UN Agencies who had means to travel wherever they wanted, to provide support that communities needed on health, routes, communication… and even electricity in some areas. Limitations of the state agencies to provide basic services to communities dilute their positional powers and make them a lot powerless when it comes to discussions on what people need and through which process they should work to get what they need. In a substantial way, lack of capacities to deliver basic services acts as a limiting factor to the exercise of the de jure power that the authorities have. The second major force in the Lake Tumba Landscape, financially speaking has been the logging companies and industrial cattle raising companies (these ones being present mostly in the southern part of the landscape). To state would seem very striking as this category has been lumped with the administrative and political structures of the State because its numbers are rather very low. In total, there were only 3–4 significant companies in the overall landscape. But despite this very low figure, logging companies can be stated to employ more than 30% of the total number of employed people within the Lake Tumba Landscape; they own most of the existing infrastructure and can literally buy anything from local administrative authorities. To give a measure of how powerful these companies are in the landscape, in some camps, it has been a surprise to find that the immigration authorities dispatched the immigration agents who are entirely dependent of logging companies and their bosses. Logging companies have been also influencing international organizations working in the landscape and in some instances, they have had very strong influence on the directions to take by conservation activities. As they can provide means of movements and means to live on to political and military leaderships, they can make and can also undo decisions taken by other stakeholders, should they wish to do so.

What would be surprising after reading the first paragraph of this chapter is that nowhere does the local civil society structures emerge as having a real power, except their distribution across the entire landscape. Even when one takes the knowledge power, this is centered on the Catholic Church, which has been building its own knowledge of the region where the Lake Tumba Landscape is located even before the independence of the country. The Catholic Church has a research center dedicated to anthropology, history, and social sciences in the middle of the Lake Tumba Landscape, located at Mbandaka. That center has been collecting data for decades and has accumulated the knowledge that no other single institution could claim. Of course, the acquisition of the first-hand knowledge forced by the physical presence on ground has also made it that the Catholic Church shares its substantial knowledge power with the Conventional Protestant denominations that have also gathered knowledge of where they are present over decades. What worries one when looking for local knowledge in the Lake Tumba Landscape besides Catholic and Protestant Parishes is the fact the administration has literally no records. There are no archives as simple as that and everyone that manages a portion of the territory takes whatever he might have found in the office. The knowledge of the churches (Catholic and Conventional Protestant) does also has some gaps, particularly in the area of natural resources. This fact is easy to understand since the churches traditionally have been investing efforts to understand communities and sociological structures as part of their own campaign for converting followers to their own creeds and not necessarily to study natural sciences. Besides that though the fact that some studies in Anthropology, history, philosophy, and sociology that were conducted through Annales Aequatoria, the Catholic Media of publication of their research in that part of the country, has been closely connected with the natural resources, as can be demonstrated by the work such as those of Hulstaert (1979).

Apart from the Catholic Aequatoria Center, there are also some research institutions present in the Lake Tumba Landscape; they include basically some higher education institutions such as the Institut Pédgagoque de Mbandaka and the Institut Supérieur de Développement Rural, the oldest higher education institutions in the region, which has been since some years now competing with newer ones, including University of Mbandaka, the Institute of Medical Sciences. There is also one institution that was essentially created by the Belgians to conduct ecological research in the Congo Basin, the Mabali Scientific Research Center. Unfortunately, however, the existing of these institutions seems to be much more of token presence, the resources available do not permit the academic communities in these institutions to do proper research. Despite that, nevertheless, some people who are affiliated to these institutions are often used by some international institutions to conduct research. This is particularly the case of the Mabali Scientific Research Center that has been benefiting from the collaboration of some institutions for years. But the research conducted through that scheme is not properly owned by the Congolese participants. They are often used only to collect data for foreign students and other researchers. While individuals implicated in such a scheme can personally benefit both materially and academically, the cooperation has not brought these institutions up to the current international scientific research standards. Indeed, the international institutions collaborating with national institutions based in the Lake Tumba Landscape own data, conduct analysis, and get the results published out in different international peer-reviewed journals without often even including the very people that collected data that they publish. This is just to emphasize the fact that knowledge power in the landscape, apart from the Catholic Church and to some limited extents the Conventional Protestant Denominations, seats with international organizations, be they biodiversity conservation or academic research ones.

When you weigh all the forms of power (positional, material, intellectual, and internal), you end up clearly identifying that the strongest stakeholders are the Churches; you might want to read Catholic Church and the Conventional Protestant Denominations: they hold a position that none other stakeholder has, they have some bits of material power and are possessors of the knowledge of the communities and environments in which they have been present for so long. Most of their leadership is also thought to provide the moral direction to communities; hence they could be said to have, at least nominally, the internal moral power as well. Of course, this cannot be said of all the evangelical and episcopalean churches that are literally springing up like mushrooms across the country; these are powerful too but they lack some of the intellectual and financial attributes of power. Indeed, their strengths literally reside in the fact that they have the messianic message that can move people beyond what they can do in normal times. So, churches combined together are the major powerful type of stakeholders but unfortunately, they use that power for different motives, including some that are not always so laudable. That said, it should immediately also indicate that not all the logging companies are that corrupt or want to corrupt; the situation a lot more complex than just pointing the fingers to the people who have the financial power and are inclined to use it for their own benefits. In fact, years of lack of true political leadership, mismanagement, and poor salary scales have left the country in a situation where each person with even a very minimal allotment of *de jure* authority would use it, in diversified manners in a very often negative way to try to extract everything from anyone. The ones with a large portion of the authority would even use massive means of coercion to get what they want. The situation in the Lake Tumba Landscape in this sense is but a reflection of what the situation is globally in the country. Looked from this angle, one would argue that years of centralized management of the country have had a very deep effect in the administration of the entire nation and that the powers that remain for provinces and other administrative and political entities are nearly nill. But, once more, the situation is much more complex than this obvious statement. In fact, there is a gradient of political and state authority that decreases as one moves out of Kinshasa, where the true center of power is located. The further one moves, the less and less state authority he (she) feels. This is what leads to the sense of chaos and each person that has a small share of the de jure authority becomes truly a small chief of his (her) fiefdom. That is why in the middle of nowhere you would often come across an administrative agent that can stop you from moving any further. Indeed, the Democratic Republic of Congo is one of those rarest countries where the immigration is located almost at any place.

Apart from the churches, logging companies have a huge material (financial) power; they can almost buy whatever they like within the landscape, including injust legal favors as well as their own way out when everything goes wrong. Indeed, it is very current to see the political and administrative authorities being plainly pocketed by the wealthiest people in the Lake Tumba Landscape and particularly the owners or the representatives of some logging companies. This is easy to understand because when the political and administrative authorities rely on individuals to either move around or simply gain something for their immediate survival needs, the authorities are weakened and become prey to the will of those who have more resources and can accommodate their daily needs. However, the situation is trickier than it first appears, with the statements above about the behavior of the political and administrative agents who try to extort any kind of benefits from the weakest it would also become clearer why some logging companies and other businesses in the Lake Tumba Landscape have been frequently using their financial powers to get easy movements and conduct their work. This though should not absolve everything that is linked to the natural propension of the strongest to use their muscles to gain what is at the stake.

After the churches, the logging companies, there is a stakeholder category that is multiple in its forms and motivations. That category stakeholder is the conglomerate made of political parties. Because there are so many political parties in the Democratic Republic of Congo, which is reflected on the situation of the Lake Tumba Landscape, a decision was made at the stage of the study design to take note of political parties without necessarily naming them unless there was something significantly important to record about one or two. At the time the study was conducted, myriads of political groups were encountered throughout the landscape but most of them were rather very proverbial in essence. But the legacy of the years of wars has made it that only political parties that once were armed movements counted really in the memories and thinking of communities. So, in that respect, there were less than five political parties that were recognized as such. Among the five, there were four that came out as the legacy of war and one was a legacy of the 3-decade single-party regime. Essentially, the power of these political parties had more to do with the effects of arms their militia possessed than any coherent program or any other form of real power. Saying so also equates to discovering another type of stakeholders that is not always taken into account when power analysis and stakeholder analysis are conducted within the context of biodiversity conservation. This massively ignored though always lurking in the minds is constituted of the poachers of all kinds. Poaching is defined as an illegal trade in natural resources, which includes illegal extraction of timber, illegal fishing and hunting, and illegal extraction of any of the natural resources. In the Lake Tumba Landscape, poaching is essentially around timber and wildlife. Poaching of timber has been partially described above and is encouraged by the industrial logging, which uses the local communities to avoid restrictions on their logging permits and other legal mechanisms such as taxes. The bushmeat trade, which is linked to logging industry in many ways, is practiced by groups of people with multiple characterizations though they all have one single motivation, which is to make money from wildlife species. In the case of the Lake Tumba Landscape, many poachers are members of the communities residing within the landscape but there are also those that come from outside of the landscape. Both internal residents and the in-coming poachers have guns with them and wield significant power that is directly linked to the possession of arms; in some cases, they can occupy extensive areas of the landscape that are avoided by the police or any other form of authority unless that form of authority becomes complaisant with them. Poachers do have power because of who they are (internal, personal positioning) and because of the network that supports them throughout their activities. Studies have documented the links between the poachers and not only the military but also the political elites in the African countries. In the case of the Democratic Republic of Congo, military elites provide arms and ammunition to poachers that hunt wildlife species to fuel the markets in towns such as Kinshasa and Mbandaka. Military elites are paid back by poachers; given the modicity of their monthly pay, they use any means that could supplement the meager salaries they are paid by the government. Talking about armed poaching on wildlife does not, by any means, devalue the impact of other forms of wildlife poaching such as snaring. The question being addressed here is not that of the impact of poaching on wildlife species but that of the power each stakeholder has within the social network in the landscape. As a group of armed people, poachers have much greater power than the power that some of the administrative and political authorities have. The fact that people vested with the public authority can become complaisant vis-à-vis armed poachers demonstrates the limitation of the political authorities to exercise a true control over means of violence. But, it also means that the armed poaching is part of the criminality, which is due to the fact that illegal trade in animals (including fish) and plants (including timber and charcoal) has become one of the largest sources of criminal earnings in the world, estimated to be worth US$50–150 billion per year worldwide (UNEP 2014).

Individually powerful people are also present in the Lake Tumba Landscape. Some of these people have their power because of the material well-being they exhibit. Of course, this is not new; wealthy people are powerful allover the world. The power of the wealthier people in the Lake Tumba Landscape is even more expressed given the near inexistence of the middle-class citizenry throughout the landscape. In a poverty-striken zone like the Lake Tumba Landscape, everyone who exhibits his or her riches is taken with great respect and counts on political and social decision-making processes. Of course, it should also be stated here that being rich is context-dependent; some people who were felt by their communities to be rich are not rich in the way people would be expected to be called rich in other countries. For example, someone possessing a truck to transport goods in Southern part of the Lake Tumba Landscape was described as powerful person because he was rich by virtue of possessing that truck. Equally telling was the fact that in the northern part of the landscape some people were described as being rich and powerful because they owned outboard motors and were selling salt, sugar, and other petty commerce goods. Other individually powerful people were people that previously held positions in the administration and politics of the country. There are many instances of people who previously worked for the colonial administration and were given deep respect for the positions they held during the colonial era. This is easy to understand in a context where old people are given respect in general. The Lake Tumba Landscape is located in a country where gerontocracy is promoted because being old is easily equated with wisdom, experience, and knowledge. Hence, the general respect, if combined with prestige of working closer to the colonizers, is accrued significantly. Indeed, working closer to the Belgians is seen as specific merit because of the hardships imposed by Belgians on their workers; one must have been good to be kept by the Belgians who imposed dire working conditions on Congolese. Apart from those who have worked with Belgians, power was also perceived as being attributed to the people that occupied political positions such as former territory administrators, former and current members of the parliament, and former or current ministers (provincial or national). Of course, this should not also be a surprise because power is felt to remain with those who have been involved in the management of their nations at one period of their lives and throughout the history of that nation.

After such a long narrative on churches, the logging companies, traditional authorities, the conglomerate made of political parties and poachers, the question that rightly comes up is where are the communities themselves? Are there any relevant Community-Based Organizations (CBOs) in the Lake Tumba Landscape? If they exist what is the power they might have? Answers to these questions are difficult to give in yes or no forms; they are much more complex. Indeed, communities are all villages, non-traditional centers, and towns of the Lake Tumba Landscape. However, the limits of their powers are rather gray to be captured in a straightforward way. In fact, a better way to respond to these questions would be to start with the last question about community-based organizations. There are several dozens of these within the Lake Tumba Landscape but most of them have been created by different international projects that swarmed the country after the multi-decadal crises. It is not surprising to find several committees of different issues in the same village in the Lake Tumba Landscape: 'committees for water management', 'committees for sustainable agriculture', 'committees for self-sufficiency', etc. All these committees are part of a situational analysis that has become more or less redundant in a sense that every intervention identifies the lack of structured communities as an obstacle not only to the development but also to a truly participative process. Indeed, this finding is a common denominator among international organizations. But rather than trying to identify what has been done previously and what local communities themselves might have as ideas, the international organizations start reinventing the wheel and feel that they are the first one to restructure communities in different ways just to discover other committees some days later. Because communities are felt to be unorganized, they are over-structured by different international actors. Over-structured as they become, they are weakened at the same time because they will have to deal with different and sometimes conflicting notions of what being an organized community mean. In a sense, the idea that in order to have communities organized to make participation effective kills the participation itself when structuring communities does mean grouping them ad infinitum. Structuring communities ad infinitum had significantly contributed to weakening them because most members of different committees are the same to nearly all of the committees existing in the Lake Tumba Landscape.

The second factor that could explain why local communities are seen as weak in terms of a political power analysis is an historic reason: all the communities in the Democratic Republic of Congo, as it also the case for the Lake Tumba Landscape, have lived a 3-decade political regime that reduced all the possibilities that everyone could seize to exercise any freedom of thinking. Indeed, the 32-year Regime of the late Marchal Mobutu has leveled down the potentialities for people to be able to take responsibilities for their own lives. The irony of the Regime of the late Marchal Mobutu has been that it was openly being promoting economic liberalism while in reality it modeled itself like central planning governance. Under this regime, all individual initiatives were crushed and denied freethinking to communities and their leaderships. Everything has had to come from above. This attitude has been kept similar to the Mobutu era until now and continues to suppress the potential for communities to really gain the power they need and deserve to become truly part of the decision-making process. But both the ended regime of the late Marchal Mobutu and one that overthrew it are deeply modeled by the Belgian Colonial Model whereby the King owned not only the geographic space of the country but also its riches and its people. The King or thereafter the presidents and other politicians were the fathers and had to think and care about everyone. When people were asked to participate in open discussions, they were given only putative responsibilities and illusory alternatives. One talks of the illusory alternative when one asks people to find alternatives to their problems while fully knowing that he (she) has his (her) own idea of the solution and is asking people to discuss just to fool them to endorse the ready-made solution that he (she) brings for them.

## 18.5 Who Wins and Loses What in the Lake Tumba Landscape?

To respond to this question, one should get back to the basics and restate the fact that landscapes, in their acception here, are primarily a paradigm through which the global community would like to conserve biodiversity while delivering sustainable livelihoods to communities that are located within these massive areas. One would argue, following Oates (1999) that mixing conservation projects and development ones might not be the best option for conserving biodiversity and that this has been tried and failed in most cases in Africa. Also, one could also argue that there is no new thing in the landscape approach if its core argument is around reconciling biodiversity conservation and needs for development because this has been already tackled (Borgerhoff-Mulder and Coppolillo 2005). While these points make sense in a certain way, the originality of the landscape approach in the context of Central African countries is not only in the objectives it sets out but also in the process of how to get these objectives achieved. This process, in its essence, is to be participative. For the landscape approach to work as they have been being sold to the general public, the participation of communities is a pre-requisite and should be at all levels, including the definition of objectives, the identification of means to achieve these objectives, how to monitor the achievement, and how to police the agreements reached.

The above reminder was to set the discussions on who wins what in the context of the general political economy within the context of the Lake Tumba Landscape. It would be difficult to try to sort out what specific gains specific churches and specific faith-based organizations, including the newly developing ones (Evangelical and neo-Pentecostal). In order to get the feeling of what the churches and other faith denominations gain, it is worth reading their gains from the perspectives of their proclaimed missions. Apart from their missiological mission, the creed of the churches and other faith-based organizations is the moral and material well-being of their members. In order to achieve these two goals, churches are keen to plead for the inclusion of the minorities and to ensure that riches are equally distributed across different social groups. If the landscape paradigm would bring greater material well-being for communities, which subsequently will ensure a good moral health for them, churches should be said to have gained in being part of the discussions and the decision-making processes that the landscape approaches will bring about. Of course, saying this does mean that the moral and material well-being of their members is all what the churches are behind in the Lake Tumba Landscape; churches have acquired properties and have been investing in diverse kinds of enterprises, including ones that are directly linked to natural resources. Indeed, some churches own extensive palm oil plantations, some are invested in artisanal timber logging (at least some members of the clergy) and practice cattle husbandries. What would they gain in their private affairs? Obviously, the gains are several folds: the first gain should be the confirmation of the churches as part of the decision-making processes; their prestige would almost quadruple as they would be seen as helping communities think independently, they should be officially acknowledged as shaping the process for a better future, they should be able to have their own opinions become part of the general thinking about the development of the Lake Tumba Landscape area and, finally, their participation in these decision-making processes will ensure that their interests on lands and other assets they possess to acquire the acceptance they need to avoid being part of the conflicts. Beyond that churches should be benefiting greatly in seeing their own activities, be they in logging or other extractive activities, be part of the general policy endorsed by all the stakeholders, including the state and other industrial companies. The endorsement by stakeholders will confer to their possession a firm legal and social status and acceptance that no one else would dare question. Finally, because their properties and lucrative activities would become generally accepted, they would have to follow the same rules and principles of the management. In this sense, the landscape approach should benefit the churches in providing them the much-needed expertise in the management of natural resources.

Of course, these gains come with some sort of tolls to be paid. In the case of churches, other faith-based organizations and other faith denominations, the toll is about losing the monopoly of the influence they have always enjoyed in many parts of the Lake Tumba Landscape. Indeed, in some areas, churches provide education, health, transportation, and imported goods to whole communities. This situation, which has come through the de facto absence of viable state structures, has brought churches to become so powerful and in some cases to abuse that situation of influence. The landscape approach will, in many ways, dilute this type of influence as it should bring more people to discuss what is good and what is not good and who should do what under which circumstance. That said, however, churches will remain among the stakeholders that should make the maximum benefits from the landscape conservation approach.

For state authorities, it has been indicated above that the positional authority of the state is still strongly felt throughout the Lake Tumba Landscape while the same authority is reduced to near-token when it comes to financial attributes. The landscape approach would be beneficial to the state agencies in the Lake Tumba Landscape because it should mean bringing people to discuss how to boost the economy of the communities. In discussing this, state agencies should benefit from having community contribute in the elaboration of budgets through a participative budgeting exercise. This exercise should ensure that official taxes are discussed and agreed-upon with communities and means to ensure that each individual pays for the agreed-upon taxes would be part of the discussion as well as how the money would be spent. This should greatly increase the financial means of all the administrative and political entities located within the landscape and will increase the financial power of the local and territorial administrative and political authorities. Through increased tax payment rates, authorities should likely gain their independence from logging companies and other wealthier people present in the landscape; they will then become able to apply and enforce the law without having many conflicts of conscience. This single fact should increase their authority over the entities they are meant to govern. Naturally, the landscape approach would require a somewhat expensive cost from administrative and political authority; particularly it would reclaim from them to greatly shift their ways of doing things. Truly, once participative budgets are adopted and funds are available, the idea of participative monitoring of the execution of the budgets will mean greater accountability for administrative and political authority. Since the administrative and political authority in the Lake Tumba Landscape, as it is the case across the Democratic Republic of Congo, are second-rated in what is known as the downward accountability; administrative and political authorities will find demands for downward accountability as being a huge price to pay. Equally, the adoption of the landscape approach would seem to be most of the administrative and political authorities as a diminution of their usually large spectrum of political power, which is often used without proper control. The landscape approach with its focus on participative process will, in fact, demand more transparent management and inclusion of voices of those who have been traditionally often left voiceless will mean many inches taken from the social spread of the administrative and political authority.

The logging companies too will win in the sense that working with communities and other stakeholders should ensure that their reputation is positively boosted. This single fact would increase their profile, not only within the country but also externally, where the final market for the timber they extract from the Lake Tumba Landscape is located. Undeniably, with ecology emerging as one of the most important concerns of the end of the twentieth century and continuing to be an area where the global world community will have serious disquietedness over a large part of the twenty-first century, preserving the World's biodiversity has become an integral part of that ecological disquietedness. Worries to preserve the world's biodiversity have become one the most stringent narratives of the modern global society and requirements for the traceability of goods such as timber and associated products are increasingly demanded by the global world's civil society. The logging companies in the Lake Tumba Landscape will hugely benefit from showing that what they do is in alignment with the foundational ideas of social corporate responsibility (see Chap. 3), which will respond to the moral demands of their clients in Europe or the wider western world. As indicated above, most of the timber extracted from the Democratic Republic of Congo is shipped to European countries whose demands for transparency are among the most advanced and the levels of democracy in these countries would mean that civil society organizations can have a much stronger impact on the commerce of goods that are produced from morally flawed principles. Apart from winning a greatly improved reputation, working through the principles of participation of all stakeholders would help the logging companies win the hearts of communities in which they operate too. Definitely, a major criticism that was voiced by 85% of the respondents of structured and semi-structured interviews organized to assess the relationship between stakeholders throughout the Lake Tumba Landscape against the logging companies was that they behaved like in a conquered land and country; logging companies did care very less about the others, meaning simply the local communities. It was declared by the same magnitude of respondents that after logging companies acquire their logging permits from Kinshasa, they do believe they have no other moral obligation vis-à-vis any other form of authority or vis-à-vis the communities. This assertion has been given indiscriminately and at the same proportions by all social groups, including, strangely enough, the administrative and political authorities. The fact that this criticism has been voiced in louder tones by the majority of the administrative and political authorities too is more than telling in several ways. Firstly, it tells how powerless the administrative and political authorities are in reality for those who are at lower administrative ranks. This is particularly the case when those at the lower administrative ranks face the power of financial interests. The branches of the executive that are at lower administrative ranks are also powerless when they are compared to the center of the power that is located in Kinshasa. Secondly, it also shows that the logging companies themselves take advantage of the situation of powerlessness of the lower administrative hierarchies and do not care about the foundational ideals of a democratic society. Of course, it would not make sense to ask logging companies to implement rules and principles of a liberal democracy in a country that is truly democratic only in its name. But logging companies being of liberal democratic countries in their origins should have a prudent way to treat the communities located in the areas where they extract the timber. Therefore, working through the principles of community participation that are promoted by the landscape as a global biodiversity conservation paradigm should bring the logging companies back to the minimal principles of a participative democracy whereby all voices are heard and reconciliation, partial in many cases and total in some of them, of divergent interests are achieved with the agreement of all. Thirdly, the very fact that the criticism of the behavior of logging companies comes in louder tones from the majority of the administrative and political authorities is also more than telling about the limits of the representative democracy (Mfou'ou 2015), in general. As it is well-known by all, a representative democracy is a system of government in which all citizens freely elect some of their compatriots to represent them in the public offices. By the fact of electing their representatives' citizens delegate their authority to the elected citizens to make laws that would govern the society. There are minimal conditions for the electoral process as such to be acceptable by all, the first one being that citizens that compose the electorate are fully informed and capable of making moral decisions. The second minimal condition is that for the system to work there is need to be in truly open societies where information for public policies is free. The third minimal condition is that the open society for the information is conducive to a true and fair competition among the candidates. Finally, candidates should be able to present true alternatives that are available for the communities to select which option they feel being the best for their own interests.

Without these minimal conditions, it is just a joke to be talking of democracy. In fact, none of the four minimal conditions is really settled in the case of the Democratic Republic of Congo. For example, in most domains, if not simply nearly all of them, the information on public policies is not open and remains opaque in many dimensions. In some instances sharing and accessing information is often equated with crime. In a system where some people use the public means, coercion, and open fraud to access political power the representative bodies and individuals holding the political power over the entire country do not necessarily respond to the will of the communities. In such a system, views other than those of the ones from the people in power and social groups that don't support the incumbents of the political positions are left to grapple with limited private means and have no resources to disentangle the coercive pressure exerted on them by the system. Elected authorities in Kinshasa are, therefore, disconnected from the true realities of people and communities that the very same elected authorities are meant to represent. That explains why even the local authorities are far from being sanguine on the decisions that are imposed on them by the pyramidal authority system. Hence, when logging companies come with their permits issued from Kinshasa and pay token prices to the local traditional authorities ([red] blankets, [red] cloth, salt, soap, and sugar) and ignore the rest of the communities, they simply trample the rights of other major stakeholders; the offices in Kinshasa do not genuinely represent most of those communities.

What is said about the flaws of the representative democracy above could be said of different liberal democracies around the world. There is nothing new these statements but there is a difference of degrees; the Democratic Republic of Congo is, by no means, comparable to other countries even in Africa. Of course, one would be called to read all these principles through the lenses of anthropological, historical, and social contexts of each country. Yet, the fact is that even when contextualized the degrees of disconnectedness between the elite and the global community in the Democratic Republic of Congo are larger than even similar experiences in Africa. That is why the participation as approach has a significant role to compensate the deficit in governance; it will bring authorities, local communities, civil society organizations, and other stakeholders closer and will help logging companies to become more and more acquainted with the realities of communities and buy confidence from communities.

As it is the case for other stakeholders, logging companies will lose in their over-displayed authority. The landscape biodiversity conservation process means that logging companies will be brought down to playing by the transparent rules rather than being on what seems to be conquered lands. Playing by transparent rules will have an increased financial cost, as part of their social corporate responsibility; and participation will certainly levy a toll on speed because, indeed, logging companies will have to truly negotiate with the communities and come to an agreement. Negotiations take time, which will be seen as loss for a category of stakeholder known for its will to do things quickly and spend very little money. Logging companies are mercantile in essence; speed and quick money is efficiency for their thinking. Of course, there should be no blame placed on logging companies when they are trying to make money; as indicated in Chap. 3, there is nothing wrong with that provided that they are also paying for the costs of externalities they generate for the global community.

Civil society organizations were described above as being weak and dispersed in their motivations within the Lake Tumba Landscape. This is not surprising for anyone who knows something about the Democratic Republic of Congo. Indeed, following the same trend as the national civil society, organizations have mushroomed throughout the Lake Tumba Landscape and, as indicated above, the majority (104, representing 69%) of the stakeholders in the landscape were made of national and local non-governmental structures. According to Fukuyama (1992), the modern citizenry is better articulated through intermediary institutions such as political parties, private corporations, labor unions (syndicates), civic associations, professional organizations, churches, associations of parents, boards of societies, and literary societies. Fukuyama (1992) indicates that Tocqueville thought that this type of associative life was necessary because it served both as a laboratory and a school for democracy. Fukuyama and Tocqueville were certainly right that given the distance between the communities and those who govern them, there is a need for something to emerge in between to care for the true needs of communities. Indeed, these intermediary structures should be the conduit through which the needs for communities should be channeled up to the top of the political decision-makers. However, what is intriguing about the proliferation of non-governmental structures in the Democratic Republic of Congo and in the Lake Tumba Landscape is that most of them have no clear thematic focus; they can shift from one theme to another without transition provided that there is funding and other resources available. In the absence of resources to create private enterprises and in the presence of limited job opportunities, creating a structure known as non-lucrative has become part of the survival strategies for many clever citizens. This very situation makes it hard to see what the multitude of the civil society organizations would benefit from the participative processes that are at the heart of the landscape approach. In some cases, it has been argued that the existence of a multitude of civil society organizations renders the participative process at best very slow because the participative process requires that all major and minor stakeholders are brought to the interests' conciliation table and their voices are given the same weight. That said, it should be noted that all the organizations will become more coherent in their approaches after meeting with different other similar organizations. Becoming more coherent and focused on the objectives, being able to assess internal structural strengths and weaknesses, which are all processes of the biodiversity conservation landscape approach, are things that would greatly benefit all of the civil society organizations. The exercise to look for what core values and the thematic areas to invest efforts will help the local civil society organizations acknowledge the fact that they cannot do everything as their objectives are to be circumscribed in one area. The exercise to go through structural strengths and weaknesses will help local civil society organizations become more realistic in what they are really capable of doing and what they cannot realistically achieve. Both exercises should be incentivized by acknowledging the extent of the problems they will have to face if they are to invest in everything possible. Indeed, seeking to become coherent in their interventions, local civil society organizations will gain by being brought into coalitions and networks of similar values, objectives, and operational models. Working in consortia, coalitions, and networks will increase their positional power and their voices would become a lot more audible than it is the case when they are spread-thin to raise funds from any organization that can provide it and do anything provided that it comes with funding. The landscape approach, through pushing hard on participation of all would, in some ways, be a conduit through which to reduce the number of actors through bringing the likes together in coalitions, consortia, or networks.

Reducing the numbers of independent civil society organizations will mean that the civil society organizations within the Lake Tumba Landscape will obviously lose some of the maneuverability in terms of agility to seek funding. Indeed, most of these structures were created in the hope to capture the international funding that comes to support the desperately poor people around the country. There are many points that can be made about the motivations of civil society organizations within the Lake Tumba Landscape but there is no space to dwell on these here. Suffice it to be indicated that the flood of international non-governmental organizations that spread across the Democratic Republic of Congo during the war has served as the big bang point for the proliferation of numerous structures to ensure that funds from the international humanitarian aid are acquired and used by communities. While some of the structures that emerged from this big bang have made great progress toward becoming real agents of sustainable development, the majority has been lurking behind in terms of structures, operational models, and management skills. Indeed, as is it the case for churches, there are those that are single-family or single-community enterprises. There is no bad thing in being socially diverse but sometimes social diversity becomes a huge hurdle for action when this goes down to microlevels as it is the case in the Lake Tumba Landscape. Indeed, there are many microlevel structures; some, such as the case of Association Mola at Lokili, do even dependent only on one single individual while others, such as the case of Association du développement des Ekonda et Ntomba, are openly about the well-being of one single tribe. The two cases that singled out here are a patent way some people use the possibilities offered by the openness of the political system prevalent in the Democratic Republic of Congo in terms of creating civil society organizations in order to get their own political agendas achieved. Indeed, there are widespread examples of local leaders using the abilities offered by the civil society organizations to anchor their political ambitions within the society. Indeed, many civilian local leaders extensively create their own non-governmental organizations to plan for future political benefits, including accessing to political positions or capturing financial resources from the national political leadership.

Political parties are, as it was the case for civil society organizations, diverse in the Lake Tumba Landscape. In some areas, one could have a sense there being more political parties than numbers of families. But, as indicated above, there are fewer and fewer political parties that can genuinely claim having an effective power, except those that are national by their very existence. Even for those that are national, political parties have a fate of being attached to their individual founder taken as the charismatic leader on whom everything depends. Hence, political parties, as structured political enterprises, have very limited power in the true sense of the word, at the landscape level. Being able to come to the table where every other stakeholder will increase the visibility of each political party at the local level, which an enormous benefit given the fact that there are many of them and communities do not know what they stand for. The second major benefit that political parties would make from their presence in the landscape and its conservation approach is the gain in the knowledge of the communities, available natural assets, and negotiation skills. Indeed, as was the case of the political authorities discussed previously in this very section, many political parties are created at the national level and lack the necessary connections with the communities they talk on behalf of. The participative approach would help them come closer to communities and acquire a deeper understanding of the true problems of the communities. Of course, coming closer to communities is likely to have high price on most of the minor political parties as the open and participative debates would likely lead people to understand the true nature of a great majority of these parties, which is mostly that they lack true political vision. This would decrease the audience of those who have no clues about what political parties should be doing and how while it will increase the weightiness of those that will be able to learn and increase their professionalism.

Traditional authorities, as discussed above, having been brought to reoccupy the moral position of authority that was left vacant by the political authorities who had seen their power reduced to very symbolic expressions in many areas of the Lake Tumba Landscape because of lack of resources to enforce the rule of the law. As such, traditional authorities would be expected to lose some of their importance as the administrative and political authorities gain back the ordinary functions of modern state authority. But this should not be the only thing at the stake for traditional authorities: the landscape biodiversity conservation approach, by insisting on a clear delineation of tasks and responsibilities, will reinforce the position of the traditional authorities as the conduit through different social desiderata will be brought up to the administrative and political leadership and other stakeholders. They will win in revamped prestige at both ends of the spectrum (political authority and communities) through acquiring space to renegotiate conditions of contracts signed by the administrative and political authorities either in Kinshasa or capitals of the provinces for local communities. This single fact will increase the legitimacy of their power in their communities. The landscape biodiversity conservation approach, as a means to distribute the power of action across different actors, will reaffirm the authority that traditional chiefs exercise de facto and make that authority become part of the *de jure* authority.

Against all the normal practices in the biodiversity conservation programs across the Central African region, the work in the Lake Tumba Landscape identified poachers as important stakeholders. If, as stated above, poaching is one of the major threats on the biological diversity, poachers who are actors of poaching should be brought to the table to provide their perspective on wildlife trade and be asked to contribute to the solution of this vexing problem. In fact, defining threats to biodiversity as was done in Chap. 12 above is only partially relevant as it accounts only on the biological side of the story. In principles, a genuine definition of threats to biodiversity should also include a sociological interpretation of the concept. In other words, a threat to biodiversity should be also gauzed using external metrics that would weigh the contribution of biological resources to local populations' well-being. In other words, this would come down to looking at the availability of resources and opportunities that people rely on alternatively to using biological resources, i.e. wildlife and plant species. When these alternative extractive or non-extractive economic sources of subsistence opportunities are absent or have disappeared, they represent a serious threat to the sustainability of local natural resources. This is because they will force communities to rely on bushmeat trade as a source of financial incomes. This definition of threat applies to such cases as lack of or limitation in paid employment, which would lead people to fall back on poaching to make incomes to support their lives though becoming able to pay school fees for one's children or being able to purchase medicines when needs arise. This broadened definition of threat to biodiversity, which was an outcome of different discussions with communities in the Lake Tumba Landscape revealed the fact that when biodiversity conservation organizations talk about poachers, communities think of that different perspectives attached to this notion should be debunked. Clearly, communities felt that armed gangs that are sent by military and political leaders are dangerous and not only for the wildlife but also for communities. Bringing this category of traders is necessary and would need the implication of all the authorities, including at the national level because this group operates in fairly similar manners as do the drug dealers and can be rightly assimilated to the mafia. However, communities perceived that there are other groups of people qualified as poachers, because they are involved in wildlife and plant trades, whose prime motives are to make a decent living through the use of biological resources available in their villages and forests. This was an important insight for biodiversity conservation organization because as Almquist et al. (1993) said, understanding how local populations perceive and use their natural environment was linked to their expectations in terms of economic development, in order to help adapt conservation and sustainable management plans to the landscape's social reality. Communities felt that they are often held back from making a dignified living when every single hunting or extraction of a tree for commerce was qualified as poaching. How local populations understand their local environment and comprehending what mechanisms they use to extract natural resources and the way they manage them would be essential in tackling the issue of poaching and bushmeat trade. In the discussions around changes and adaptation to new situations, there was a great deal of new insights and new perspectives because the communities felt that threats to biological diversity were also threats to their own sustainable livelihoods. These discussions helped to understand what these populations see as present and future threats to their livelihoods. To them, all commerce on bushmeat is not and should not be called poaching because that contributes to their livelihoods and it is part of their culture. Indeed, over the discussions, some interesting ideas on how to deal with the bushmeat crises were presented, including the resumption of colonial laws on periodical hunting, a clear law indicating species to be hunted and those that would be exclusively protected, and working toward the certification of the bushmeat to ensure that the wild meat that reaches the markets is not generated by the mafia but by communities with genuine rights to extract some specimens of wildlife for their own consumption and for a legalized trade.

With this perspective, community members that hunt as their principal livelihood activity should become part of the discussions; they profit from being the opportunities to share their own views on the bushmeat trade and how they can contribute toward a solution that would factor the interests of biodiversity conservation and their own livelihoods. In essence, the biodiversity conservation organizations would also benefit from the presence of hunters to discuss the levels of wildlife species off takes and reach agreements on what can be hunted, when and how. This is a true gain for biodiversity conservation. Another gain for the same categories of stakeholders is that, when the agreements on species off takes and methods to hunt and how to certify bushmeat are in place, biodiversity conservation organizations would collaborate with local communities to sort out the problems with bushmeat trade mafia. This would be helpful in several ways, particularly in identifying areas where these mafia-like traders in the bushmeat are likely to hide before embarking on massive decimations of wildlife species as they often do.

What will the communities win in this whole gamble? Indeed, as indicated above, the portrayal of the stakeholders that emerged from this analysis is that communities are often the left outs of the whole current biodiversity management system in the Democratic Republic of Congo. Saying so is far from going back to the victimization narrative that is prevalent within the international organizations about local communities across Africa. To the contrary, this statement is an indication of lack of a sound participation process and lack of democracy in natural resources in the country. As explained above, laws and regulations governing natural resources in the Democratic Republic of Congo are, in many ways, relic of the colonial era and, despite efforts to adapt them to the current conditions, have been too state-centerd. By state-centeredness it should be understood that communities are not given political means to be part of the decision-making processes, which then has implications on their share of the outcomes of the extraction of the natural resources located in the areas they live in. Hence, a major benefit of the landscape approach to the local communities will be to help them come to the table of negotiations and argue for the implementation of clauses of laws and regulations that are beneficial to them all. A specific gain that they would garner from this process is the implement of the clause of the current forestry code that is about the local communities' forests. This is a factual advancement in the tradition of land laws in the country but it needs to be followed by concrete actions. For the concrete actions to come forth local communities need to be part of the discussions to ensure that among other functional land units, 'community forests' (see Chap. 4 above, for legality of this nomenclature) are given the due value they deserve.

Despite the major gain, they will make by being able to ensure that 'community forests' are given the due value they deserve and become a material reality, communities would also have to give up some practices that are destructive of their own environments. This is so by virtue of the law on natural forests, which is clear that certain activities cannot be done anywhere except in specified areas. Indeed, the fact of physically delineating functional land units that is preached by the landscape biodiversity conservation approach means that communities and their activities are fixed in some areas. This would require enormous shifts in practices of activities such as agriculture, opening of new village sites, etc. Communities should be brought to remain in one area, for example, even after some of their members have died of an unknown disease; they would not be allowed just to take their household utensils and vacate the area as they wish. Equally demanding for communities will be the fact that the landscape approach would also delineate areas whose use would be a function of times of the year and with specified tools to be used to extract biological resources. This is, for example, the case of fishing zones in some portions of the lakes (Lake Maindombe and Lake Tumba) positioned within the Lake Tumba Landscape where communities agreed to sequence the period of fishing and specified types and sizes of fishing gears to be used. Communities should constrain themselves from violating rules they have come up with over the negotiating table. Undeniably, people that made communities throughout the landscape were also asked to come up with a properly agreed-upon law enforcement mechanism that would ensure that everyone feels bound by different agreements.

## 18.6 Political Economy, Landscape, and Total Economy for Landscape Communities

Beyond these logically acceptable limitations for communities, the total economy of the landscape means that, at the end of the game, communities in general will win more than the losses they incur. Indeed, taking a broader look at that the total economy of the landscape one would have to factor things that communities draw directly from the biological resources that are essential for the local communities and the broader human community as whole. This is to move from the normative economy to the positive economy of communities. Indeed, normative economy and gains it comes with, as was described above, gives who should win what if the process would flow normally where the positive economy deals with facts and makes an effort to put these facts in a theoretical framework that makes sense. For local communities, the positive economy is made of such essential items as fuelwood for energy, water for drinking and sanitation, fish and wildlife meat for protein, thatches for construction, and other non-timber forest products. For the national and international communities, in terms of the normative economics, facts that would make part of a sound theoretical approach are diverse benefits yielded from biological resources and include financial incomes generated through taxes levied on natural resources, accessing the extracted goods via the market system as well as multiple sorts of ecological services.

For the local communities, for example, the valuation of such items as the fuelwood used for cooking energy by each family residing within the Lake Tumba was estimated taking into account that there were nearly 5,000,000 residents within the landscape. This population is clustered in 800,000 households, taking into account the national average of 8 people per household. Taking the average of 2.5 KWh to cook one full meal, the energy from the fuelwoods that is used for one meal for the entire landscape on daily basis is 2,000,000 KWh; and because households at the Lake Tumba Landscape had indicated that they would cook at least twice a day, the energy used just for cooking food is, on average, 4,000,000 KWh per day. The financial equivalent of this energy has been generated using the social price of the National Electricity Society, which is set at a flat rate of 2.65 American dollars (USD) for a lump of 100 KWh (SNEL 2016). Once the 4,000,000 are divided by units of 100 KWh one reaches 40000 units of 100 KWh per day. Multiplying these 100 KWh units by the price gives the enormous amount of 106,000 USD per day or 3,180,000 USD monthly being supplied by the fuelwood energy to the communities that reside in the Lake Tumba Landscape. In the current situation of the country, there is no alternative source of this cooking energy for more than 95% of the people in those communities and that situation is likely to remain still for long. So, gaining to protect the forests for fuelwoods is an essential part of the sustainability of livelihoods.

Similar huge numbers emerge when one wants to translate the benefits of using freshwater directly by communities of the Lake Tumba Landscape. Indeed, if we take the usual mean of 20 l of water per person per day, the total amount of water needed for the 5,000,000 people of the Lake Tumba Landscape is 100,000,000 l/day for the totality of the communities. The price social price of water practiced by the National Water Company is 693 Congolese Francs by $m^3$ (ACP 2015), which can be rounded to about 0.6 USD. Since 100,000,000 l equate 100000 $m^3$, the total cost covered by sources of water located within the landscape is about 60,000 USD daily, which would amount to an equivalent 1,800,000 USD monthly. To give the magnitude of what these figures represent one should bear in mind the fact that the current budget of the Democratic Republic of Congo is about 7 billion USD only and that the global national product (GNP) was about 500 USD per person per year in 2015.

The above numbers are what would have been spent if the energy and water were bought for communities to use. In order to provide indications of what communities actually make as money from the use of biological resources in the Lake Tumba Landscape, a look at the study by Colom et al. (2006) provides some clear indications and magnitudes of benefits that communities already make; and these gains would be greatly increased if they would manage these resources in a sustainable way. To get a sense of that reality, examples of fishing, hunting, and the five most important non-timber forest products are given here. More than 80% of the communities located in the northern part of the Lake Tumba Landscape reported being occupied by fishing most of the year. This part of the Landscape, including the town of Mbandaka, is home to approximately 1,800,000 people. Taking the same average number of people per household as above (6 people per household), the northern part of the Lake Tumba Landscape has 300,000 households on average. According to Colom et al. (2006), 80% fishing households in this part of the Lake Tumba Landscape reported generating incomes ranging from 11 USD to 44 USD per week per household from fishing during the dry season when fish off take is rather at its maximum. If one applies this range of income generated by fishing to the 80% of households (640,000 households), the amount of financial income being generated by each of these households is between 44 USD and 176 USD per month. Cumulated, fishing brings in between 28,160,000

USD and 112,640,000 USD monthly into the region. While it might be tempting to use these scores across the year to obtain the final annual contribution of fishing into the local real economy, it would be wise to restrain from doing so because fishing is very seasonal and figures being presented are just to show the magnitudes of the contributions of each biological resources to the well-being of the communities in the Lake Tumba Landscape. That having been said, however, the figures above give a sense of how important fishing is in the Lake Tumba Landscape. Clearly, if taken throughout the year, these figures would indicate that figures that appear on the report by Ingram et al. (2010) are rather very minimal for the entire Congo Basin. For comparison reasons, The Guardian (2011) reported that the total annual military spending of the Democratic Republic of Congo was 239,000,000 USD, which can be converted into an average monthly spending of 19,916,667 USD. Clearly, the average monthly expenses of the national Congolese army are lower than the lower limit of 28,160,000 USD generated by fishing alone in the northern part of the Lake Tumba Landscape while the lower limit of financial income generated by fishing is not that far from the amount of total monthly spending for the army.

Hunting is the other important activity in terms of generation of financial income in the region. In the same region of the northern part of the Lake Tumba Landscape, Colom et al. (2006) reported that hunters declared being able to generate about 22 USD per week for the household. The same calculations as in the case of the fishing would lead to a total average income of 88 USD per household per month, which can be converted into 56,320,000 USD per month for the total number of households. With the exception of fuelwood and water, important non-timber products collected in the Lake Tumba Landscape, include the following products in decreasing order: caterpillars, mushrooms, honey, cola nuts, fruits of *Dacryodes edulis*, fruits of *Annonidium mannii*, and leaves of Marantacea species. To this list from Colom et al. (2006), could be added some other products that (Ilumbe 2006) described as being identified as strategic resources and vitally strategic by the communities. These products include the palm oil trees (*Elaeis guineensis*), zinzer plants (*Aframomum melegueta*), African Elemi (*Canarium schweinfurthii*) and fifteen other plants. According to Ilumbe (2006), 15 if the 16 plant species qualified to be vitally strategic were for traditional medicine uses, which they combined with other uses such as production of fruits (3 species), production of caterpillars (2 species), and providing construction materials (3 species). The prominence of the traditional medicinal species among the vitally strategic species calls for some thoughts, which go beyond the scope of this chapter. But, still communities have a clear sense of what product has what importance in the long list of available forest products. Should be added to this list because of their over-emphasized importance are two Raphia plant species (*Raphia sese* and *Raphia gentiliana*) which Inogwabini (in preparation) has described as being part of the most economically used species in the same region.

It should be stated that all these products are seasonal. Edible caterpillars, in their diversity, are highly prized as an alternative source of protein; they are sold all over the Lake Tumba Landscape when they are available. According to the Food Aid Organization (FAO 2014), around 150 g of caterpillars are sold in exchange for 1 USD. Communities in the Lake Tumba Landscape reported assembling on average between 10 and 15 bags of 35 kg every season per household; they would be expected to get 350–525 kg per season per household. These figures say that, on average, one household makes between around 2300 USD and 3500 USD per season, which runs between June and August each year. This is an enormous economic benefit particularly compared to the state employment that cannot pay more than 200 USD monthly (or a total net income of 2400 USD). Of course, the work involved in collecting caterpillars is a collective effort for the whole household and should not be compared with the financial productivity of one single individual but it still remains that the work invested in collecting caterpillars is just a fraction of the work done by those who are employed and the financial income generated by caterpillars is much more appealing than seeking for poorly paid jobs. Back in the early 1990s when statistics were available, it was estimated that the region of Bomongo and Lukolela, which are part of the Lake Tumba Landscape, produced between 2,100 and 2,800 metric tons of caterpillars per year (PNUD and UNOPS 1998). Another important product of the Lake Tumba Landscape that contributes significantly to the economy of communities is palm oil. Indeed, there are naturally grown palm trees; people also cultivate palm trees to generate palm oil, which product is used for cooking locally and is also '*exported*' to major towns of the Democratic Republic of Congo and the neighboring countries such as the Central African Republic and the Republic of Congo. For example, and to show how important the production of the palm oil is in some areas of the Lake Tumba Landscape, according to discussions with local communities, 90% of households in the territory of Bomongo own traditional palm oil distilleries. With an estimated population of 183,982 people (Mbimba 2015), this would lead to 27,597 distilleries, if the mean household size is equal to 6 individuals. These individual palm oil distilleries produce, on average, 250 l/year. If one considers the price of 11 USD per 25 l (Colom et al. 2006), which is in relative agreement with the figure of 13 USD per 20 l in other areas of the Democratic Republic of Congo (Van Hoof 2011), the total amount of financial income generated by the communities of Bomongo alone is 3,035,703 USD annually.

It was not the purpose of this chapter to go through each non-timber forest product and give its economic value. Rather the inclusion of what this means to communities in the discussion of the political economy of the landscape was deemed necessary to show that the landscape approach as a means to achieve conservation biodiversity would be more beneficial to communities than any its opposite, which the creation of fully protected areas that would exclude any use of any resource. By the same token, in accepting to work within a priority conservation landscape, the communities would not only be making profits but would also ensure that the resources are available for longer time. Indeed, one of the outcomes of the landscape approach is the delineation of space of authority, space of rights and space of duties. In the space of rights, one of the most important rights that local communities would be gaining is the exclusivity to access certain resources of the landscape while prohibiting those who do not have that right to access the same resources. This right would be used against mafia-type poachers and could make hunting for local economy and subsistence of communities more sustainable.

There is another important stakeholder, which is the group of the people who trade all kinds of goods in the Lake Tumba Landscape. At the first glance, it would be tempting to consider this group as made of businessmen and businesswomen but the reality is that they invest very little in the landscape and their activities are mostly the exchange of 'imported' commodities with natural resources. Hence, it is more appropriate to call them simply merchants. This is a group that is difficult to describe because it includes a variety of actors; most of its components reside outside of the Lake Tumba Landscape. These are merchants, people that come in to sell goods such as salt, soap, clothes, working utensils, etc. and, in exchange, buy natural resources such as smoked and fresh (and sometimes living) fishes, smoked meat, caterpillars, fuelwoods, charcoals, timber logs, etc. that they take back to major towns to be sold. With this description of what they do, the merchants and other types of traders, have a much greater impact on the natural resources of the Lake Tumba Landscape than some of the stakeholders depicted above. But, the fact of being located outside of the landscape made it difficult to assess its powers and interests in the same manner as it was the case for other stakeholders. However, using surrogate methods enabled the examination of what they would exert as influence. These surrogate methods were also included in the structured and semi-structured interviews and focus group sessions. Through these surrogate methods, the study has been able to assess the reflected images of the merchant stakeholders group. Rather than asking what communities and people felt about their own group, the reflected image method consists of asking people how they perceive the influence of any other group, in which ways they do perceive that influence.

The reflected image method is constructed from the image methods used in physical spectrometry and consists of assembling dispersed elements of different wavelengths to figure out who the person is the negative of the picture is. As it is the case in traditional photography, it is often the case that when a group of influence is absent, feedbacks from other groups are mostly bad or negative feedback. This corroborates what the allegory of the optical photography is about. Indeed, when a picture is taken with a traditional camera, the image in the black box is reversed (inverted) because of the location of the object in a certain distance away from the lenses. As in the traditional photography, the work from the point of taking the picture onward is to produce a true image that resembles the true person. The reflected image method has, therefore, the obligation to construct a serene image of the absent stakeholder, which is made of its positive as well as negative influence through several steps described here. First, rather than just asking about the perceptions (*what do you feel that X does do?*), the method insists also on the '*how*' the felt the influence of the powers (*how does X does do what you feel?*) can (be it negative or positive) is felt. Second, and apart from the how, the negative or the positive influence and powers are felt, the reflected image also demands that described behavior and activities are supported by concrete examples to illustrate the descriptions being given of the absent stakeholders (*can you give an instance where X did do what you felt?*). Another important step in this method is the construction of the relative irreplaceability of a given stakeholder, which summarized in the question: '*can you tell what stakeholder X positively does that you think it would be difficult for any other stakeholder to do*?' The relative irreplaceability step is an important one because it does ask specifically for actions that are positive and would construct a better image of the absent stakeholder even if the general tendency would be to dwell on the negative side. Of course, it is possible to encounter statements such as nothing, barely zero, not so much, etc. But, despite these negative answers, one would very often end up getting something, even minor, that would somehow brighten the image of the absent stakeholder.

Finally, as would be expected, the method takes one to assemble a large amount of views from different stakeholders and overlays them to produce a consolidated image before checking it out in a proper focus group session as a way of triangulation. Ideally, as one would do in normal photography, one should compare the image with the physical person; this means that the generated reflected image of a given stakeholder who is unable to attend the discussions should be brought back to the stakeholder in question to see what his (her) own opinions about the image he (she) reflects outside of his circles are.

Through the above method, it has been identified that most of the merchants trading in natural resources goods from the Lake Tumba Landscape invested very little in the region. They were thought (to their agreement) that they held a significant power both financially and positionally. Their financial power was rather obvious because they were the only group bringing in goods and commodities that were difficult to obtain in the Lake Tumba Landscape; these goods and commodities being primary necessities for communities, the traders had monopolies in some places. In many cases, they were fixing prices of natural goods such as fish, palm oil, manioc tubers, etc. In some areas, the traders imposed barter back as a means to exchange goods with local communities. This situation was deeply anchored in their ways of interacting with local communities because of difficulties of transport; the merchants were also, in many cases, the owners of means of transportation. Because of this, the merchants were a very strong group of stakeholders because the transport has immense influence on extraction of natural resources, including timbers, bushmeat, over-exploitation of fish stocks, etc. Positively, however, the merchants were seen as the only group that brought services that other stakeholders were unable to provide such as bringing medical products, soaps, salt, and clothes in remote areas. Positionally, they because they were providers of primary goods in these remote areas, traders were very close to the local political establishments; they had close ties with the military and security apparatuses in the Lake Tumba Landscape and could use that influence to sneak through any good. Indeed, what this group of stakeholders are able to do was exemplified in cases of captures by security services of full protected species such as bonobos. In these instances, traders were able to buy their ways out the legal nets and freely move out the Lake Tumba Landscape.

The above statement indicates why traders as a group of stakeholders should be part of the decision-making consultation to enable the landscape approach to become fully operational. In fact, the landscape approach cannot work if there is no agreement on rules and principles; it cannot work if some stakeholders are held above the rule of law. Clearly, however, the traders will lose by accepting to play by the same rules and principles; they will not be able to openly influence legal decisions in their favor.

In contrast, they will win enormously in becoming seen as playing fair and with transparency, particularly when it comes to prices they practice in buying natural goods such as fishes, fuelwoods, palm oil, and timbers.

## 18.7 Uncovering the Obvious

The political economy analysis of the Lake Tumba Landscape has come to find out that women constituted a separate group of stakeholders. This was so even though women were represented in all other groups of stakeholders. This is because, as it has been identified by Kisambu (2016), even when women where enlisted in these other stakeholders' groups, they have very limited participation in the decision-making meetings such as village assemblies and village councils. Obvious reasons for this situation, as is widespread known across Africa (MacCormack 1982; White 1982), include combined effects of illiteracy, timing of village meetings, the burden of domestic work (cooking, fetching for water, child-care, collecting fuelwood, etc.) and, principal among them, weight of the customary practices (Kisambu 2016). Additional burden to the status of women in the region of the Lake Tumba Landscape came from the toll that whole communities had to pay for the cycles of wars that happened in the country over the last two decades: a major consequence of these difficult times is that many men died and left single women to carry the load of whole families on their own shoulders. It was upon uncovering this obvious yet un-reflected upon situation that the socioeconomic appraisal of the Lake Tumba Landscape and the Rapid Rural Appraisal were designed in such a way that women would be treated as a separate group of stakeholders. In the socioeconomic study that was conducted by Colom et al. (2006), women were interviewed separately from men using a separate set of gendered questionnaires. These specific questionnaires were administered by women research assistants to ensure that women in the Lake Tumba Landscape would be open to express problems that are of particular importance to them but could not be conveyed in public areas where usurpation of the by male is often the modus operandi and where the dynamics of exposed and veiled strengths of powers are inherently present.

Both the socioeconomic and the rapid rural appraisals uncovered the fact that women had major stakes on land right issues; they were the single most important stakeholders' group in the use of the non-timber forest products and were also part of the other regular work such as fishing, and palm oil production. Unfortunately, despite the widespread acknowledgment of the critical role played by women in the overall economy of the Lake Tumba Landscape (which parallels the global situation in the Democratic Republic of Congo) and the potential impacts of their daily livelihood activities, women have been often barred from the involvement in the land administration, the governance of natural resources as was reported from Ghana (Torvikey 2016). Once more, exclusion of women can be essentially tracked back to the legacy of the traditional customs, as they were reinforced by the colonial education system (Yates 1982), and the fact that the Landscape as a biodiversity conservation approach has been placing a substantial emphasis on including the voices of women in the processes that lead to decisions on land rights, usages of the forest products (be they timber or non-timber), the prospects for sustainable livelihood, etc. is a major win, even though it is not a final win in itself. Hence, apart from the generic awareness campaign about the role of women in the global economy of the Lake Tumba Landscape, several tools were used to ensure that women's voices were heard and accounted for in the discussions and the decision-making processes. The landscape's socioeconomic teams worked on ensuring that women became effective members of the village sustainable development and village community conservation platforms. This was made possible through women's specific fora on sustainable development and community conservation program whereby women debated their own specific problems before being brought by their representatives to the village level forum. From these fora, women representatives to the village assemblies were elected by their peers. To ensure equitable repartition of voices in the village committees, it was decided that women would have a representation of 50% of these committees' memberships. Efforts were also made to craft into mandatory procedural requirements that village committees' meetings would start only if the quorum of 30% of women's representatives were present.

The above procedural measures were, in essence, what has been also tried in other circumstances (see e.g. Kisambu 2016) but true innovation from the landscape perspective was around the fact that women were given leadership not only in matters where their expertise was acknowledged by the overall community but on issues that were of the global importance for communities in general. In a sense, it would not be fair to say that women were given the leadership because the fact that many women were brought to the forefront of the social issues because their husbands and sons died throughout the wars has also brought them to take the lead in many issues that traditions saw as the attributes of sole men. However, these matters to which the leadership of women was expanded included, for example, issues such as negotiations on the assessment of needs in land (see Chap. 19 in this volume), fishing regulatory procedures, economic analysis, etc. Of particular interest for biodiversity conservation organizations was the acknowledged influential role of women in the commercial bushmeat trade, which has been also described in other African contexts (Ellis Historically, these areas have been often left to very few people to debate on and often these few people were, in their majorities, men. So the fact of insisting that women should be part of the leadership in these areas was a stroke on the traditions of the region where the Lake Tumba Landscape is located. Apart from that stroke, the landscape approach took a path that would easily qualify as being affirmative in its actions through a more inclusive programming, not only focusing on women-specific issues but also by including empowerment of all as part of milestones to the success of biodiversity conservation. Indeed, specific studies (e.g. Inogwabini 2016) to enlighten not only the plights but also the potential of women to lead communities in the Lake Tumba Landscape were conducted to ensure that women perspectives and the influence that their works has on natural resources, the environment, and the economy of the Lake Tumba Landscape is decently and sufficiently factored in the decision-making process. These studies also looked at the influence important issues such as policies and practices of land uses, agriculture, extraction of timber; commercial fishing, commercial bushmeat trade, etc. have on women's abilities to live as they would want through examining relationships between deforestation, land usages, agricultural practices, and the women's participation in community's livelihoods.

All the above steps have increased the women's social capital in several ways though the landscape as a biodiversity conservation approach would not be expecting to wash away traditions that have been anchored in people's mindsets for centuries. However, changes that the fact of seating in a meeting room with other stakeholders, including traditional authorities, were felt immensely by women. Indeed, looking at this from a perspective of a peasant woman who had never been offered a place on the table where issues of collective importance are discussed would give one a chance to see how deep the change was. Also being offered a conduit via which women's perspectives on sustainable livelihoods was also a major gain not only for women but also for the entirety of communities.

## 18.8 Overall Who Wins in the Political Economy of the Landscape?

It is really tempting for one to state that the political economy of the landscape approach is a win-win game. But this is too easy and alerted minds would ask the question whether the theoretical backgrounds of the win-win analysis are not a simple game of words or wishful thinking in reality. Indeed, as indicated at the introduction of this chapter, you cannot win on all fronts; there are things you win when negotiating your interests and there are things you have to give up. To repeat Professor David Johnson, the Rolling Stones were at least once right: 'you cannot always have what you want'. This means when you acquire x, you will not get y—all have a cost. This means that the whole question is how much one wants to lose to get what? There is a need to weigh the relative value of things one desires and one should sacrifice those things that have less meaningful outcomes. Of course, this is the analysis of opportunity costs, which is defined in this context as the identification of the next best alternative to the current situation wherein communities have needed to improve their conditions. In that context, every stakeholder has something to lose to gain a better alternative in his (her) activities within the landscape. While all stakeholders will win something, the major winners, at least in the ideal conditions of the implementation of the landscape approaches, however, appear to be in two: the first one is the biodiversity and the second is composed of local communities. Biodiversity conservation wins because rather than being an annex to the core discussions of development, its needs are brought to the table of discussions and are at nexus point of the entirety of multiple scenarios: being brought at the front of the agenda is a major win for something that has always been the least considered in the development debate. Local communities win in significant ways because they gain power to be heard and to influence the economic and political decision-making processes in addition to being able to harvest the natural resources legally and in sustainable manners.

Finally, despite the fact that 'you cannot always have what you want', however, it can genuinely be argued that everyone wins in the landscape approach. Indeed, everyone wins in confidence and trust, which as Stiglitz (2015) says, offer the foundations for new forms of contracts and facilitate democratic processes and [smooth the processes of making inclusive] laws and are necessary for social stability. But confidence and trust are also functions of sincerity (without rude and condescending behavior) and transparent relationships. In fact, the insistence of the landscape biodiversity approach on inclusive participation in the decision-making process is the only path to fixing the confidence between local communities and the elite society that has been broken because of the opacity on the management of natural resources across the country.

## 18.9 Limitations of the Political Economy Analysis in the Context the Landscape

The neatness and the strengths of the political economy analysis should not preclude us from seeing some of its major limitations in the context of the landscape. The most important of these limitations is around the fact that the relationships between different entities analyzed in this context are mostly made of informal networks. Indeed, even though there are regular frameworks, stakeholders, and institutions discussed above are often defined in very soft boundaries; and sometimes these boundaries are even difficult to draw and when they are made apparent, they could be blurred very easily given the porosity and the complexities of the social makeup in the region of the Lake Tumba Landscape. For one and according to Fritz et al. (2009), political economy analysis is also a function of historical legacies, and prior experiences with reforms, social trends and forces (e.g. ethnic tensions), and how they shape current stakeholder positions and actions. For one, the historical legacies of the region are that of iron-handed management style, which is a legacy of the King Leopold II of Belgium, combined with years of lack of democracy. Both have suppressed the expression of communities. Secondly, the experience of reforms in the Democratic Republic of Congo has been essentially that of the Zaireanization, which was launched in 1973 with the motive to conquer the economic independence of the country. Unfortunately, despite the laudable objectives of the political action, measures announced to restore the control of the economy to Congolese ended up ceding most of the country's economic assets to a few *acquéreurs* (recipients) essentially from a group of privileged high-ranking party members, government officials, and local elites and notables who took over major plantations, ranches, and large commercial business enterprises formerly owned by the Belgians. These measures literally blew up the economic situation of the country and left a somewhat queer taste of whatever comes with the notion of economic reforms. Indeed, the landscape approach has been put into questions by communities in the Lake Tumba Landscape because men and women of these communities could, at best, not see how that would be different from other laudable experiences that failed in the past. These communities were rather very sceptical to see the participative process ignited by the landscape paradigm would lead only to tumults, disorders and […] scramble to loot whatever was left in forests at even faster rate by those who are politically influent and those who are at proximity to the ultimate sources of the political power, which is what Young and Turner (1985) observed decades before. The other important historical experiential story that communities have of reforms is that of the structural adjustments that were imposed on the country by the Breton Woods' institutions (the World Bank and the International Monetary Fund), which were based on the neoliberal approach. A summary of that experience has been succinctly provided by Heidhues and Obare (2011) when they concluded their study by saying that 'the neoliberal approach, on which Structural Adjustments Programs were purely founded, while theoretically sound, was fraught with pitfalls and failures to effectively address challenges to economic development in Africa'. Brief, the memories of constituencies in the Lake Tumba Landscape conserved the negative impact of their history with reforms and changes. That negative story was the single most important fact that set hurdles very high to jump and make the political economy analysis a less sensible tool. Thirdly, the social trends and forces shaping the communities were not so different than in the past: the politics of the country remained relatively the same. Democratic processes through which voices of the least educated, least influent and the least economically wealthy can be brought up to the political decision-making tables are still in their infancy and in some areas have been experiencing major drawbacks. These failures, inherent to any apprenticeship as they may be, continue to erode the confidence that communities would have in the institutions supposed to represent them. Even ethnic tensions, which many felt like died with the eclipse of generalized wars between 1965 and 1990, resurrected and were increasing not least because of the diminished resources. This was particularly the case of what happened in 2009 when the Dongo conflict also known as the Enyele rebellion broke out around disputed natural fish pounds. Severe other natural resources-linked conflicts are on-going in 2016 in some other areas of the Democratic Republic of Congo, including the Bantu–Pygmy conflict in the new Province of Tanganyika, formerly part of the Province of Katanaga and the Clashes between two villages of Mbwete and Tudimala in the administrative territory of Gungu in the Province of Kwilu, formerly part of the Province of Bandundu.

The relationships between entities in the context of the Lake Tumba Landscape were, in their essential, informal and the distribution of powers was rather difficult to disentangle because of unwritten rules, processes, and structures of the communities. Because of that and for the political economy analysis carried out above to work, there was a need to formalize some of the relationships, using the existing legal frameworks as well as introducing new ideas through a process to structure communities in different functional cells of the interests. That process was carried out throughout the Lake Tumba Landscape and has been described in chapters that follow, principally Chap. 20. Indeed, structuring communities in different functional cells of the interests has become a general trait of different aid programs (health, development, etc.); so, there is really nothing new here. The only novel thing in it in this context is that the process comes from a detailed Political Economy Analysis, which scientifically grounds what has been taken-for-granted in the factuality of the conservation practice.

## References

Agence Congolaise de Press (ACP) (2015) Les tarifs de la Régideso ne couvrent pas les charges d'exploitation. http://acpcongo.com/acp/les-tarifs-de-la-regideso-ne-couvrent-pas-les-charges-dexploitation/. Accessed on 14 July 2016

Almquist A, Deshmukh I, Donnelly-Roark P, Frame G, Pitkin B, Swartzendruber F (1993) African biodiversity: foundation for the future. The US Agency for International Development, Washington DC

Borgerhoff-Mulder M, Coppolillo P (2005) Conservation: linking ecology, economics, and culture. Princeton University Press

Chambers R (1983) Rural Development: putting the last first. Longman

Colom A, Bakanza A, Mundeka J, Hamza T, Ntumbandzondo B (2006) The socio-economic dimensions of the management of biological resources, in the Lac Télé—Lac Tumba landscape, DRC segment: a segment-wide baseline Socio-Economic study's Report. Submitted to the World Wide Fund for Nature

Food Aid Organization (FAO) (2014) Caterpillars in diets Congo (DRC). Focus on food habits in the DRC Basin: insects as source of protein in local diets. Agriculture and Nutrition: Lessons from the field and good practices. http://www.fao.org/fileadmin/user_upload/nutrition/docs/policies_programmes/Good_practices/12.FSNL_Factsheet_Caterpillars_in_DRC.pdf. Accessed on 18 July 2016

Fritz V, Kaiser K, Levy B (2009) Problem-driven governance and political economy analysis: good practice framework. World Bank, Washington, DC

Freeman RE (1984) Strategic management: a stakeholder approach. Pitman, Boston

Fukuyama F (1992) The end of history and the last man. Free Press—a division of M^c^Millan, Inc

Grandin B (1988) Wealth ranking in smallholder communities: a field manual. Intermediate Technology Publications

Heidhues F, Obare G (2011) Lessons from Structural Adjustment Programmes and their Effects in Africa. Q J Int Agric 50(1):55–64

Hulstaert G (1979) Sur l'emploi des plantes médicinales chez les Mongo. Revue de Recherches Scientifiques 1:61–75

Ilumbe BG (2006) Inventaires multi-ressources préliminaires dans la zone à gestión communautaire de Bobangi, Landscape 7, Province de l'Equateur. Rapport du volet ethnobotanique. Typescripted Report Submitted to the Innovative Resources Management and the US Agency for International Development, Kinshasa

Ingram V, Ndoye O, Iponga DM, Tieguhong JC, Nasi R (2010) Non-timber forest products: contribution to national economy and strategy for sustainable management. In: De Wasseige C, De Marcken P, Bayol N, Hiol Hiol F, Mayaux P, Desclée B, Nasi R, Billand A, Defourny P, Eba'a Atyi R (eds) The forests of the congo basin—state of the forest 2010. Observatoire des Forêts d'Afrique centrale (OFAC) and the Congo Basin Forest Patnership (CBFP), pp 137–154

Inogwabini BI (2016) Congo basin's shrinking watersheds: potential consequences on local communities. In: Rao P, Patil Y (eds) Reconsidering the impact of climate change on global water supply. Use and Management, IGI Global

Kisambu N (2016) Mainstreaming gender in Tanzania's local governance. In: International institute for environment and development

MacCormack CP (1982) Control of land, labor and capital in the rural southern Sierra Leone. In: Bay EG (ed) Women and work in Africa. Westview Press, pp 35–54

Mbimba M (2015) Territoire de Bomongo: Fiche du territoire. Cellule d'Analyses des Indicateurs de Développement (CAID). RepubliqueDemocratique du Congo, Kinshasa

Mfou'ou JM (2015) Regards croisés sur les societies civiles Africaines à partir des cas du Ceméroun, Burundi, Bénin, Côte d'Ivoire, République Démocratique du Congo, Gunée Conakry et Tchad. Cahiers du Mapinduzi 4: Les sociétés Civiles en Afrique, pp 92–103

Mitchell RK, Agle BR, Wood DJ (1997) Toward a theory of stakeholder identification and salience: defining the principle of who and what really counts. Acad Manag Rev 22(4):853–886

Oates JF (1999) Myth and reality in the rain forest: how conservation strategies are failing in West Africa. University of California Press

Pretty JN, Guijt I, Thompson J, Scoones I (1995) Participatory learning and action. IIED Participatory Methodology Series

Programme des Nations Unies pour le Développement (PNUD) and United Nations Office of Project Services (UNOPS) (1998) Programme de national de relance du secteur agricole rural (PNSAR) 1997 – 2001: Monographie de la Province de l'Equateur. Ministères de l'Agriculture et de l'Elevage, du Plan, de l'Education Nationale et de l'Environnement, Conservation de la Nature, Forêts et Pêche. République Démocratique du Congo

Société Nationale d'Electricité (SNEL) (2016) La Tarification SNEL. Available at the society's webpage: http://www.snel.cd/dl/LE%20TARIF%20PUBLICATION2.pdf. Accessed on 14 July 2016

Stiglitz JE (2015) La grande fracture: les sociétés inégalitaires et ce que nous pouvons faire pour les changer. Les liens qui libèrent

The Guardian (2011) Military spending: how much does the military cost each country, listed

Torvikey DG (2016) Strengthening women's voices in the context of agricultural investment: lessons learnt from Ghana. In: International institute for environment and development

United Nations Environmental Program (UNEP) (2014) UNEP Year Book 2014 emerging issues update: Illegal Trade in Wildlife. UNEP, Nairobi

Van Hoof F (2011) Changer l'agriculture congolaise en faveur des familles paysannes: Des dynamiques paysannes dans les différentes provinces de la RDC. Alliance Agri Congo

White FE (1982) Women, work and ethnicity: the Sierra Leone case. In: Bay EG (ed) Women and work in Africa. Westview Press, pp 19–34

Yates BA (1982) Colonialism, education, and work: sex differentiation in colonial Zaïre. In: Bay EG (ed) Women and work in Africa. Westview Press, pp 127–152

Young C, Turner TE (1985) The Rise and Decline of the Zairian. State. University of Wisconsin Press

# Chapter 19
# Assessing the Needs in Lands in the Lake Tumba Landscape

**Abstract** The most important issue identified by communities was that of land ownership. The chapter discusses the understanding by communities of land, land ownership, and land rights. Using a participative land need assessment, the study identified agroforestry lands, hunting lands, fishing basins, utilitarian conservation areas, and culturally access-prohibition areas as five functional land units that communities had in their minds. The study found that it was difficult to extrapolate figures from means of land needed by households to project total land needs in the landscape. Traditional land rights are not necessarily correlated with community population sizes; extrapolating would necessarily lead to problems because some communities would lose and other gains. This recommends that land need assessment is conducted on village-by-village approach, which is prohibitively expensive. But, positively, communities would be willing to give up some of their lands if there are no other activities such as logging in their lands and provided that the government keeps the link to the traditional land ownership, factoring the history of the land occupancy, land use-related culture, land layouts, present dynamics, and future needs.

**Keywords** Land needs · Land rights · Traditional land rights · Land functions · Hunting lands · Fishing basins · Culturally-prohibited lands

## 19.1 Introduction

In the discussions that were held throughout the Lake Tumba Landscape on the landscape approach, almost all communities clearly indicated that for them the most important issue that had hoped to be sorted out by the new approach would be around the land ownership and the rights of communities to possess the lands that previously belonged to their ancestors. Indeed, the conflict between the local communities and modern State authorities after the agreements between traditional chiefs and Henry Morton Stanley had been used by King Leopold II to confiscate the lands that were previously the properties of different communities (see Chap. 4) emerged an unending conflict. Hence for these communities, access to land and possession of rights that would not be trampled over by the governmental authorities or other stronger



B.-I. Inogwabini, *Reconciling Human Needs and Conserving Biodiversity: Large Landscapes as a New Conservation Paradigm*, Environmental History 12,
https://doi.org/10.1007/978-3-030-38728-0_19

entities was the most awaited outcome of all the participative processes. Identifying resources, whether strategic or vital was all good but these resources are located somewhere; what communities wanted was to access this somewhere and at any time of the year without being constrained to relinquish that right of access by any other legal subterfuge. Because of the vital importance given to the access and possession of rights on lands, the landscape as a conservation biology paradigm to achieve biodiversity conservation has its successes pending on the successes it would achieve in proper land use planning. That land use planning process has to be participative, with agreed-upon definitions, protocols, and processes that would lead to a common understanding of who wins what and where (see Chap. 13). Indeed, viewed in terms of the weighted relative importance of each win by the communities, there was only one highest win, apart from being given the space to get the voices expressed on the negotiation table. That relative highest conquest is access and possession of land rights.

This chapter discusses the definitions of what community members had in their minds as land, land ownership, and land rights. It also presents some results of the participative land need assessment that were conducted by communities with support of the technical staff from different biodiversity conservation and sustainable development organizations that worked in the area on the subject of how to bring the landscape approach from its nice theoretical approach down to the realities of being implemented on the ground.

## 19.2 Defining Land and Land Needs

According to the Oxford Dictionary of Geography (Mayhew 2009), world-widely, land is defined as the part of the earth's surface that is not covered by water and in ecological studies, this corresponds to the terra firma (or the solid ground), which is the dry land that does not get submerged even after flood events. However, in legal terms, the land would equate to a field of activities and does imply a relationship between people and the lands they own, in which relationship is mediated by the human actions. It is in this sense that communities of the Lake Tumba Landscape have been describing land to include more than just the solid crust of the planet; they say the land issue as including whatever space they might use to extract resources to live on. Indeed, for communities in the Lake Tumba Landscape, attributing functions to available lands should be disentangled at two different levels. The first level was that of community lands and, at the second level to attribute individual functional units to members of the communities within these communal lands.

A first difficulty that should be discussed before going any further with this narrative of the process that took place was the fact that, despite de facto possession, communities have lost rights of ownership on lands where they dwell none of these communities accepted the idea of vacant land. Hence for communities, the idea of talking about vacant land was improper because they claimed unused lands did not mean vacant lands; lands have always had been owned by communities even if they were not put to any use (Inogwabini 2014).

## 19.3 Approaches to Land Need Assessment

There were both theoretical and practical assessments carried out through the Lake Tumba Landscape. The theoretical approach consisted of taking numbers of individuals residing in the landscape and disentangling them in different categories and attributing to each individual the same amount of land. But these methods were preceded by a participative exercise to define the typology of functions to be attributed to each piece of land. From the local communities' perspectives, there should be many categories of land units but all could be brought down to 5 types of land units: (1) agroforestry lands, (2) the hunting lands, (3) fishing basins, (4) utilitarian conservation areas, and (5) culturally access-prohibition areas. Agroforestry lands were areas needed by communities for their fields as well as collecting fuelwoods and other petty natural products to satisfy their immediate needs. The hunting lands and fishing basins are areas that are left without massive destructions of habitats to allow communities to hunt and fish; these zones should be left intact in their habitats. The utilitarian conservation areas are zones of forests where some strategic products such as trees of caterpillars, palm trees for collection of thatches, and other specific products should be preserved for the benefits of entire communities. The culturally access-prohibited areas are these areas whereby traditional taboos only some activities can be executed in and by specific type of people at specific periods of the year. Additional to these land units acknowledged by communities, there were two additions in the laws of the country: (1) conservation areas and (2) logging concessions. Conservation areas are any area left for the biodiversity conservation in any category of protected areas as defined by the International Union for Conservation of Nature. In some areas, these would include the community conservation areas, which were an addition made by the landscape approach to ensure that where there are no proper protected areas and no culturally access-prohibited areas, communities should always set aside some areas to serve the purpose of biodiversity. The logging concessions are, of course, areas left for extracting timbers, under circumstances and conditions laid out by the laws of the country.

## 19.4 Theoretical Land Needs Assessment

The theoretical approach was obviously far removed from the reality but was, nonetheless, a necessary part of the process because it helped identify the thresholds beyond which allocating lands to individuals and communities would not be feasible. In its basic methodologies, the theoretical approach to assess needs in lands was preceded by a rapid field assessment to collect data about the mean sizes of field and mean distances between major access routes and the '*latest*' field in the area. Firstly, mean-field sizes were used to calculate the crude needs in lands for the whole landscape. The mean size was 0.75 ha, within the margins of 0.1–2 ha. However, the analysis was not done to capture individual land needs per se but factored the notion of households and the community lands. Indeed, children, old people, and people with severe physical handicaps do not use any land for agriculture. However, several people in the same household can have their own independent land-related activities. For example, the socioeconomic study in the southern Lake Tumba Landscape indicated that the mean number of fields per household was 1.6 fields, within the margins of 1–5 fields. Secondly, mean distances from major access routes were used to create buffers around roads to project the potential areas to be covered by the first type of land use unit, which is the agroforestry land. For the entirety of the Lake Tumba Landscape, the mean distance between the 'last' field and the main access route was about 5 km (5.2 precisely), within the range of 0–12 km.

With these metrics, it was possible to use standard geographic information systems tools such as ArcGis to generate a default land cover of land needs. Indeed, with 800,000 households with 1.6 fields each, it is easy to calculate the total number of 1,280,000 fields; and if each has to be of the mean size of 0.75 ha, the total land needed just for agriculture would amount to 9,600,000 ha or 9,600 $km^2$. This represents just 1.2% of the total landscape area (80,000 $km^2$). But this should not be seen as all that is needed by local communities and should be read with the fact that 60–65% of the Lake Tumba Landscape is composed of swamps. If that fact is factored in, the entire remaining terra firm land area of the Lake Tumba Landscape is 28,000–32,000 $km^2$. This means that the above figure of needs in lands for communities represents a much higher figure of 3–3.4%. As a planning exercise, the need for lands should factor the trends in population increases that are, as we have seen above, rather higher in the region. Projected until when the Lake Tumba Landscape would become home to more 1,000,000 households, which is around 2021, if the population increase rate remains 3%, which is likely to be the case, the land needs would be higher than they appear. Indeed, with a constant increase of 3%, in 2021, there would be about 6,150,000 people in the Lake Tumba Landscape, which would equate to about 1,025,000 households. These figures would lead the needs in land to amount to 12,300 $km^2$. These figures indicate that an increase in populations by 3% implies an increase in the needs of land by about 8%. So the difficulty of land needs assessment is that it also has to factor needs of people who are not born yet.

The difficulty to plan for needs of unborn people, which is part of the definition of sustainability, is to be coupled with other forms of lands, which fall in other types of activities described above (the hunting lands, fishing basins, utilitarian conservation areas, and culturally access-prohibited areas). Furthermore, the exercise should also take out the lands allocated for towns and other major human settlements. In so doing, the exercise should also factor the development of these towns and their potential expansions in the foreseeable future. Most of these other needs in lands are difficult to measure without being able to delineate them in the field.

## 19.5 Theoretical Land Needs Assessment and the Reality of Claims Over Land Rights

As stated above, there are numerous problems with the theoretical land need assessment. However, the most important of these problems is the conceptual problem of trying to allocate equal size areas to each individual or households. Indeed, the reality is that traditional land rights were not acquired in these terms; there were villages that had more land than others and these differences were not necessarily proportional to the populations of each village. As an illustration of this reality, Kondjo (2009), supported by a project known as 'Capacity Building for the Sustainable Management of Forests in the Green Heart of Africa' (a project funded by the World Wide Fund for Nature Program in the Democratic Republic of Congo), was tasked to produce participative maps for the claimed lands of Bodzuna, Embirima, Mpelu, and Nkala in the southern part of the Lake Tumba Landscape. He worked with local communities from each village who brought him throughout the areas and showed him where the limits of each territory owned by each community as a legacy of the ownership by the community's ancestors. He then made a tracklog of the entire perimeter for each area being claimed by each community.

Kondjo (2009) calculated the surface within the perimeter of land claimed by the community of Bodzuna to be 9,436 ha; Embirima laid claim on 13,564 ha, Mpelu claimed rights over a total surface of 10,412 ha and for Nkala the calculated surface amounted to 8,809 ha. These villages had respectively 29, 23, 31, and 11 grand families. Left aside the definition and the mean size of what grand family is these figures show how difficult it is to extrapolate figures from means of land needed by households to project total land needs in the landscape. Indeed, ratios of the total land traditionally owned over numbers of grand families are respectively 325.3 ha per grand family, 589.7, 335.8, and 800.8 ha per grand family for Bodzuna, Embirima, Mpelu, and Nkala. A clear biodiversity conservation lesson here is that traditional land rights are not necessarily related to the population sizes for communities. Hence trying to extrapolate would necessarily lead to problems because some communities would lose and other gains.

The same work by Kondjo (2009) identified different land uses per community. Forests occupied 48.8%, 63.7%, 56.8%, and 48.1% of the land whose rights were claimed respectively by Bodzuna, Embirima, Mpelu, and Nkala. These proportions of land had been allocated by communities for hunting, collecting non-forest timber products, and extracting timber for community's activities such as construction. However, it has been observed that some communities also practiced artisanal logging and sold the wood logs to industrial logging companies. Equally important proportions of lands, essentially composed of savannahs, respectively 48.4, 30, 41.6, and 42.4% for Bodzuna, Embirima, Mpelu, and Nkala were allocated to cattle husbandry both small-scaled and large-scale ones. In the particular case of Nkala, 50% of the 42.4% were leased by the community to an industrial cattle husbandry. Agriculture, which was used in the theoretical land need assessment as the prime factor in allocating land, shared smaller percentages of the magnitudes of 2.6%, 5.3%, 1.3%

and, exceptionally higher, 9.3% of lands for Bodzuna, Embirima, Mpelu, and Nkala, respectively. Human habitations and village installations (houses, toilets, and other infrastructures) occupied even smaller percentages of the land claimed by communities, ranging from 0.2% to 3%, respectively, 0.3, 1, 0.3, and 0.2% in Bodzuna, Embirima, Mpelu, and Nkala.

The above figures call for certain observations. The first and obvious one is that land needs are far from being simple mathematical models. This statement does not imply that mathematical models are inept to picture the realities behind land needs but rather that if one would need to use a model that model should be sufficiently robust to capture variations that are linked to history, culture, present, and future needs. In fact, taking the average of 0.75 ha per person would have led to hopelessly smaller estimation of needs in lands in this part of the landscape. Indeed, with a total population of 1,762 people, communities in the whole area around Malebo would be estimated to need only 1321.5 ha while their real current claims total 42,221 ha, when figures above are added together. This means that modeled tabulations of the land needs above would represent only about 3% of the real needs and claims.

That is why mathematical models, in order to be robust enough need to factor the history of the land occupancy, land use-related culture, land layouts, present dynamics, and future needs. History is an important parameter because the land occupancy results from the dynamics of the history of given territory. Indeed, some smaller communities may lay claims over large expanses of land simply they happen to have occupied and marked a land as theirs when other communities were not present in the area; the precedence in the chronology of occupancy is a major factor in this sense and can be justified by is what Inogwabini (2016) has called the right of the first occupants. Secondly, communities that are smaller, in comparison with their current neighbors, may have been large enough in the past when they occupied the current land. History may have seen people from these currently smaller communities diminish either through epidemics or emigration for several reasons (education, search for paid jobs, etc.). Indeed, discussions with the local communities around the Malebo area (in the southern Lake Tumba Landscape) had clarified why such a tiny community of 175 people as Nkala had far more land than other neighboring communities by indicating that Nkala was once a densely populated village but its population decreased over years because many people died for reasons unidentified to them.

Culture too is an important parameter to factor while assessing land needs. As one would have read from the figure above, significant proportions of community-claimed lands are allocated for cattle husbandry. This picture is quite different from the experience in the northern part of the Lake Tumba Landscape where cattle husbandry is rather rarer as an activity. In fact, communities in the southern Lake Tumba Landscape use close to 40% of the entire land they claim for cattle husbandry which is not even mentioned in the report by Akwah et al. (2006) on the land use planning in the northern part of the landscape. The cultures of the two regions are very dissimilar in some of the essential livelihood activities. In the area around Bobangi where Akwah et al. (2006) organized several participative exercises on land needs, land use, and community land management communities insisted on community fishing zones that were to be managed in parallel with communities hunting zones. The fishing zones are hardly even mentioned in the southern part whereas cattle husbandry areas are absent in the northern part. This is just one illustration to show that culture, in the sense of habits and ways of using the space and resources, influence land needs.

Clearly, the example above contrasting fishing zones versus cattle husbandry zones in the northern part and the southern part, respectively, is not only a good example of how cultures influence land use and land needs but also a good one for how the land layouts do influence cultures and, subsequently, land needs and the definitions of land functional units. According to data extracted from Akwah et al. (2006), communities in the northern part of the landscape, particularly people residing in the area from Mobenzeno to Mikungu (along the Congo River) estimated their needs in hunting zones to be around 16,000 ha (about 21% of the total area of 75,000 ha of the Ngiri area). These hunting zones, of course, are inclusive of other activities identified in the case of the southern Lake Tumba Landscape and include collecting non-forest timber products and extracting timber for community's activities such as construction. Comparing the figure of 21% for the northern part and that of a calculated average of about 54% of land allocated for this category of use in the southern part, the differences speak by themselves for the populations of relatively similar sizes. The same differences also clearly emerge when comparing the needs for cultivable lands for the 17 communities located between Mobenzeno and Ilanga, which were estimated, using the same data from Akwah et al. (2006), to be were estimated at 20,000 ha (about 27% of the total areas) and the average estimate of the four villages around Malebo, which is around 4.6%. Completely absent in the northern part was the category of cattle husbandry land. Conversely, the identified category of fishing zone, for which an area of 12,500 ha (17% of the total area) which was a need expressed by the 10 semipermanent villages of the northern part of the Lake Tumba Landscape at region was totally absent as a need in Malebo. These differences show adaptation of communities' activities to their environments as well as the land layouts. In fact, communities in the northern part of the Lake Tumba Landscape are, as described by Harms (1989) (see also Chap. 20), communities of Bobangi, Likoka, Libinza, etc. of the northern Lake Tumba Landscape have fishing as their principal livelihood activity; this has been on-going for ages and at least since the nineteenth century (Harms 1989); communities of Bateke of the southern landscape are known to be good traders, hence have cultivated their interests in cattle husbandry for commercial reasons.

## 19.6 Combining the Theory and the Reality of Claims of Rights to Make Sense

Given the above-described discrepancies between the simplistic theoretical approaches that give very smaller areas for communities and the reality that seems to provide larger figures, the need to find solution that would be not only a suitable solution for all stakeholders but also a sensible and morally justifiable one emerged. The morality was based on the fact that local communities had the rights over natural resources found in the locations where they happen to be established whether they had the jure right or not. As it has been argued in Inogwabini (2016), basic human needs of local communities should be fulfilled for reasons of justice, fairness, and right to life. But, as it has been also argued in this volume, the most important resource for these communities is land and waters that land stores. Looking at the ranges of needs and claimed lands, the most obvious conclusion is that there cannot be a 'one-size-fits-all' standard allocation of land that would apply across the entire landscape.

The second logical conclusion is that measuring fields is not a fair measure to express claims from communities. There are obvious reasons for this to be the case. Firstly, the nature of the agricultural practices in the region is that of shifting culture, which implies a lot of movements from one location to the other. People move their fields at least every two years and when the nutrients of the soil are totally tarnished, communities take very long time to recultivate abandoned field. As such, the needs of lands expressed by communities are logically in alignment with this factor, which cannot be factored by single measures of actively cultivated areas, which presided over the theoretical methods. Indeed, for the theoretical methods to provide estimates that are closer to the true needs, they have to factor the space needed for moving one's field to leave the cultivated space to recover. The same logic of shifting occupancy does also apply when it comes to fishing and hunting. Fishing is always shifted from one location to the other depending on seasons and the radiuses of these shifting occupancies are rather very large. The magnitudes of these shifts have been reported to vary from several hour-walk distances to day-walk distances. In terms of geographic spread, these estimates are reasonably beyond 40–50 ha annually per person. Hunting too does move depending on times of the year as well as the abundance of wildlife species. People would walk long distances before they are able to spot any species or even any sign of a living species. These facts make it difficult to assign a buffer around the main communication route to delineate fishing and hunting zones through the easiness of the approach that may look very appealing at first glance. Shifting occupancy also means mobility of people; in some of the particular cases of fishing and hunting, people would move into the forest for relatively long periods to fish and to hunt. For fishing, this corresponds to the periods such as dry seasons when catches per unit effort are higher. During these periods, fishermen (fisherwomen) and hunter would invest the most efforts they can to capture as many species as possible before coming out for selling the products they have amassed during these periods.

The theoretical conservation biology question from whole this situation shifting occupancy (shifting cultivation, shifting fishing, and shifting hunting) is then: with dynamic land use what are the appropriate methods to calculate needs in land for communities in order to stabilize their effects on the wild habitats? Obviously, there is no single model. In the Lake Tumba Landscape the approach applied was a multi-model one. First, it was a combination of measuring field and calculating rates of fallows and land needed to care for the necessity for cultivated land to recuperate some significant portions of its initial fertility. Secondly, the definition of hunting or fishing zones had to be used and make calculations of these needs as public goods rather than proceeding through individual needs. Thirdly, the hunting and fishing zones were to be calculated also using the needs of communities in off-takes of wildlife species; this meant a detailed knowledge of the ecology of species of interests.

Materially, the above combinations led to projecting an area of 2.8 ha for agriculture by a standard household, which is made of 8 people, as described above (in this chapter and Chap. 18). The figure of 2.8 ha comes from multiplying the average measured field size with 4, which is the period during which the ploughed and harrowed lands are left without being sown to regain their fertilities. It was postulated that during that period each household would need to move to a new piece of land to sustain its own livelihood's activities. This adjustment alone will bring the land needs of local communities at 2,240,000 ha of the total of the Lake Tumba Landscape area, which is (once more) equal to 8,000,000 ha. This amount represents 28% of the total area. This is huge difference when compared to the meager 1.2% calculated above. This figure should add the land (forest) needed for hunting and fishing. The mapping of the communities by the mapping of Kondjo (2009), Akwah et al. (2006) have come up with an averaged figure about 440 people as composing what could be fairly called a community, which is made of an assemblage of villages that share the same culture, respond to the same traditional form of rule and apply similar principles while using the same natural resources which they withdraw from one or multiple joined land units. As indicated above, the hunting or fishing zones are reliably calculated using communal lands (forests) rather than being presumed to be individually owned. In this case, the estimated 5,000,000 of the Lake Tumba Landscape would represent roughly 11,350 communities. With an estimate of 156 ha for mean hunting zone per community, inclusive of fishing zones in the northern part of the Lake Tumba Landscape, this would lead to 1,770,715 ha representing 22% of the total landscape area. Put together, the needs for communities would, therefore, equate to more or less 50% of the entire Lake Tumba Landscape, albeit with the fact that some areas would combine different functions.

Regardless of what may be argued about the accuracy of the estimates, the value of this exercise resides in the fact that it shows how complex the exercise to assess community land needs is. There are multiple hidden factors that need to be brought to light and incorporated within the analysis. Of course, the land need assessment is far from being the only exercise in attributing functions to different land units in a given landscape. Yet, the process to identify needs of communities is equally important as, for example, identifying areas for mining or timber extraction. Once these needs are known then a mapping exercise has to be conducted to separate areas that wield other types of value such as biodiversity areas and industrial extraction of natural resources.

## References

Akwah G, Mogba Z, Tchunza G, Yoko A, Punga J, Mucici PM, Isolumbu J, Akiak J, Konso MR, Bokondokondo F (2006) Rapport synthèse des ateliers préparatifs d'affectation des fonctions et usages des espaces dans la zone à gestion communautaire de Bobangi (Mission du 13 au 27 Mai 2006). Type-scripted report submitted to World Wide Fund for Nature, Democratic Republic of Congo

Harms R (1989) Fishing and systems of production: the precolonial Nunu of the middle Zaïre. Cahier des Sci Hum 25:147–158

Inogwabini BI (2014) Conserving biodiversity in the Democratic Republic of Congo: a brief history, current trends and insights for the future. PARKS 20(2):101–110

Inogwabini BI (2016) Congo Basin's shrinking watersheds: potential consequences on local communities. In: Rao P, Patil Y (eds) Reconsidering the impact of climate change on global water supply, use and management. IGI Global, USA

Kondjo KH (2009) Cartographie participative des terroirs communautaire de Mpelu, Bodzuna, Embirima and Nkala (villages aux alentours de Malebo), province du Bandundu. Projet de renforcement des capacités pour la Gestion Durable des Forêts dans le Cœur Vert de l'Afrique. Type-scripted report submitted to World Wide Fund for Nature, Democratic Republic of Congo

Mayhew S (2009) A dictionary of geography, 4th edn. Oxford University Press, Oxford

# Part IV
# Using the Data to Strategize and Manage the Landscape

This part of the book is essentially about how to use the data collected throughout the Lake Tumba Landscape and presented in part 3 to manage natural resources of the landscape for the benefits of both the biodiversity and human communities. As would have been already clearly stated in the first part of this book, the landscape approach is a way to reconcile the needs for local communities to live a decent life through the use of natural resources, including biological resources and those that would ensure the perpetuation of biological diversity throughout the landscape. Hence, this part of the book deals with technical ways of making that reconciliation possible on ground. The fourth part of the book uses data and other forms of knowledge collected throughout the landscape and assembles them in a way that deliver on both ambitions.

The first chapter in section (Chap. 20) starts with a general landscape planning process, which is the backbone of every possible achievable activity to ensure that communities and biodiversity benefit in the framework of the landscape approach. As a planning exercise, Chap. 20 starts with a brief situational analysis, even though this has been given so much attention in all previous chapters where the habitats, different landscape species, and threats to biodiversity were described. The brief situational analysis is given not only as a recapitulation of what has been already said but also as part of the methodological approaches on how to write management plans. Indeed, without knowledge of the situation, management plans would have no value at all; they are to address issues identified during the analysis of specific given situations. In presenting the situation of the Lake Tumba Landscape, the situational analysis gives some important details that were not presented earlier and tries to link them with specific objectives of the management plan. These management objectives were a combination of the national Congolese objective, the livelihood and development objectives for communities, and the objectives that biodiversity conservation organizations such as the World Wide Fund for Nature were pursuing either in the Democratic Republic of Congo or in Central Africa as the region. The chapter then moves on how to achieve these objectives, with what means and in what time span. Objectives, means, and the timing are linked through the logics of the intervention that is provided in the same Chap. 20. Although there are no tables and pictograms, the points on the logics of the interventions use the language of the logical framework, which is in some cases combined with the language of theory of changes and some principles of system thinking theories. Indeed, the landscape approach as a whole comes down to thinking about biodiversity and human communities as being part of a huge system whereby to ensure the survival of one component there is a need to improve the relationships between parts of each component of the system.

The next Chap. (21) presents different approaches to setting areas aside for biological diversity conservation. This chapter is equally important not only in its insistence on the fact that to conserve biodiversity throughout the landscape, there is need to get protection of the landscape species, which is in alignment with the traditions in creating protected areas, but also in the methods and different and alternative scenarios to get to the same results. Obviously, the chapter argues that there is no need to have fully protected areas across the landscape, lower IUCN categories such as category VI are sufficient if proper management mechanisms are put in place and agreements with communities are reached. The chapter also argues that besides the traditional protected areas, mechanisms such as Ramsar Convention should be exploited to reach the same biological diversity conservation targets. Indeed, the line of arguments being put forward in this whole volume is that multiple purposes and multi-disciplinary approaches are needed to make biodiversity conservation work. Principal among the argument is also the fact that human communities, made of men and women, should not be seen only as part of the problem with conservation but also equally as part of the solutions to problems biodiversity conservation faces. Chapter 22 that follows the needs to set aside lands for the long term sustainability of biological diversity in the Lake Tumba Landscape is about making law enforcement work. The chapter is about how to use scientific methods to identify and deploy conservation resources efficiently. Beyond that, however, the chapter adds on the possibilities of making law enforcement differently using the local and traditional law enforcement mechanisms. As such, Chap. 22 is a complement of the preceding Chap. (21) in that it argues along the same lines and argues that the best conservation efforts should not be only limited in paying law enforcement services but also be invested in getting binding support from communities. Finally, part four ends with exploration of how means can be identified to resources the efforts to conserve biodiversity across the landscape.

# Chapter 20
# Planning for the Management of the Landscape

**Abstract** This chapter presents a tool-kit of a planning process for priority conservation areas. While it is heavily based on the context of the Lake Tumba Landscape, it embraces broader and general principles of planning for conservation in multiple-use areas. These principles include conducting a situational appraisal (geography and the biological diversity, human ecology), economic analysis (logging, oil and other economic activities), setting global priorities, identifying resources (human resources, training needs, and financial and logistical needs), and identifying means to reconcile divergent demands (actions for sustainable uses) and exit scenarios. This very scheme was used for the planning process in the Lake Tumba Landscape, and a lesson to be learnt from that process was that using charismatic species (bonobo, forest elephant, and lion) residing in the landscape as the measure of sustainability revealed to be a powerful tool because people easily identified the existence of these species with their own history in the region. Another lesson was that greater participation in the planning process diffused rumors and ensured better relationships with different stakeholders, including local communities and state authority.

**Keywords** Situational appraisal · Increasing human populations · Logging · Oil extraction · Landscape global conservation priorities · Human resources · Training needs

## 20.1 Introduction

This section chapter presents a tool-kit of elements of the planning process through which conservation organizations must go through when working in a priority conservation area. Although it is linked to my work in Lake Tumba with the World Wide Fund for Nature, its aims are rather broader and are targeting the generally agreed-upon principles of how to plan for conservation in multiple-use areas such as landscapes. Indeed, to most people, except for their large sizes in the context of Central Africa, landscapes could be easily classified as protected areas of Category VI within the IUCN nomenclature.

 
B.-I. Inogwabini, *Reconciling Human Needs and Conserving Biodiversity: Large Landscapes as a New Conservation Paradigm*, Environmental History 12,
https://doi.org/10.1007/978-3-030-38728-0_20

## 20.2 The Situational Appraisal for the Implementation of the Landscape

The situational analysis conducted at the beginning of the project in 2005 and which has driven the process of writing the master document for the Lake Tumba Landscape can be summarized in the following areas: the geography and the biological diversity of the Lake Tumba Landscape, the issues related to the rampant poaching in the area, rapid human population growth, extensive and uncontrolled logging, and extraction of other potential resources that can be found in the Lake Tumba Landscape. Other issues of importance and that would have legitimately been part of this section have been treated either in the section on the political economy of the landscape or in specific sections treating different aspects of the landscape such as the Chap. 5 on the habitats, which does also include a long section on the human ecology of the communities residing in the Lake Tumba Landscape. The separation of such sections has been dictated mostly by reasons of fluency of reading and the coherence of the arguments.

## 20.3 The Geography and the Biological Diversity of the Lake Tumba Landscape

The Lake Tumba Landscape segment in Democratic Republic of Combo is 80,000 km$^2$ and has been described in large details in Chap. 5; this is a very brief summary, which serves as an introduction to the situational analysis that will take the reader to the next important section on priority making for the Lake Tumba Landscape. So, briefly, this 80,000 km$^2$ large area encompasses a bulky assemblage of habitats, albeit all of the lower tropical altitudes. The most dominant feature of the Lake Tumba Landscape is the presence of massive areas of open waters (the Congo River, and two large lakes). The other important feature is the presence of large swamps, which are less dried most of the time. Biodiversity-wise, as illustrated in Chap. 5, the biodiversity of the area comprises primates (bonobo, Angolan pied colobus, Allen's swamp monkey, black mangabey, Thollon's red colobus, red-tailed monkey, de Brazza's monkey), forest elephant, forest buffalo, top predators (leopard and lion), two crocodile species (*Crocodylus cataphractus* and *Crocodylus niloticus*), hippopotamus, and duikers (bushbuck, sitatunga, blue duiker, black-fronted duiker, yellow-backed duiker, etc.). The fish diversity is rich and includes some regional endemic species.

## 20.4 Rapid Human Population Growth

Mirroring the general trend of the country, the Lake Tumba Landscape is known by its rapidly increasing populations. Just to give an order of the magnitude, the town of Mbandaka, located at the center of the Lake Tumba Landscape, currently has about 1.2 million people (INS 2015) whereas back in 1996, the same town counted about 220,000 inhabitants (PNUD and UNOPS 1998). Some of the increase in population in Mbandaka is due to people migrating from villages to towns, which is another general trend for the country. However, the Mbandaka and other large human settlements in the Lake Tumba Landscape are also characterized by burgeoning internal birth rates. According to PNUD and UNOPS (1998), the internal birth rate in the province of Equateur where Mbandaka is located galloped up from 2.06% per year between 1970 and 1956 to 2.6% between 1970 and 1980. Finally, it jumped from 2.6 to 3.4% between 1980 and 1994. Other administrative entities located within the same landscape had also seen the same increases in their respective populations. For example, the populations of Bikoro almost doubled in the same period, going from about 185,000 in 1994 to about 367,000 inhabitants. In most of the administrative territories of the northern Lake Tumba Landscape, the most alarming increases were that observed in the territory of Mankanza where the population more than quadrupled jumping from 37,000 people in 1996 to 170,000 people in 2015. The same situation was also observed in territories such as Bomongo (about 47,000 people in 1996) and Lukolela (72,000 in 1996) where populations respectively increased from 184,000 to 291,000. In the southern part of the Lake Tumba Landscape, the demographic situation is not that different. In fact, none of the administrative territories of the southern Lake Tumba Landscape had less than tripled its populations between 1994 and 2014. Indeed, the population of Inongo went from 207,000 to 621,000 inhabitants (3% increase), the population of Kutu almost tripled from 320,000 people to 804,000 people (2.5%) whereas that of Bolobo more than quadrupled from 60,000 individuals in 1994 to 250,000 people in 2014. Almost the same thing happened with the population of Yumbi that went from 60,000 people in 1994 to 220,000 inhabitants in 2014. Only the population of Kiri doubled from 118,000 people to 283,000 people (2.3% increases).

## 20.5 Rampant Poaching

With the demographic descriptions given above, combined with the fact that the majority (about 85%) of the total population, as reported by Colom et al. (2006), of these people are dependent on natural resources for their livelihoods, it is evident that the intensity of hunting and fishing would be higher. Poaching is known to be one of the major causes for loss of biodiversity. It is encouraged by easy access of the zone and easy availability of ammunition and automatic machine guns consequence of the social and political crisis that swept the country between 1991 and 2002.

## 20.6 Logging, Oil Extraction and Other Issues of Importance to Biodiversity Conservation

Concessions for logging exist in the heart of the landscape, from which is extracted the *Mellitia laurenti*, a black hardwood highly priced by the logging industries. The potential for extracting oil had been tested in the past. Indeed, ESSO explored the Congo River and confirmed the presence of petroleum reserves at Mbandaka, but exploitation was never initiated. With increased development needs, the government will initiate the extraction of this petroleum stock. High and unmonitored migrant fishing rates place a burden on fish populations. Bushmeat hunting is rampant; poaching activities are insidious due to the availability of weapons and ammunition and the physical location of the landscape at the crossroads of river routes. The military base is present at Irebu and hosts unpaid and lowly disciplined soldiers that travel long distances to hunt for large mammals such as elephants and hippopotamuses. Despite the intensity of hunting and fishing, however, there are still embarrassing levels of poverty. In fact, poverty adds to excessive strain on natural resources through unplanned and unsustainable subsistence agricultural practices. About 60–65% of the landscape is in a swampy zone, making the landscape one of the most extensive wetland zones in the country.

Based on the above assessment, the driving philosophy of the landscape approach is to first take immediate actions that would alleviate the loss of the biodiversity in the landscape and then move the consolidation of those actions by providing long-term approaches and alternative solutions to the most intractable conservation issues that include points that are described below.

## 20.7 Setting Global Priorities for the Lake Tumba Landscape

The Lake Tumba Conservation had an umbrella vision, which was stipulated as follows: 'Sustainable natural resource management measures to conserve the Lake Forests are in place through a concerted land-use planning that attributes appropriate functions to each land portion and water unit included in the landscape. All charismatic species residing in the landscape segment such as bonobos, forest elephants, lions, and their habitats are protected in perpetuity.' The visioning exercise was undertaken through a participative process whereby many stakeholders met at Malebo in 2009 to discuss about the course of action to be taken in view of all the work that was needed to be done if success was to be expected of the collegiate work being carried out for more than 3 years in the region. The global issue addressed by the exercise was an assemblage of themes that would promote sustainable economic development and the alleviation of poverty to benefit local communities and the national community globally. Given the importance of poverty seen in the contextual analysis above, conservation activities would not be positively perceived if they were to prevent people from accessing resources that would help them mitigate further impoverishments. This vision was the guiding principle for all the consultations and the planning processes and presided over the specifications of operational objectives, which were of four global themes that were the paramount objectives of a consortium of conservation organizations operating in Lake Tumba.

### *20.7.1 Forests Protection*

In alignment with priorities set out by the Congo Basin Forest Partnership, the first global aim of the landscape paradigm (Chap. 1 in this volume), to protect the forests of Central Africa, the top priority action was to develop an integrated and participatory land-use plan. This was to go via several mechanisms, including, among the first priorities, the establishment and maintenance of viable, representative networks of protected areas in the world's threatened and most biologically significant forest regions. The Lake Tumba Landscape was to contribute to that milestone by assisting the Congolese Institute for Conservation of Nature (ICCN) to become more present within the landscape. This was because historically, Lake Tumba Landscape has never been a conservation priority area; there was just one small protected area, Mabali Scientific Reserve, in the region. Mabali Scientific Reserve has always been managed by the ministry of scientific research. With that in mind, plans were to help the creation of two natural reserves in the landscape. In the end, the two were created and named, respectively, the Tumba-Lediima Nature Reserve and the Ngiri Triangle Nature Reserve. The target to have the Congolese Institute for Conservation of Nature operates therein by 2010 and 2012, respectively. Still, at the time of the planning, another global target objective on forest protection was to improve management across the world's production forests, through a combination of credible certification and a stepwise approach to improved forest management. In alignment with that global objective, the Lake Tumba Landscape projected to develop, complete and implement the management plans through a participative process for each use zone within the landscape. The targets were 2 reserves, 2 community-managed areas and 2 extractive zones. It also pledged to lobby for the establishment of sustainable forestry management within at least 2 logging concessions in the Lake Tumba Landscape and logging companies are in the process of timber certification by 2011.

To develop an integrated and participatory land-use plan, there was a need for a broader conservation vision that emerged as part of the overall conservation paradigm shifts that begun in the late 1990s. This period coincides with when the conservation world noticed that protected areas alone would not help protect the world's biodiversity in its entirety even if natural reserves and other protected areas were providing the backbone of the wildlife conservation (Bajracharya and Dahal 2008; Madhusudan and Raman 2003; Karanth 2001; Ghimire and Pimbert 2000). This shift in paradigms meant that conservation organizations were forced to envision large-scale planning and to integrate other stakeholders in the process of designing new conservation schemes. That need was particularly acute in tropical forests of the world, which happened to be dually holding large components of biodiversity while also sheltering the poorest people economically speaking. Furthermore, in the case of the Democratic Republic of Congo, the land law in the Democratic Republic of Congo, as it is the case in most of the Central African countries, was conducive to numerous conflicts not only of usage but also between different social groups. Therefore, to ensure that conservation action is aligned with the economic needs of the country dictated by the legitimate seek of profits and the urgency to feed millions of people in this bourgeoning demography of the country, only a participatory planning process that allocates different zones to different modes of use was thought to be the best way for conservation to achieve concrete results that would last for a long term.

The second most important priority to protect forests of Central Africa was to establish new protected areas. As part of the global land-use planning process, one of the key functions for biodiversity conservation is protected areas wherein assemblages of the biological diversity unique to the zones should be protected. The inclusion of the establishment of new protected areas in the Lac Tumba Landscape Segment was dictated by two major elements: (1) the need to help the government of Democratic Republic of Congo achieve its commitment to increase the portion of its national territory included in the Protected Areas Network to 15% of the total and (2) the fact that most of the habitats in this part of the country had never been included in the Protected Areas Network (Inogwabini et al. 2005a). Therefore, as a contribution to both the land-use planning process as a model for setting conservation priorities and to the government of the Democratic Republic of Congo, establishing new protected areas has been selected as one of the top priorities of the Congo Basin Forest Partnership in its action in Lake Tumba Landscape.

Creating new protected areas was a captivating measure among others especially for the species of conservation concerns present in the landscape. Discussions indicated, right at the front, that there would be one protected area for the bonobos and another specifically for the chimpanzees. To achieve the conservation of some of the landscape species, particularly bonobos and chimpanzees of the Lake Tumba Landscape, it was agreed that these species would need specific conservation measures. Hence, it was a priority for all stakeholders to collaborate with and support the DRC Conservation Authority and local communities to establish new protected areas in the Lake Tumba Landscape. That support was to be multidimensional and should include such basic elements as collecting field and formatting that in such a way that they should support the decision-making process by providing all necessary data and documentation to help the gazettement process for new protected areas in the Lake Tumba. The support should also be in term of aiding the Congolese Institute for Conservation of Nature to prepare technical papers for gazettement; this was part of the capacity building too because of the limited experience the Congolese Institute for Conservation of Nature as a former colonial institution has in building constituencies on such issues as creating protected areas. Indeed, as indicated above and somewhere else (Inogwabini 2014), most of the protected areas (parks and reserves) that the Congolese Institute for Conservation of Nature inherited were either created in the colonial era or through the imposition of the Mobutu's regime: one stroke of the pen was enough to create a new protected area and drag people out of the lands their ancestors occupied for millennia. With the general push for human rights and democratic governance, times have changed and no single person, including any sensible die-hard preservationists, would want to see people being dragged out without their own prior informed consent to cede their lands for biodiversity conservation efforts. So, the Congolese Institute for Conservation of Nature needed to nurture a new type of mentality which would need more negotiations and discussions of trade-offs for its conservation policies to be implemented without causing so much anger in people who would feel being the losers of the whole gamble. Apart from these technical supports, there was also a need to provide material support to the Congolese Institute for Conservation of Nature to organize field visits to negotiate limits with local communities and to help bring stakeholders together for negotiations leading to the creation of new protected areas. Plans were also that the support to the Congolese Institute for Conservation of Nature should not be ended once the government of the Democratic Republic of Congo published legal decrees creating the protected areas in the Lake Tumba Landscape. The support should be continued to help the Congolese Institute for Conservation of Nature render these new protected areas a reality on the ground. But this was a separate priority, which is detailed below.

The third priority to protect forests of Central Africa was to maintain protected areas. Establishing new protected areas can be become a conservation achievement only when protected become operational and are maintained throughout their existence (Inogwabini et al. 2005b). Yet, the experience in the Democratic Republic of Congo, as in other countries in the Central African Region, is that many protected areas exist only on paper because of lack of resources (human and financial). Therefore, it has been recommended that creating new protected areas should be accompanied by some serious actions to materialize those new conservation zones on ground and make them become viable conservation entities. This priority includes materialization of new protected areas' limits, building operational infrastructures and, most importantly, building human capacity through training that empowers the protected area managerial authorities to be fit to its duties.

With the protected areas created, it was necessary to make them operational, viable and part of the communities accepted assets. To make the protected areas operational, the assessment conducted of the capacities of the Congolese Institute for Conservation of Nature and different consultations held led to the conclusion that the Congolese Institute for Conservation of Nature had very limited capacities to reach out outside its traditional areas, which are composed of older protected areas. Hence, there were several critical interventions needed to operationalize the newly created protected areas. One of these critical interventions was to support the Congolese Institute for Conservation of Nature install key law enforcement personnel in newly created protected areas. This, it was agreed, should be preceded by identifying and training the core management personnel. Because of the length of the time it would take to train protected areas managers and other support personnel, it was agreed to select the top management team from the existing junior staff of the Congolese Institute for Conservation of Nature and send them to rapid training session in order to move to the higher levels of management and come back with the expertise to manage the newly created protected areas. As far as the other support and paramilitary personnel were concerned, it was agreed that they should be recruited locally for the communities to see that they are being part of the management of the protected areas; the locally hired paramilitary and support staff were to be trained locally or at least at the nearest possible point from the concerned protected area. Training would not be only on paramilitary but should also include the soft side of how to deal with communities as well as the history of the communities living either within or near the protected areas. Indeed, as indicated in this volume, it was perceived that the best way to get communities accept more protected areas in the Democratic Republic of Congo was to have loose protection IUCN categories, namely category VI, which would allow communities to extract some resources for their own subsistence. Apart from building the capacities of the human resources, it was also felt that building the material capacities would be important to allow the Congolese Institute for Conservation of Nature operate as efficiently as possible. Thus, the second type of priorities included building minimum infrastructures to host the headquarters of these new protected areas in appropriate sites. Of course, minimum infrastructures were to cover both buildings and other operational equipment such as communication (Internet, HF radios, etc.), and office equipment (computers, printers, etc.). On the softer side of the interventions needed for these newly created protected areas to become effectively operational, it was felt of the utmost importance to help the Congolese Institute for Conservation of Nature conduct annual assessment of the biodiversity and other indicators of success; this should have to start by acquiring the baseline data, the tools to analyze and the capacities to format the data, interpret its meaning for the management of protected areas. The creation and the use of a change tracking tool was part of this softer capacity building exercise. Unfortunately, given the financial weakness of the Government of the Democratic Republic of Congo, the softer support would have also to cover salaries and rations for law enforcement to become effective. To curb this situation in the long-term perspectives, the parties agreed to help the Congolese Institute for Conservation of Nature establish linkages between protected area and payment of environmental services; this exercise would lead to a wider exploration of other funding sources, which would ensure sufficient and long-term funding availability to manage the newly created protected areas.

Finally, the protection of forests of Central Africa needed priority action around the rational Management of forest resources. There are no ways that all the forests in Central Africa, and particularly those found in the Lake Tumba Landscape, will be entirely locked for conservation. Indeed, increase in human populations and the genuine demands for development, which human communities of the forested areas of Africa express, have made that the philosophy of basic natural needs that Rousseau and his followers promoted are no longer holding for communities that aspire to better standards of living; people need more than just food and shelter. These additional needs will be met only using the natural resources of their countries, provinces, and villages. As Inogwabini (2010) stated, in fact, with all new developments, involving massive movements of people of different cultures and in communication technologies, nothing is left that could be called ‘immaculate culture’, not even in previously remote areas or even previously unshakable cultures such as those described in Conrad's hearts of darkness and, generally, Thoreau's Thorstein Veblen. Therefore, the best way to avoid dramatic losses of biodiversity in land units that will be allocated to extractive activities such as logging would be to influence the management logging concessions within the landscape with a special focus on timber certification and overall sustainability. This is because one of the three pillars of the landscape approach is embedded in the rationale that human communities residing in large conservation priority zones should use the natural resources of their ancestral lands in a sustainable way in order to improve their livelihoods (Sayer et al. 2006). Therefore, any conservation sound program in the landscapes should include human aspects that include both access to resources (land rights, water basin rights) and participation in the management decision-making processes. Because >60% of the landscape is of the Landscape in a swampy zone, a special focus on livelihood activities depending on water will prevail in the implementation of the actions directed toward local communities and their uses of the natural resources found in their ancestral lands and water basins. The livelihood component of the project has its own set of objectives, which are an integral part of the overall project and can be summarized in the following lines. Firstly, there was a need to build a space for productive dialogue, access, and permission to pilot negotiated interventions, and ensure key local and national level stakeholders are informed of planned objectives and approaches. Creating the space for constructive dialogue was not only innovative but also very demanding in a country with a long history of centralized power and where communities have always been brought to accept decisions made for them only using force, sometimes even a brutal crack. But there was no way to go otherwise. As Inogwabini (2007) indicated, biodiversity conservation can be reconciled with development in Africa. But only if we acknowledge that sovereign, democratically elected governments must decide the course to take. The problem is not too much authority but a lack of authority, i.e. there are insufficient tools for good governance. Reconciliation of the fulfillment of basic human needs and the conservation of biodiversity depends on good governance, and this depends on the overall health of democracy. However, the health of a democracy is measured only through how much people participate in the decision-making process and how the openness of the debate on the public affairs and the availability and sincerity of information that shapes that debate is. It is only through the open and sincere debates on natural resources that people would trust the process and accept its outcomes. Secondly, it was agreed that debates involving all stakeholders should propose strategies for resolving conflicts and tensions relating to the governance, policies, institutions involved in natural resources, and particularly forests. The debates should include community groups and the local civil society and local structures of the government as well as other national stakeholders to discuss and evaluate these governance structures and policies and their implications for the lives of the local communities as well as the impacts of the governance structures and policies on the national economic profile. Thirdly, the single most important aim of these discussions and negotiations was to formulate and pilot a jointly owned action plan to reconcile conflicts and tensions relating to policies, institutions, and communities, including a strategy for a complete reconciliation. Finally, the discussions should lead to a viable and sensible mechanism to evaluate participatively and disseminate knowledge of the impact of the action plan piloted on ecosystems and livelihoods for key stakeholders and policy-makers.

Apart from these country-centric processes to conserve biological diversity, there was a great need to have a consensus around the conservation of forests across Central Congolian Lowland Forest Ecoregion. As indicated above, when landscapes were created in Central Africa, there was no legal framework to support the idea. Lacking the legal carpentry in places meant that landscapes that were crossing multiple countries would be managed differently according to national legislation. This would have meant having a single cross-boundary landscape with different standards and moving at different speeds. Also, lack of a common understanding and compatible legislations across Central Africa would increase the risks of leaking in activities such as poaching, illegal logging, and trans-boundary traffic of natural resources. To mitigate these risks, it was felt that trans-border agreements should be put in place and these should be implemented for trans-boundary landscapes as well as the global governance of forests across Central Africa. In the case of Lake Tumba Landscape, which is part of the larger Lake Tele–Lake Tumba Landscape, the institutional, policy, legal and financial frameworks should be in place to support development of conservation constituencies for biodiversity conservation and sustainable development.

## 20.8 Freshwater Should Not Be Ignored in the Planning Process

Another priority for the Congo Basin Forest Partnership was that additional hectares of freshwater habitats are more sustainably managed in priority river basins and ecoregions. This priority and the simplicity with which it is stipulated hides the reality that freshwater as a separate ecosystem has been always among the poorly known and very rarely included in the conservation plans of Central Africa. Indeed, when the process that led to agreeing on the concept of landscapes was launched, it was very apparent that there were only a few Africans with good knowledge of freshwater ecology and even expatriate colleagues with the knowledge of freshwater biodiversity could be counted with the fingers of one's two hands. This was contrary to the situation with other biomes and biota of the region. Worst still was the fact that most of the universities and higher education institutions in the Congo Basin since historical times had not included curricula on freshwater in their programs. Hence, there was no way to expect bringing in staff freshly from the universities to help with collecting data and formatting these into edible knowledge. It makes perfect sense to infer from these realities that the priority set by the Congo Basin Forest Partnership on the regional freshwater objectives was, in essence, not clear about the work it would have taken to make certain that priority freshwaters are identified and delineated in their different microhabitats and levels of fragility they are exposed to. Of the landscapes that were designated from the process launched at the Libreville Meeting, only one was specifically selected because of its potential to protect freshwater habitats. That one landscape was the trans-boundary Lake Tele–Lake Tumba Landscape, of which Lake Tumba is the Democratic Republic of Congo. The designation of this trans-boundary landscape was justified by the fact that swamps around the two lakes were connected though they are also separated by the Congo and Ubangi rivers. While there was a long-term presence of an international conservation organization on the side of Lake Tele (Congo Brazzaville), there was nearly nothing on the side of the Democratic Republic of Congo, despite the existence of the Mabali Scientific Research Center at the shores of Lake Tumba since the late 1930s. Indeed, the Mabali Scientific Research Center and its tiny Mabali Scientific Reserve, as it was the case with other scientific research stations over the Democratic Republic of Congo, were no longer operational in the true sense of the word. When the landscape concept was adopted and Lake Tumba included among the conservation priority areas, the center was run by unpaid personnel with outmoded knowledge and were conducting barely no research though they kept on with obsolete methods and keep very useful old research records on fish species, water levels in the lake, time-series data on weather, and a substantial collection of plant species and some anthropological records. In short, Mabali was more of a relic of history than an institute that could provide any up-to-date knowledge to inform the participative planning that was needed to come up with an agreed-upon management scheme for both the lake and the adjacent swamps.

Given the lack of knowledge, the first milestone for the freshwater objective was, of course, to complete baseline studies of freshwater biodiversity. Then, and then only, the collected and formatted information would become usable for planning purposes of the landscape by 2009. The information should, in priority, help identify and delineate different water basins and help draft their co-management plans by 2010. Of course, rather than being limited to drafting co-management plan, the efforts were thought to be able to help different stakeholders adopt the co-management plans and initiate sustainable fisheries whereby different stakeholders agree on rules to implement sustainable use of freshwater resources.

## 20.9 Species of Conservation Concern and Others Should Be Conserved

The landscape approach being a conservation tool, the Congo Basin Forest Partnership also had a priority to ensure that species of conservation concerns were conserved and preserved from unduly harm in the landscapes of the Congo Basin. For the Lake Tumba Landscape, the priority was to practice active conservation of charismatic species in the landscape with the aim to globally stabilize or increase populations of priority species and to safeguard their critical habitats. The four landscape-species in the Lake Tumba were the bonobos, the chimpanzees, the forest elephants, and the lions. The Lake Tumba Landscape holds the largest population of bonobos remaining in the wild and an un-described population of central chimpanzees located within the swampy forests of the Ngiri Triangle (Inogwabini et al. 2007). It does also hold a sizable population of forest elephants, and African lions and a large cohort of other large mammals, birds, and invertebrates. The bonobo population of the Lake Tumba Landscape represented (at that time) perhaps 20% of the world's total and was estimated to number between 5,000 and 7,500 individuals. With this priority, the landscape's species target was to, at least maintain the population at these levels or increase its sizes. But that has also meant that the population status of bonobo is fully determined across the Lake Tumba Landscape and that a bonobo management plan is submitted to the government, adoption and implemented by different stakeholders and then turned into an action plan.

The chimpanzees in the Lake Tumba Landscape were in the Ngiri Triangle but were not fully documented; although the relative indications of the abundance of the species are available in terms of encounter rates, surveys conducted in that area could not provide estimates of populations due to small sample size (Inogwabini et al. 2012). For the species conservation to apply to this species, the first and foremost important activity was to fully describe the species and its habitats to identify appropriate conservation measures. Equally difficult to estimate was the population of elephants in the southern Lake Tumba Landscape though relative abundance was also available in this case. Similar reasons for sample sizes prevented the calculation of a reliable estimate. However, given the urgency and the knowledge of the ecology of the species in other similar circumstances across the continent, it was agreed that for the relic forest elephants in the southern Lake Tumba to survive a conservation plan landscape should be written and approved by stakeholders and implemented by 2012.

Because elephants do move on long distances and need a large niche to live in, it was also agreed that the World Wide Fund for Nature should help the authorities in charge of biodiversity conservation in the Democratic Republic of Congo to strategize the creation of the corridor between Lake Tumba Landscape and Salonga. This process to put this corridor in the application should have been completed and adopted by the same span of time to allow the elephant migration. Finally, despite there being some basic knowledge on the population of lions in the southern Lake Tumba Landscape, the plan to conserve different landscape species emphasized the need for great expert's knowledge for lions before envisaging further steps to come up with a species conservation plan for lions. Despite that decision to wait for more ecological knowledge, however, there was an agreement to start with a program to raise the awareness of communities in relationship with the conservation status of the lions and work out some measures to prevent further escalation of the conflict opposing the presence of the species and humans.

Finally, conserving these species was essentially felt to be done in two directions. The first of these was to work with the Congolese Institute for Conservation of Nature within newly created protected areas. Naturally, the second direction was to seek better ways of conservation populations of these landscape species that were residing outside of protected areas. This included deployment of a variety of activities that encompass the preservation of viable numbers of individuals in combination with the involvement of local communities in the planning, the implementation, and the management of conservation activities tied to their own livelihood. A cross-cutting priority was to communicate conservation activities, and lessons learnt has always been the forgotten child of the conservation activities. In fact, very often field biologists come to the field and collect data and depart to their home countries where most of the analysis is conducted and where scientific and grey papers are published. Local communities do not have access to peer-reviewed journals and do not hear back from the field biologists; and even if these papers would be available, they would be in a format that is difficult to digest by most of the members of the communities and even if the format was simplified, the languages in which the layman's literature is less often the language that local communities speak. So, it was felt that communication with local communities, local and national authorities should become one of the priorities of the landscape approach. If all the stakeholders must participate in the process of making decisions, they were equally entitled to receive the same information to guide their decisions. Indeed, sharing the information is part of the corporate social responsibility and it is located at the heart of the free and informed consent for communities. For the biodiversity conservation organizations, it has become obvious that to make a difference and deliver long-lasting results, local communities in ca. 200 villages and permanent camps in the Lake Tumba Landscape have to buy-in priorities jointly established by all parties. The best way to achieve this was to go through a strong communication program that would share results and lessons learnt not only with the decision-making bodies, authorities, and the international conservation community but also, and principally, with local stakeholders. The information should be formatted in such a way that it can be palatable by all but not only by field biologists doing scientific research on ground. Indeed, there was a need to communicate scientifically to ensure that the research being conducted in the landscape is scientifically sound and is being conducted using the world-class standards. But this does not imply not reformatting the same information as to be consumed by all stakeholders. Discussions over the issue of sharing knowledge had also led to the conclusion that sharing information should be done via a cohort of channels and should not be limited to the written papers and posters. This is to factor the fact that most of the populations in the landscape had very minimal levels of literacy and had no means to buy radios of television sets. Of course, even if they had means to purchase radios and television sets, the national radio and television broadcasting network coverage was so limited that communities might not even be able to use them.

Hence, means such as community meetings, churches, conservation plays coupled with feedback sessions, arts, and schools were envisaged. These would ensure that parties solve the epistemologically general disquietness surrounding the question 'whose knowledge is it? (Chambers 1983)' which is often asked when people collect data in the field world and return back in the western world, get degrees, acquire author rights and, in some cases, have patented products from the knowledge generated from the work they conduct jointly with communities. Indeed, in ecological research, as well as other domains such as anthropology and other social sciences, scholars design research (questionnaires or other experimental methods) and collect the necessary data, which they format, analyze, and get published. Often, publications are limited to thanking those who helped collect the data and sometimes they are not even given the credits they deserve. The epistemological question '*whose knowledge it was*' remains one of the most important issues in knowledge management. While the scholars who design research and write papers should be recognized as genuinely contributing to the advancement of science, the work they often do is to unearth the data that have been buried beneath the general social knowledge possessed by communities. With this view, the data should be owned by both those who unearth it and make sense of it; without their efforts, the knowledge may have remained buried and hence become unusable in many respects. But the same knowledge, which may be in forms that are different from the scientific formats, has been kept alive by the communities even though it might have been underearth for so long. Hence, it is decent to affirm that it belongs to the communities that preserved it throughout millennia and are often the first ones to want to share it with other cultures and the wider community; in that sense, they too are the owners of the very same knowledge that gets published in languages and formats that they might not be able to chew because they have not been trained to use such types of information. Thus, the format of using community meetings, churches, conservation plays coupled with feedback sessions, arts, schools, etc. as envisaged in the Lake Tumba Landscape is just doing justice to the other owners of the knowledge: it is to give them back what they deserve even if their names do not necessarily appear on scientific journals.

## 20.10 Clearly Laid Logics of the Intervention

As should be expected, the logics of the intervention in the landscape was linked with the global objective being pursued by the landscape, which was (to recall what has been already said above) to establish sustainable natural resources management throughout the Landscape. This global theme promotes sustainable economic development and the alleviation of poverty to benefit local communities and the national community globally jointly with the conservation of biodiversity while thinking of future generations. Briefly stated, the intervention in the Lake Tumba Landscape is a project on the sustainable uses of different natural resources of the landscape. The above global vision needs to be broken down to specific objectives in order to become operational and tied to specific activities, which should be linked to specific milestones. These specific objectives were identified through the planning process whereby the majority of core stakeholders were present. Core stakeholders are people and agencies who had direct interests in the landscape, particularly those that had investments or the direct authority of the management of the resources. These included state agents, logging companies, conservation and development organizations, and the heads of villages or other entities concerned with the existence of the landscape. Of course, given the size of the landscape, meetings were organized by sectors and in some cases, they were not attended by the same core stakeholders. Specific objectives that came out of the objective-definition meetings included (1) development of an integrated and participatory land-use plan, (2) rational management of forests, (3) active conservation of charismatic species in the landscape, (4) management of natural resources by local communities, (5) establishment of protected areas in the landscape, (6) maintenance of protected areas created in the landscape, and (7) communication of conservation activities and lessons learnt. In essence, the objectives defined the measures of successes and should be the ones to be used to gauze the success or the failure of the program. Indeed, once they would be completely achieved, they should open the gate to the exit of the international conservation organizations from the landscape or at least international conservation organizations should be playing a backbench role (see below) and leave the operations of the landscape evolve through a democratic process whereby national institutions and communities would be at the front line.

Core stakeholders (as defined above) and international, national, and local non-governmental organizations, members of churches, and other interest-groups, located both inside and outside of the landscape also met to discuss results specifically expected from the above specific objectives. The process was that this wider group was provided with the available information to craft sensible results. To make the process viable and sensible, smaller groups were constructed around themes defined in the specific objectives and different people were asked to join the group they felt was the one in which they would contribute the most. The smaller groups had then to come up with specific milestones to achieve each specific objective. The principle milestone for objective (1) (development of an integrated and participatory land-use plan) was to have an approved land-use plan for the landscape, which should come through (a) the identification and demarcation of functional land-use units. Functional land-use units followed the predefined Central Africa Regional Framework to include protected areas, community managed areas, and extractive zones. Protected areas were defined in this framework in alignment with the IUCN's (2007) nomenclature. This broad definition is that a protected area is an area of land and/or sea especially dedicated to the protection and maintenance of biological diversity, and of natural and associated cultural resources, and managed through legal or other effective means. As it should be expected from a broad definition, this large definition of protected area by IUCN needs to be broken into details to make sense in operational terms. This is why IUCN also has adjoined the definition with a categorization, which was also used in the discussions, about what stakeholders wanted to see, implemented in Lake Tumba.

The IUCN categories including, as all the conservationists know: Category Ia (strict nature reserve/wilderness protection area managed mainly for science or wilderness protection, Category Ib (wilderness area: protected area managed mainly for wilderness protection), Category II (national park: protected area managed mainly for ecosystem protection and recreation), Category III (natural monument: protected area managed mainly for conservation of specific natural features, Category IV (habitat/species management area: protected area managed mainly for conservation through management intervention, Category V (protected landscape/seascape:

protected area managed mainly for landscape/seascape conservation or recreation), and Category VI (managed resource protected area: protected area managed mainly for the sustainable use of natural resources). The most interesting feature of the discussions around zoning, and particularly when it came to protected areas, was that all stakeholders were aware of the history of conservation in the Democratic Republic of Congo. This history has been described by Inogwabini (2014) as running in parallel with the story of alienation of land and natural resources which began in early colonial times and impregnated with a legacy of undemocratic laws promulgated in the time of Leopold II that still govern land rights and the conservation of biodiversity. While all acknowledged that conserving biological diversity was good, they also grasped the need to make the management of protected areas effective and reconciled with the needs expressed by communities. At the end of the discussions, it was rather a good surprise that the state agents that participated in the debates concluded that the best way to make that reconciliation happen was to create some protected areas within the Lake Tumba Landscape but all of these new protected areas should fall within the IUCN Category VI, with proper zoning to reflect the reality of management.

The other accepted functional land unit was that of Community Managed Area, which was still a new concept in the conservation scenes of the Democratic Republic of Congo. Indeed, the idea in the concept is that the community owns *de jure* the land that they have been using de facto as per the traditional land ownership principles and sustainably use the resources of that land for their own betterment. The idea germinated around the history of CAMPFIRE in Southern Africa and was promoted to combine the sustainability of resource uses and the legality of ownership. The concept was both widely wildly endorsed by communities as they felt that it could come to solve the problems they have been facing since long historical time on the land ownership. In its essence, it linked to the Community-Based Natural Resource Management (CBNRM), which has been declined by Bond et al. (2006) as an approach to the management of land and natural resources which is relevant too, and has the potential to provide solutions to some of the problems found within the communal lands of Southern Africa, where the majority of people live with, and depend on, natural resources. As concrete milestones, stakeholders agreed that for the landscape project to be successful in Lake Tumba Landscape it should at its end have several community-managed forests and water basins officially adopted and each of these should have identified demarcated and respected limits. By being officially adopted it was meant that each of these community-managed areas should have land and water basins over which rights of communities are clearly indicated in official documents that are endorsed by the authorities and converted into laws, which laws should clearly identify allowed activities clearly identified in a participative process whereby communities would be leading the selection process and the enforcement mechanisms of the agreed-upon actions.

Of course, a more philosophical reflection on the community-managed land area would be around the question 'if communities are to acquire a communal land, what would happen with historically learnt lessons around the tragedy of the commons?' The tragedy of commons is an economic concept that has become popular after it was published for the first time by Hardin (1968). Hardin's theory envisaged a resource 'open to all' in which each individual receives significant surpluses from using the common resource while paying only community-managed by very thin price of the use of that resources. Each individual cares little and pays little in terms of the externalities. Even when the extraction of that common resource exceeds the capacity of the environment to provide it, if the costs incurred are too little and benefits become larger and larger, individuals will still be motivated to continue to extract the same resource further and further because they received all the benefits and only minimally contribute to the collective efforts to maintain a healthy environment that would continue to provide the same resource in quantities and qualities that all want. The tragedy is in the fact that each individual is induced himself to increase his own benefit without limit in an environment where resources are certainly limited. Because of this, everyone works toward the ruin of the entire system, which is due to the fact that each individual pursues only his own interests when everyone believes in the free usage of goods commonly owned.

The tragedy of the commons would apply in the case of land communally held and where every member of the communities would be allowed to have access to the land and pay very little while working hard to ensure a continual betterment of his (her) livelihoods. Indeed, this happens when the exploitation of resources such as fishes and wildlife is commonly used without any regulations and no one holding the mandate to ensure that exploitation is matched with the payments of externalities caused to the global community. This is a vexing question only to those who have limited knowledge of the actual workings of the traditional land rights across Central Africa. In fact, land has always been collectively owned; it was however managed by community leaderships whose roles included ensuring access to the people who were the right bearers in the community. The extraction of resources was also known to be monitored by the same community leaderships and not all the people were allowed to access the community forests as it is currently the case. Indeed, for many Congolese, the current regime that gives the ownership of all the lands and forests to the state is the one that makes it difficult for communities to ensure that resources are managed in a sustainable manner. Previously, killing elephants, for example, was a communal activity and could only be organized at the demand of the community leadership. The other example that can be used to show how the community-owned land was previously used is the case of access to the land itself: each family was known to own its portion and to access that portion; if someone was not the bearer of the right, he (she) would have to ask the permission from the community leadership. So, there were mechanisms to ensure that different uses of resources owned by the communities would not lead necessarily to a tragedy. Indeed, Feeny et al. (1990) were right when they reported that there were accumulated evidences that the tragedy of commons does not always ensue when good are collectively used and that private, state, and communal property are all potentially viable resource management options. The suggestion by Feeny et al. (1990) that a more complete theory than Hardin's should incorporate institutional arrangements and cultural factors to provide for better analysis and prediction was directed to the right target.

The question that might rightly be raised is where do all the major changes that have affected African countries lead us if there is a need to bring back the communal management of resources which prevailed in historical times to the current situation where movements of people are a major pattern, where the traditional authority has been crumbled and is crumbling still further to leave the space to new tools and techniques, new forms of legalities, and legitimacies? This is a major question and would need more than the space allocated to it in this volume. But what can be said at this point is that for many communities across the Democratic Republic of Congo, and particularly those in the Lake Tumba Landscape, giving the *de jure* authority for communities to manage their resources would prevent communities from being sold out by the state or other owners of large portions of the lands when there are multinational companies trying to acquire lands for their own activities. Of course, the idea of landscape is not to oppose development against a nature conservation, not to oppose communities against the industry but to bring them to work jointly to ensure that interests of each stakeholder are taken into account in the planning for the management of natural resources. In this sense, affecting functional lands to the sole usage of communities does act as a solid buffer against the abuse of any instances and would prevent conflicts over lands in the long run. Doing so would, if combined with the notion of Social Corporate Responsibility discussed below, lead to a more just country in the Democratic Republic of Congo in ways it distributes its natural resources across different social strata.

The third category of functional land or water unit was the zone of sustainable extractive activities. Extractive activities were defined in broader terms and included mining areas, logging concessions as well as areas from which offtakes of fishes were the principal activities. In this sense, lakes included in the Lake Tumba Landscape were included within the category of sustainable extractive zones. The discussions about what sustainability meant in those areas came to the single conclusion that each extractive activity in the landscape should bear the cost of social corporate responsibility (CSR) throughout the execution of its activities. According to Steurer (2010), CSR aims to better integrate social and environmental concerns into business routines on a voluntary basis. In the views of the stakeholders in the Lake Tumba Landscape, CSR was not simply about giving back some portions of revenues from activities in the area to communities or support the development of the area through building infrastructures (roads, schools, health facilities, etc.) but it was also covering the area of caring about environment, taking into account needs of local communities to extract resources in logging concessions, for example, as well as ensuring that traditions and cultures were protected. But there was also a sense of the weakness of the extractive activities to pay attention to duties imposed by the CSR in the absence of the government's intervention. Indeed, SCR was felt to be able to function properly only in areas with advanced democratic management of natural resources. This was far from being the case in most of Central Africa and particularly in the Democratic Republic of Congo. Concrete milestones that were agreed were around logging companies, which were expected to truly engage in the certification process and ensure that local communities residing within or near these logging concessions are aware of their rights and duties while they are part of the CSR-certification process too.

Finally, on the zoning process, the last milestone has been for each functional land unit to have its management plan approved and some initial implementation of activities laid out in the approved management plans for critical zones. Critical zones were defined based on their importance in contributing not only to the conservation value of its area but also on how critical resources drawn from each area were for communities at different scales: local, provincial, and national. Of particular importance was the management plans for each new protected area to be created; these management plans, organizational and minimum management capacities to be put in place and the availability of minimum operational funding were felt to be the backbones of a sustainable management for each new protected area.

Finally, as indicated above, the landscapes in Central Africa are part of the biodiversity conservation paradigm with element preservation. As such, all the planning processes would have to include a component of biodiversity conservation. In the case of Lake Tumba, different stakeholders agreed to put emphasis on identified key conservation species residing in the landscape. Key conservation species were defined as species of conservation concern as identified by the IUCN red list. These included great apes (bonobos and chimpanzees), forest elephants, and the lions. As described above, bonobos, chimpanzees, elephants, and lions were described present in the landscape during the data collecting phase; their respective populations were estimated and their distributions mapped out, with key populations of great apes (bonobos and chimpanzees) identified. As milestones for species were identified, the process led to pledging to actively conserve protect 65–70% of bonobo and chimpanzee populations. The adjective active here meant that the conservation action would be as close as possible and the monitoring of the population would be providing estimates at a yearly frequency to gauge the impact of the conservation investments on the sustainability of the species. Active conservation action also meant that for each species, there should be a species management plan. As for the elephants and lions, the planned action was first to identify all the key populations within the five years following the agreement and craft the management plan for each species.

With the above specific objectives and milestones, the approach to achieve both objectives and milestones was molded through a logical analysis of what the problems were, what logic intervention was needed to solve the identified problem, target groups to achieve the objectives or milestones, indicators of successes (or failures), assumptions and risks of the actions, the means of verifications for achievement of milestones and the who should conduct the action. Clearly this is the language of log frames but in addition to the log frame of the landscape, the process also included the definition of the levels of expected impacts of each activity on the landscape collectively.

Problems were essentially different in nature but could be summarized in the broader categories that follow.

Firstly, there were problems around human communities. Local communities were either very minimally or completely non-structured. This situation created another problem which was that because of the lack of organization (structured), their involvement in the decision-making process was reduced to the very minimal in the natural resources. Because they were not part of the decision-making processes, they could not advocate for their situation of lack of rights over lands and water basins where they had exercised their activities. Lack of rights impacted the long run livelihood activities that would enable the same local communities to improve their living conditions.

Secondly, there were problems over the planning and legal framework. Indeed, as it was the case across the country, there were no land-use plans and a very weak conservation strategy for key conservation species residing in the landscape. Landscape species only could be said to have some legal protection as they were fully protected species and were under the aegis of the national conservation law. Lack of plans both for land use and for species conservation has led to destructive forest and water uses anarchical logging practices, and rampant poaching of conservation target species all over the landscape.

Thirdly, there were fundamental problems of lack (or limited) of knowledge not only of the biodiversity of the landscape but also of its diversified cultures and human communities. Lack of knowledge led to the lack (or the inexistence) communication scheme for communities and decision-makers and because of that lack of a formal communication scheme, no coherence could be found in the awareness rising for communities on conservation activities.

For obvious reasons, the third set of fundamental problems, particularly the one on the lack of minimal knowledge, came first among the priority actions. It was a simple fact that without the minimum knowledge possible, there would be no further action. Hence, logical actions were designed and had to be implemented first and foremost. Under the assumption that the employment would be warranted to trained people, the first logical intervention to solve the issue of lack of knowledge was to train a pool of competent Congolese to undertake different sorts of assessment across the landscape. After the training of these Congolese experts, it was felt that the second logical action would be to deploy these teams to conduct biological and socioeconomic assessments with the assumption that conditions for these assessments were peaceful and that the necessary financial resources would be available. If resources were not available, it was required that international conservation organizations leading on the process would work to raise funds for the funding to become available and help throughout the process. Indeed, this requirement, as well as the assumption that funding would be available, crossed all the sets of logical actions to solve the problems that were identified to impede the normal implementation of the landscape as a conservation paradigm.

## 20.11 Actions Planned to Achieve the Global Biodiversity Conservation Priorities

To achieve these global priorities and to be aligned with the logics of the intervention described above, a strategic planning process has been conducted and was based on different experiences, not only the one that World Wide Fund for Nature could genuinely claim to have accumulated throughout Central Africa but also from different other organizations, including international development agencies, churches, local communities, and the national governmental representatives. A deep situational analysis, whose short summary is given at the introduction of this chapter, helped guide the planning group to lay out clear priorities to address different threats identified at the landscape level. Actions planned were in alignment with identified priorities and were, in essence, three in their nature: collecting the data, which was an essential part of helping the development of an integrated and participatory land-use plan. Needless to say that all data-collecting activities were to start before anything could be done and they were all intricately interlinked. Hence, the first activities that were launched in the Lake Tumba Landscape were around training teams to conduct the biological assessment including birds, fishes, large mammals, and vegetation and habitat in different biota of the landscape. Staff members were recruited from young graduates from Kinshasa under the conditions described in the section on the capacity building below. They were brought to the Mabali Scientific Research Center where they were trained in the basic ecological surveys, including compass reading, use of maps, measuring perpendicular distances, reading instruments such rain gauzes and thermometers. Because there was very little expertise in the social sciences, to conduct socioeconomic surveys that would inform the land-use process on local needs in land and other natural resources, the landscape leadership had to recruit some external capacity to combine both the generic socioeconomic study and to conduct a somewhat in-depth stakeholder analysis. The second type of activities to help the land-use planning process included a formal spatial depiction of baseline data including socioeconomic features, biological information including species and habitat distribution. That exercise of spatial analysis aimed at establishing a different zonal plan that shared similar attributes within the landscape. With the data handy and formatted in pictograms that would be easily readable by stakeholders that were dissimilar in their background and professions, the third type of activities in the process of land-use planning was to establish platforms enabling the inclusion of different stakeholder visions in zoning. But, unfortunately, as a legacy of long years of a strong centralized political regime, communities had a history of being less organized. The history of the Democratic Republic of Congo has been that political leaders think they have the right to think about what people need the most in the place of communities themselves, which has inhibited the abilities of communities to mobilize their energies around issues of common concern. So, to move on with the planning process, years were invested in helping structure local communities and local social organizations to become a cohesive force that would conduct the assessment of the needs of the general public and become the conduit through which all the identified social desiderata would be conveyed to the rest of the stakeholders. Structuring communities in a cohesive unit that would unite forces to gain their voices heard is easier said than done; in practice there were other activities needed to enable the social structuration become a reality. These included community capacity building activities such as helping them formalize the creation of local committees for the Natural Resources Management and training their respective leaderships in the advocacy to move their causes through and facilitate the process for the legal formalization of these committees. The final activities in the land-use planning process would, then, follow and were meant to include the production of maps of macro-zonal based on biological and socioeconomic assemblages and the discussions of different stakeholders. Indeed, macro-zones were defined as more than just maps depicting biological and social attributes; they had to have an element of political acceptance by the majority of stakeholders. A macro-zone map, in this sense, was a political creation but a political creation that should come into being only after a democratic process that had included the ideas of all social categories. This was and is still the most difficult set of activities. Indeed, as stated above, it is a truism to say that the Democratic Republic of Congo has a very limited democratic experience. In the vacuum of democratic experience, bringing people together with people with either political or economic powers to identify issues of common interests was but a daunting demand.

After convening, vetting, and modifying the plan with different stakeholder groups in the presence of the national authorities, the idea was to adopt a global land-use plan for the entirety of the landscape. It was interesting to see how difficult it was to bring the national, provincial, and territorial authorities to agree to be treated not only as partners but as equal partners in different meetings. Once more, the democratic culture in the country was not sufficiently ripe to accept the idea of comanaging natural resources jointly with communities and other stakeholders. In my sense, the most difficult situation that landscapes across Central Africa will have to face before they become a true conservation paradigm that truly protects species, habitats, and cultures of different communities will be the advancements of different countries toward truly open and democratic societies. It is very clear that with macro-zones being political creations, they have the nature of regimes in place in their backgrounds. The idealistic idea that participation will solve all the issues of the natural resources management in the Democratic Republic of Congo would remain wishful thinking as long as long the society would not truly embrace democracy. Participation alone does not wash away all the lethargies that were brought about by long centuries of undemocratic management of societies and other resources. Indeed, when people were brought together, it was interesting to read how their reactions were guided by who was in the room; people acted just like when school kids are aware that the headmaster is somewhere around and would come and stop or crush the discussion. Participation, in this case, should not be seen as the magic panacea; the dynamics of power do play greatly in the participation process. Despite that though, the plans were to complete and adopt a management plan, which meant it should be endorsed by the government of the Democratic Republic of Congo.

Apart from the land-use planning process, the second important set of activities aimed at making the management of forests become a rational practice. It was plainly agreed that not all the forests in the Democratic Republic of Congo would be locked into protected areas. This agreement was rooted in the fact that there are more biodiversity and more habitats outside of protected areas. With that view, it was agreed that the first set of activities in the area of rational management of forest would include the identification logging companies that are operational in the landscape and assess their willingness to embrace the certification process. Among the criteria to be probed in detail, it was agreed to check the willingness of the logging companies to apply the core requirements of the corporate social responsibility and the level of implementation of the biodiversity conservation norms by the selected logging companies. The assessment was decided not to be limited within the administration of logging companies but it should be extended to include open testimonies of other stakeholders and principally the local communities residing near or inside logging concessions. Once the two logging concessions were selected, it was envisaged to hold negotiations with them to initiate the process of sustainable forest management. In fact, at the time of the beginning of the landscape design process back in 2001–2002, none of the logging companies in the Democratic Republic of Congo was being certified by any international mechanism. Of course, there were some companies with goodwill but the countrywide political anarchy that ensued from a half-decade war could not allow the proper implementation of international regulations. So, everything has to start from zero; one needed to build even the legal architecture to ensure that they were compatible with the changes in the international scenes. This is how the new forest code was issued in 2002; its main aims were multiple as it was regulating things that are wildly separated such as land rights and forest management. In this perspective, initiating a sensitization program on the new forestry code for local communities and logging companies was also identified as an important activity to make certain that all players are playing by the same rules. After these negotiations and an entrenched understanding of the then-new forestry code, it was felt that there would be a long learning process for all parties to come to a common understanding of what was needed for the timber certification process. Hence, with the experience gained from working with the two selected logging companies, the next aim would have been to work with logging companies, at the broader scales, and local communities to ensure the sustainable management of forests is being promoted by all actors. Because of the importance of the water for the Lake Tumba Landscape, it was also agreed that management plans for logging concessions should include specific methods and provisions to preserve the quality of habitats and species diversity in the watersheds of the southern part of the landscape through the implementation of national regulations on water basins and watersheds. As part of this specific protection of watersheds, it was agreed that conservation organizations would provide technical support to the Government of the Democratic Republic of Congo to classify 650,000 ha Ramsar site within the Lake Tumba Landscape. Indeed, given the limitations of existing legislation on how to manage landscape, which is a new entity not only with the conservation paradigm but also in the management of natural resources, it was felt that the legal vacuum could be bridged by invoking legal instruments that the

Democratic Republic of Congo has already signed off. This is the case of the Ramsar Convention, which was signed by the Democratic Republic of Congo back in 1996.

## 20.12 Management of Natural Resources by Local Communities

The overall goal of the landscape approach has been defined as a way to have everyone, including communities to become part of the sustainable management of the natural resources with the goal to maintain the biological diversity in their areas. Hence, the management of natural resources by local communities was one of the major priority objectives. But that has to come only through a concerted effort and via a participative process; it has to be built on the local realities while accounting for the general context of the Democratic Republic of Congo, Central Africa as a region and the global environment. Concerted efforts needed to be invested in priority activities, among which the first one was to conduct a Rapid Rural Appraisal to identify local problems and proposed solutions. This activity was the only one to be planned from the offices in the area of support to be provided to communities. Avoiding planning other activities from the expert's perspectives was deemed valuable as it meant to resolve the problems inherent to the top-down approaches that have marred development programs. Indeed, when Oates (1999) says that most of the conservation programs that deviated to become development programs have failed, the reasons for that failure can be easily attributable to the 'top-down' approaches, which had brought solutions from the international experts to fix the lack of development in these areas. The issues raised by Oates (1999) are, in many of their aspects, not the problems of mixing conservation with local sustainable development but rather these are of the way sustainable development issues are identified and solutions proposed. In this respect, the landscape as a conservation paradigm would clear the space for the issues raised by Oates (1999) to be clarified and new methods adopted to ensure that conservation efforts are helping communities to carry on with their sustainable livelihoods.

The above said, however, there were activities for which communities needed support from conservation and development entities operating within the Lake Tumba Landscape. These activities were crucial because they were either related to the environmental and/or conservation laws for which communities had no other alternatives but to comply with their implementations. The other set of activities that needed the expertise from conservation and development agencies operating within the Lake Tumba Landscape was the one where the knowledge of the resources was needed in order to help communities acquire a weighted view on the potential for the available resources and plan accordingly. Indeed, one of the problems that most conservation practitioners have always encountered while discussing with communities about natural resources is how wide the margins between what people think about what their forest hold as natural resources and what really is there are. This is not to infer that communities do not have knowledge of the potential natural resources they have in their areas; indeed, in many cases, they know better than anyone else but the question of quantification of those resources is always a difficult one to answer to form perspectives presented by communities.

Crucial environmental and/or conservation laws activities on which communities needed support along the Rapid Rural Appraisal were the support to the implementation of community forests as it appeared on the Forest code of 2002 (see Chap. 4 above), support to the development of Congo River Basin Initiative, establishing linkages with creation of new protected areas in the Lake Tumba Landscape. The provisions of the Forest code of 2002 were, indeed, already there and the communities needed to take advantage of them and be prepared to fully implement them whereas the Democratic Republic of Congo had already committed to being an active participant of the Congo River Basin Initiative and openly declared to increase its network of protected areas from 10 to 15% of its national territory. Hence, communities needed to be prepared to negotiate their stake of responsibilities and their share of benefits that would ensue from these legally binding commitments. The activities to produce quantitative knowledge on the resources available to support a sustainable livelihood planning process included the inventory freshwater fish species in the Congo River, Lake Tumba, and Lake Maindombe and adjacent swamps and undertaking fish stock studies in critical water basins of the landscape segment such as the Congo River, Lake Tumba, and Lake Maindombe. Both diversity inventories and quantification of stocks of fishes were necessary to inform the sustainable offtakes and the community management processes and plans. This type of data helped identify platforms for stakeholder participation and the creation of community forests and freshwater basins. Platforms were also important in establishing collaborative agreements between fishing communities for Lake Tumba and Lake Maindombe.

In addition to these applied research studies on management, there were other needs in knowledge on sustainability, which included the data necessary to support the zoning process of Lake Tumba and Lake Maindombe. In this case, the scientific data were collected on breeding sites (Inogwabini et al. 2009), which then were used in a model delineating potential fishing-exclusion zones with the participation of local communities. Other data on the sustainability of the landscape's resources were collected on freshwater habitats and water quality with the view to establishing a water quality monitoring program and assessing the effects of different activities on water quality and fish resources. This category of data was necessary to come up with an establish watershed management program, which also had to make links between water quality, freshwater resources, and poverty alleviation mechanisms.

Finally, there was a need to conduct a study on alternative sources of proteins and promote sustainable hunting and fishing practices. This study was to be, by its very nature, a participative one; it was to be a study whereby researchers would have to retrieve the lost knowledge on food cultures and habits; knowledge on the selection and conservation of seeds as well as that of traditional husbandry. This study was to concentrate not only around known sources of traditional food types, their usages, how they could be accessed, and comparing their availabilities with the current situation but also around new types of food, coming from other cultural areas and how these novel culinary goods were perceived within different communities in the Lake Tumba Landscape.

With all these data, the main objective was to initiate pilot livelihood programs in areas of high conservation values. This was what has been done in the area of Malebo where bonobos, elephants, and lions were found. Needless to repeat, these species are of paramount importance for biodiversity conservation given their conservation status, as depicted by the IUCN Red List. The idea to establish a Community Conservation Program was aimed at attenuating difficulties that communities face and which force them to rely on hunting and practice the slash-and-burn agriculture, which all deplete the biological diversity. To address the criticisms that Oates (1999) rightly fingered out, it was important to establish a link between conservation targets and the Community Conservation Program.

But the activities for the Community Conservation Program were to come from the essential Rapid Rural Appraisal. As has been outlined by Chambers (1983), Rapid Rural Appraisal is an interdisciplinary endeavor and needs time, which should be not too short but also not too long and thinking through about the most appropriate methods or the best composite of methods. Rapid Rural Appraisal also demands efforts to avoid placing the wrong targets at the center of the inquiry and proceeds, to use the words of Chambers (1983) by combining the search and use of existing information and identifying and learning from key informants. Key informants include such categories as anthropologists who have worked in the area under study, social workers, group leaders, university students, etc. Briefly described, key informants should include categories described in Chap. 18 as the bearer of stakes. As a composite approach, the Rapid Rural Appraisal should not be closed method but rather a mixture combining direct observation, structured and semi-structured interviews, and group interviews. The appraisal should also include both formal and informal discussions with different stakeholders to identify the right priorities for the residents of the Lake Tumba Landscape in their diversity. In addition to these traditional techniques, the teams working on the Rapid Rural Appraisal also included open sessions of questions and responses. These open sessions were demanded by communities as the means for them to triangulated the information and ensure that the voices of the less powerful and, consequently, voiceless in the communities were also being taken into account.

After the use of multiple combinations of the above approaches, communities in each sector of the Lake Tumba Landscape came up with their own priorities. It is important to note here that these priorities varied depending on cultures, history, and ways different communities perceived the future of their areas. Of course, this is not surprising given the fact people would see only the façade of things that they face. Hence, for example, communities in the northern part of the Lake Tumba Landscape residing along the Ngiri River identified the mechanization of the palm oil processing as the priority while in the southern part, communities residing along the Malebo–Lediba, the reopening the road Malebo–Lediba emerged as the priority number one. The road Malebo–Lediba is the national road number 2, which is described in Chap. 6. This difference can be understood as a differential in the cultures in the two areas. Indeed, as described in Chap. 5, for people in the administrative territory of Bomongo, wealth is equated with the abilities for one to produce palm oil, which is not only the principal commercial product but also a natural constituent of food. Communities in the south of the Lake Tumba Landscape are known to be good traders and, as indicated by Inogwabini (2010), their very name 'Bateke' takes its root from 'Koteka' (to trade). Indeed, the traditional true name of this population is the Tio but because they were considered a mainly mercantile people during the 17–19th centuries, they have come to be known by the pseudo-name of Bateke derived from the verb '*koteka*', a Kikongo word meaning 'to sell' (Ndaywell 1998; M'bokolo 1992, 1995; Vansina 1973). Hence the road to ease the commerce can only be prioritized. The second priorities for the two regions were also different and can also be tracked down to the respective cultures of the region: in the northern Lake Tumba Landscape, communities prioritized sustainable offtakes of fish for commerce while in the southern Lake Tumba the second most important need was around acquiring good manioc seeds and being trained in good agricultural practices to ensure that the acquired seeds would produce more manioc roots. Indeed, Bobangi, which is a community of Ngiri River, as well as other communities such as Likoka and Libinza of the same zone, fishing has remained the principal activity in their villages since the nineteenth century (Harms 1989). The communities of Bateke are known to be good cultivators of manioc.

The community Conservation Program was to provide support to these priorities identified by communities to ensure that people would continue to use natural resources in a sustainable way while conserving the biological diversities in their areas. The first priority of the communities around Malebo was the road, which is often negatively viewed by the traditional conservation paradigm because it is felt to bring in poachers and increase depletion of large mammals and other biological resources. As is the case in other sites, roads built to move timber and other goods from forests to the lakes and rivers pose significant threats to wildlife species (Bennett 2002) and existence of roads act like magnet for different communities willing to trade biological goods with modern commodities and encourage new human settlement villages, facilitating, therefore, human demographic growth and movement in relatively remote areas, and also lead to an increase in commercial hunting (Wilkie et al. 1992). To support communities in their endeavor to improve their livelihoods through commerce by fixing the road, negotiations were held with communities to agree on the measures to be implemented to avoid the potential for the efforts to reopen the road to become a threat for wildlife species. Throughout this process, communities working with the conservation organizations in the southern Lake Tumba Landscape agreed on the fact that the road should be reopened using labor-intensive methods, which meant that only residents in villages would be paid to manually fix the road. It was also agreed that each village would create a conservation law enforcement team in charge of checking goods being transported through the fixed road to avoid hunters using the road to carry trophies from species of conservation concerns, which were listed through a participative process and to avoid exceeding the quantities agreed upon. The agreed list included bonobos, elephants, lions, buffaloes, and a cohort of monkey species. Furthermore, there was an agreement on hunting seasons and hunting tools that would be acceptable by all stakeholders. Through the agreement on hunting tools, it was agreed that net hunting and other techniques such as snaring for small mammals would be acceptable but gun-hunting would only be conditionally accepted. But even with the conditional acceptance of gun-hunting, it was agreed that this should be limited to known reasons and individuals and bringing in more guns should be prohibited. Equally prohibited and totally was the case of automatic weapons. Communities would ensure that mechanisms of community surveillance would help deter those who had these types of hunting tools and report to the traditional authorities. Indeed, a year after the work on the road, communities in several villages arrested and confiscated bushmeat along the road and several guns were also confiscated and given to local traditional authorities. Figures of arrest wildlife traffickers were of the magnitudes of 4–7 per month but then decreased sharply. In the first 3 months, five automatic guns were seized too. Whether the decreases in arrests were symptoms of the impacts of people being aware of the practice and avoid entering the area of whether they have created different circuitries to operate within the area was not assessed though it would have provided a great opportunity to compare traditional law enforcement mechanisms with community law enforcement.

## 20.13 Active Conservation of the Landscape—Species in Lake Tumba Landscape

The other conservation priority of the Lake Tumba Landscape was to ensure that charismatic conservation species were actively protected. This should not come as a surprise as the landscape approach is a global paradigm to ensure that biodiversity is maintained while the needs of communities are also squarely addressed. Just to repeat what is said in Chaps. 6 and 11 through 14, species of conservation concerns in the Lake Tumba Landscape included bonobos and lions located in the southern part of the landscape (Chaps. 6 and 12), chimpanzees located in the northern part of the landscape (Chap. 11), and elephants located across both parts of the landscape (Chap. 14). Equally included in this category was the guild of diurnal primates (Chap. 13) because of the diversified ecological roles that diurnal play in the ecosystems wherein they occur. Furthermore, because of the potential for humans to contract diseases that are borne by primates as well as the observed and acknowledged drastic declines in the populations of primate species, local communities agreed to include monkeys in this list though in some villages the lists included specific species but not the entire guild.

In order to reach the goal of conserving the charismatic species of the Lake Tumba Landscape, one of the most important issues to solve was that most of the populations of these species were located outside of formally protected areas. Even though creating new protected areas in the Lake Tumba Landscape was a priority, one would not think of transforming the entire landscape (80,000 $km^2$) into a series of protected areas. Hence novel thoughts were needed to do the job. One of these novel thinking was to establish conservation partnerships with traditional authorities and major private sector enterprises in areas holding large numbers of conservation target species such as bonobos, forest elephants, and lions. This was particularly appropriate in the southern part of the Lake Tumba Landscape where the traditional authority was still very strong and even stronger than the legally established modern power. Indeed, Inogwabini and Leader-Williams (2012) have found that in the southern Lake Tumba Landscape, respect for the traditional authority correlated strongly with the presence of the bonobos in villages where the species was documented. Because of that Inogwabini (2010) has suggested that the organized and strong traditional system of authority in the southern Lake Tumba Landscape should become the body designated to enforce conservation law, which would be possible given the current administrative and political decentralization promoted in the country, which encourages effective devolution and should clearly indicate the role that traditional authority should play in the gamble of conserving charismatic species. The road example given above was, in essence, a trial of this type of creative thinking.

But despite the impracticalities of creating new protected areas across the entire landscape, communities needed to possess rights to expel intruders from the areas adjacent to their villages and needed to be able to locate species of conservation concerns. Briefly, communities needed to possess forests in which these charismatic species were located in order to have a stronger voice and accepted authorities over the species. Hence, there was a need for the conservation organizations to work with local communities to initiate the framework of the creation of Community-Based Natural Resources Management Land units in zones holding sizable populations of bonobos, chimpanzees, and forest elephants and where protected areas would be unwelcomed. With this in mind, the establishment of a Community Conservation Program in these areas harboring large numbers of conservation target species in areas outside of any formal management units would be used as a trade-off mechanism. And, to ensure the sustainability of the Community Conservation Program, there was need to establish a sustainable funding mechanism, which was seen, in the case of the southern Lake Tumba Landscape, as being partly in the habituation of bonobos for ecotourism to support the community conservation action in areas where this is possible (see Chap. 21 of this volume).

Because the communities expressed worries about their health in relationship to bonobos and other diurnal primates, the focus on conserving charismatic wildlife species needed to be done with something beneficial to humans. These worries were explainable by the fact that outbreaks of Ebola were associated with primates and that some lay reports literally stigmatized bonobos, chimpanzees and other primates. That ‘something beneficial to humans’ also needed to be different from pure sustainable development activities such as distributing seeds, reopening, and fixing roads. Indeed, as the traditional chief of Bobangi once told us, sustainable development begins with good health. Because of that, conserving charismatic wildlife species also included the establishing of a wildlife health monitoring system for habituated groups of apes in order to monitor potential effects of habituation on the populations’ survival as well as potential cross-species transmission of diseases. As depicted in Chap. 9, the cross-species transmission of diseases is possible and has been documented to be possible on several occasions.

The wildlife health monitoring, as part of the strategic objective, to actively conserve charismatic species was combined with another important activity, which was to develop a biological monitoring system for bonobo, chimpanzees, forest elephant, and forest buffaloes to gauge effects of the project's conservation action. For many years, conservation efforts have been invested in conserving species across Africa but measuring the effects of these efforts has been slow to come through because of the costs inherent to biological monitoring, lack of human capacities to conduct proper monitoring exercises in most of the Central African countries, and limits of methodological frameworks.

In the context of the Lake Tumba Landscape, the monitoring of charismatic landscape species was designed to be simple yet robust enough to account for changes in the populations. Beyond combining simplicity and robustness, the methods were also designed to be multi-resolutions and a combination of several approaches. For example, the research presented in Chap. 10 on the use of presence–absence has been guided by the will to identify the easiest and cheaper methods to assess the populations of bonobos across the southern Lake Tumba Landscape. Also, as portrayed in Chap. 14, the dung accumulation method was used to monitor the population of elephants in the Malebo area. Through the dung accumulation method, new elephant's dung piles are counted and marked on a preset frequency; this gives the indices of the presence of the elephants in a given area. Furthermore, numbers of new dung piles are counted and compared with preceding periods to establish the increases in relative abundances of elephants per period. Of course, this method does not give the population's densities but it provides, with some robustness, a precise sense of increase and/or decrease in populations of elephants. Another example of a simple yet robust method of monitoring of wildlife that was tried with relative successes in the context of the Lake Tumba Landscape was the recording of calls of diurnal primates. In fact, this method has been tested by Hall et al. (2003) to estimate the distribution, the abundance, and the biomass of primates within the Kahuzi-Biega lowland and adjacent forests in the Eastern Democratic Republic of Congo. But, as it has been simplified and applied for the monitoring purposes in the context of the Lake Tumba Landscape, the method was essentially of the same approach as the dung piles accumulation described above. It consisted of logging the every day's calls per species and then comparing the total number of calls heard in a given period with those heard in any another period. In addition to both methods were added the metrics of the distances from the main roads and human settlements, times of the calls, etc. These were a way to factor the theoretical ecological factors that influence the distribution of species as well as the species' ranging ecology.

Finally, the simplest method for monitoring wildlife species would have been a simple forest reconnaissance whereby encounter rates are measured and compared over time. Yet, the simplicity of the method would have provided a much more robust approach if the encounter rates were correlated with rates of dung piles deposition or call frequencies. Once these correlations were established between $\tau$ = signs/distance and densities ($d$), an analysis using the conceptual framework of linear regression models would be used to fine-tune estimates of mammals across a gradient of habitat types. This would have offered a better framework to gauze the results from different efforts invested in the conservation of landscape charismatic species.

In addition to ensuring that people do not blame wildlife for diseases they contract (see the section above), the situational analysis (see below) also led to the identification of elephants being felt as major destructors of the livelihoods of many members of the communities. Hence, to ensure that elephants received the degree of the protection that is conferred to them by their legal conservation status, the other important activity contributing toward ensuring the active conservation of the species was the documentation of the degree of human–elephant conflict in the Mbanzi region in order to propose palliative solutions. The documentation of this conflict was conducted through a multidisciplinary approach; it combined both social science methods and quantitative methods (Inogwabini et al. 2014). With the results of the study of the human–elephant conflict, it was possible to propose some tangible measures through the stakeholders' platforms, which was part of the Community Conservation Program for conservation target species. Participation was, indeed, key and measures included such inputs as financial compensation for individuals who have lost their agricultural products but with a trade-off to moving agricultural fields away from main permanent elephants' trails or near tree species on which elephants relied as food for long periods of the year.

Of course, to ensure that conservation activities to ensure charismatic landscape species were protected, an important activity was to build an operational infrastructure for conservation in areas of high conservation value to help support the conservation action in the field. This is an activity that is often neglected but the simple physical presence in a given area acts as a mean of dissuasion. In fact, a study conducted in the Eastern Democratic Republic of Congo in the National Park of Kahuzi-Biega has indicated that species such as elephants and gorillas were found in larger numbers around the park's headquarters (Inogwabini et al. 2000). Hence, the dissuasion role played by conservation infrastructures should not be neglected when planning for large landscapes such as the Lake Tumba Landscape. Of course, this should be viewed not only as an asset to help the action but infrastructure does also act in itself and contributes to providing protection to species that need it.

## 20.14 Communicating Conservation Activities and Lessons Learnt to the Global Society

Many communities in the Lake Tumba Landscape declared that they had no idea what the whole issue of conservation was about. This declaration can be explained in many ways; the first of these possible explanations being the fact that the Lake Tumba Landscape became a priority conservation area only after the Yaoundé process. Apart from the Mabali Scientific Reserve, located near Lake Tumba, there was no significant conservation geographic entity within this large area. Hence, compared to the other landscapes of the country, which are all delineated around preexisting national parks or reserves, communities in the Lake Tumba Landscape have never been exposed to the campaigns of awareness raising on issues related to biodiversity conservation as those in the Eastern Democratic Republic of Congo where most of the old protected areas are located, with the exception of the Salonga National Park. Apart from this layman's explanation, it is also a global trend in the global community of the Democratic Republic of Congo that communicating results of conservation work to the wider audience has been and is often felt to be deficient. This is mostly due to the means of communication used; indeed, scientific results of studies are reported in scientific peer-reviewed journals whose languages are often so technically sophisticated to the point where even the intelligentsia of the country, with the exception of those who have been trained in this area, might not be able to digest the contents of the findings. Furthermore, access to these journals, sometimes, can be prohibitively expensive to the point that even those who can read and digest their contents are unable to have easy access to the knowledge they contain. Hence, it was identified that a tool to gather consent from communities and to ensure proper implementation of the Community Conservation Program was to develop a communication strategy for conservation activities being implemented in the Lake Tumba Landscape. The communication strategy in question was to go beyond the conventional distribution channels of the conservation news and biological research publication networks. In this sense, the communication strategy should aim at broadening its impact via the use of different communication media, including folk talks, comics, songs, paintings, and talks around the fire. It should also extensively use different communication offices within the network of the partners and increase its partners' portfolio to include nonconventional conservation entities such as television and radio networks, to promote a fluent communication of key achievements of the Lake Tumba Landscape. This should come after initiating a community sensitization program for all the landscape's communities and use the best available communication channels not only to influence people's attitudes toward biodiversity conservation in the Lake Tumba Landscape but also to gather people's views and lessons learnt to be shared with other stakeholders, including state agencies. The communication, in this sense, should be a two-way loop rather than the conventional one-way pontification methods that believe to deliver the truth to communities. This is why Chap. 18 on the political economy of the landscape insisted on open discussions and open session; it is also why in this chapter (see above), the Rapid Rural Appraisal also included sessions of questions and answers. These were important to ensure a two-way flow of information, concerns, and celebrations of key events. Indeed, a two-way flow of information is a prerequisite to building the confidence between stakeholders whereas holding information simply laminates the self-assurance people should have in a process that ties their future together. Of course, communication was also felt to be critically important beyond the categories of primary stakeholders and should embrace the communication of the results of work done within the Lake Tumba Landscape to national authorities and to the regional framework of the Commission des Forêts d'Afrique Centrale.

Communication between communities, state representatives, and enterprises of the private sector, non-governmental organizations, churches, and influential individuals residing in the Lake Tumba Landscape was more than just a trick to appease the minds of conservation biologists. It was rather at the heart of the accountability mechanisms. When people question the transparency of conservation and development activities and their long-term perspective outcomes, this question often boils down to lack of fluent and constant communication from actors. Thus, communication plays a critical role beyond the information it provides and contributes significantly to the internalization and the acceptance of both the action and its results while also ensuring the ownership of the outcomes of the conservation actions.

## 20.15 Reconciling Divergent Needs: Actions for the Pilot of Sustainable Uses

Based on the review of outcomes by the multidisciplinary project team and stakeholders at each site, action plans will be formulated, considering ecosystems, livelihoods, communities, and institutions with the objective of reconciling conflicts and tensions. This will employ iterative, participatory methods to achieve explanatory depth while exploring potentially contentious issues and identifying the most promising approaches in a locally sensitive manner. Draft action plans will be developed for each site and presented to users, local community groups, and representatives from stakeholder groups and target institutions. Following refinements concise action plans, together with proposed approaches and activities, will be published for each site and disseminated to raise awareness, especially among the local communities. The pilot for sustainable uses of natural resources in different land-use units will aim to reconcile differences among institutions, policies and processes, and in particular the most appropriate reconciliation strategies identified and methodologies employed. Based on contacts and rapport established with local institutions (civil society, community-based and government), NGOs and CBOs during previous activities, action plans will be discussed in meetings and focus groups and where appropriate community members and stakeholders requested to participate in the decision-making process regarding implementation, monitoring, and evaluation. Opportunities to address and reconcile specific conflicts or tensions between local and regional, national and international institutions, and policy identified will be discussed with stakeholders, raising awareness of the issues and identifying the most appropriate approaches to reconciliation or conflict resolution. Constraints and opportunities associated with advocating community participation in decision-making, strong leadership, mediation, consensus building, collective decision-making, and participatory conflict resolution will be discussed and evaluated, while future avenues of conflict resolution will be established. This activity will be undertaken at each site to elucidate differences in local conditions, customs, institutional arrangements, or the conflict/tension demanding reconciliation that may affect the suitability or effectiveness of approaches. Where complete resolution is not possible within the timeframe of the project, a suitable strategy to achieve the desired outcome will be outlined and promoted. The process and impact will be evaluated and appropriate communication media developed to help others formulate strategies in support of action elsewhere in the region, and potentially globally.

## 20.16 Resources Needed to Achieve Project's Mission

The most important intractable issue in implementing conservation programs in the whole Central African Region is finding enough resources to achieve concrete action. The first of these rare resources is human resources, which had not been trained adequately and in sufficient numbers. This is even more so for large areas such as landscapes which are different from traditional conservation niches and had brought changes in the classic conservation paradigms by introducing livelihood as a second leg of the conservation tripod. Obviously, in order to deliver high standard results on such large-scale zone and diversified activities, the needs for financial resources increase dramatically.

## 20.17 Human Resources Needed

Finding adequately qualified personnel with both managerial and biological skills has been one of the most vexing problems of biodiversity conservation over the last three decades of conservation efforts in the Congo Basin, and particularly in the Democratic Republic of Congo. Indeed, despite efforts invested in training capable people, results have been so limited that the discussions of capacity building are always part of the global plans, raising the question of whether capacity building is not a mere verbiage in the overall biodiversity paradigm across the region. Despite this feeling, however, it should be recognized that the capacity building exercise, as it has been practiced in the Democratic Republic of Congo, was essential to train low-level law enforcement para-military personnel in the first generation. Then came the second generation of the wardens that were mostly trained in managing the law enforcement personnel composed of park guards. At that time, it was felt essential to train this type of personnel while the bulk of the biological expertise, as well as the one on planning, were relegated to expatriate personnel. When the time came to train the Africans in biological research, planning and other high level performances, the task of capacity has revealed to be daunting. This has been so for several reasons acting cumulatively: despite the efforts being invested in some cases and by some conservation organizations, some skilled people who were trained in earlier periods were given high level jobs in the administrations of their countries or even in large conservation organizations, leaving field biodiversity conservation with less skilled people. Also, some people who were meant to go abroad for specialized training remained in their host countries and never agreed to return back to their countries of origin. But also, in some cases, some organizations did not see the value of transferring skills to nationals for reasons difficult to decipher properly. Most importantly though, the magnitude of the work needed to cope with protected areas that are sometimes larger than some countries needed a workforce several times larger than normal, some square kilometers protected areas. This would certainly be the case for even larger landscapes like the Lake Tumba, which is 80,000 $km^2$, a territory larger than Switzerland and Belgium put together and which presents considerable logistical challenges. A preliminary assessment conducted on the personnel needs indicated that for the landscape project to run at its minimal speed of right operational levels over 15 personnel with technical capacity to plan and execute the work of assessing biodiversity, delivering on social and anthropological assessment and implementing socioeconomic policies to get people to agree on the idea of landscape would be needed. Clearly, the inclusion of the category of the social and human scientists is a sharp contrast with the traditional thinking around biodiversity conservation whereby responsibilities were often left to the hands of biologists. Biologists, however capable they might be as a group of lads, they are good for what they have been trained to achieve in the context of biodiversity conservation: assessing biological values of given areas, study the links between different components of the very areas, and follow trends of these components and interrelationships over time. Clearly, biologists cannot address the issues of humans and communities over the resources. And, as we all know now, most of the problems that biodiversity faces are caused by humans and their communities. To address the cause roots of the declining biodiversity in landscapes, the first group of causes is to look at the human responses to different stimuli toward their immediate environment. That is why, landscape as a conservation paradigm should place equal emphasis on human and social sciences too.

The lower-limit capacity given above would be needed in addition to a support team composed of administrators, finance team, drivers and sailors, night watchmen, and cleaners. As was the case for the human and social scientists in the preceding paragraph, a word needs to be said about the administration and financial management team. The previous experience in the Democratic Republic of Congo has been that parks were managed by wardens who were better trained in law enforcement than in any other biodiversity conservation jobs. The result of this has been very limited professionalism in administration and financial management. This situation could be said to be the case for most protected areas in the Central African Region. It has become a widely accepted idea that management of activities of biodiversity conservation should become as professional as the management of any social and political entity; this means to have clear objectives and be more accountable on both these objectives and the means provided to achieve them. In order to be at that level of management, landscapes should be managed by more than just field biologists. Indeed, in order for field biologists to do their work properly, they would need capable managers both administratively and financially. That is why the capacity building the Lake Tumba Landscape had in its early days was to include the training of Congolese expertise beyond the field biologists and included the capacity building in administration and finance. But given the prospects of zoning the landscape into different functional units, the expected levels of needs in expertise were felt to be at least four times higher than this lower-limit levels and should be quadrupled over time as training for different groups of expertise would be completed. This is why a specific framework of capacity building through high level training as well as training for low skills laborers was needed. This is discussed below.

## 20.18 Training Needs for the Project to Operate

As stated above, one of the major constraints in implementing conservation projects across Central Africa is the lack of capable personnel. Most people have very approximate knowledge of the work they are called to provide and the education system across Central African nations is very weak to respond to the modern demands of qualified personnel in this new area. To contribute to the solution of this stringent problem, the landscape project was also meant to organize training sessions both to boost its human resources capabilities of delivering high quality conservation products and the creation of the national elite for the management of the Congolese gigantic natural resources and conserve the national gorgeous biodiversity. Indeed, reporting on training and numbers of trained personnel was a requirement of the donor (USAID CARPE). But the records reported back were composed of almost everything, starting from exercises on mapping reading all the way to academic training with higher degrees (Masters and Ph.D.s). In the case of Lake Tumba, plans were to work on a degree-leading training process whereby skilled graduates from the national university system would be channeled through training to obtain research-based degrees. In addition to that philosophy, the project was working on sending its staff to nonacademic but highly technical training sessions in new techniques such as Geographic Information Systems (GIS), Management of Protected Areas, and Management of Community own lands. As a matter of fact, some of the project staff benefited from training to obtain their degrees, some others were trained in GIS while another group was trained in the management of parks.

Countries where the staff members were sent for either their academic training or technical training included Belgium, Sweden, Tanzania, and South Africa. Sending staff out for training is an indication of a far broader problem with the quality of the national higher education institutions. In fact, this is a widespread regional problem: universities and other research institutes in DRC and Central Africa have not adapted their courses and research programs for many decades; even programs in the field of biology remained very theoretical and remote from the real needs of the country. While universities continued to teach courses like theoretical molecular biology (albeit with obsolete programs), very few institutions have, for example, no appropriate courses in tropical forest ecology. Clearly, theoretical molecular biology may seem nice on paper but does not fit well within the current setting of the priority needs of the country and the region. Nonetheless, fewer institutions had thought to construct something in the area of natural resources management. Unfortunately, the quality of training in these institutions is so weak that when one recruits some of their graduate students, he needs to start from literally nothing; some of the young students we have had to recruit did not even know how to use maps and compass to navigate and how to use computers and spreadsheets; these—needless to say—are the very basics you would expect from someone with a degree in the management of natural. Years of structural adjustments imposed on African countries by multilateral donors, particularly the International Monetary Fund (IMF), had meant for many African countries to stop investing in their social structures. One of these structures that suffered from the financial austerity years imposed by structural adjustments was the education systems and particularly higher education; for years research was not funded and professors were poorly paid to the point where most of them had to combine numerous paid commitments to make months' ends meet. While universities in the Western world were investing in research and adapted their curricula to new research findings, universities in the Democratic Republic of Congo were hindered by a lack of resources and continued to use very obsolete course programs. In the end, this situation has meant that most of the students in the generation of structural adjustments have very limited capacities to take over the lead in research, particularly in areas as new as the management of forests, biodiversity research, climate change, and ecology in general. It is to curb this situation that the UNESCO has opened and supports a regional school on the management of forests. This school is based at the University of Kinshasa and serves Central African countries. But being a postgraduate training, the problem with this type of school is that it is working on a level where many gaps that the lower education levels left open may not be filled in, no matter how good the students and educators are. There is a need to start from the foundations. This is a decade-long exercise, which in itself requires of Central African nations more commitments and tenacity to build a brighter future globally. Of course, training that limited pool of staff did not reach the optimum level that was needed; education is such a long process that its fruits do need some lengths of time before they can be noticed to affect the lives of people in countries.

In the meantime, given the fact that it would take a long time before the pool of Congolese experts is ready to hit the ground, it was envisaged to bring in different sets of skills from abroad either via the Volunteer for Environment, which was an innovation we brought in, and having graduate students from western university come to conduct their field research in the Lake Tumba Landscape. The Volunteer for Environment worked with some successes and had brought in mostly people from Belgium. Also, graduate students started working on their doctoral thesis in the landscape by 2010. The combination of these schemes was meant to both, increase the profile of the site and ensure that the needed high quality data are being gathered. Beyond these two obvious reasons, the combination of these schemes was also an essential part of capacity building for Congolese students as it aimed at sharing experiences and exposing the Congolese to the international research culture.

## 20.19 Financial Needs for the Project to Operate

The backbone of all the planned action is the availability of financial resources without which all the beautiful planning processes will sum up to being nil because no concrete activities will be deployed on the ground. Estimates for financial resources that are below are not the most important piece however. The key thing to learn here is that for the first time, communities were brought together to discuss with a given conservation organization what would be needed to ensure that agreed-upon activities, including law enforcement, and the community livelihoods could be implemented. The process, however, was guided by estimates of financial resources needed to operate conservation activities developed and published by several authors (Bruner undated; Blom 2004; Bruner et al. 2001; Wilkie et al. 2001) in other contexts. Combining these estimates and the participation of communities as well as that of project personnel, a budget giving ideal figures and minimum operational costs was developed. Besides that, needs in funding were assessed by a different identified functional zone of the landscape. All lead to the conclusion that current funding levels of conservation projects across Central Africa are insufficient and would need to be increased if expectations are to be met. The overall estimate for the period of five years was about 12 Million American Dollars (or if precision need be: US$11,943,754.00) as a minimum threshold to implement all conservation and livelihood activities. This amount is a bare minimum as taking a ratio of US$200 $km^{-2}$, as other studies (Eagles et al. 2002) have suggested, would lead to an estimated financial effort of 16 Million American Dollars per year for the right level of management of Lake Tumba Landscape. Given the fact that resources to be donated by the government of the Democratic Republic of Congo will almost never reach this level of funding, different stakeholders agreed to use different scenarios to get to the minimum levels first. These scenarios included fundraising externally by international conservation organizations, recourse to ecotourism, and other formulation such as creating a landscape-specific trust fund, taping on the new initiatives such as the climate change mechanisms, e.g. funds allocated to Reduce the Emissions of Greenhouse Gases due to Degradation and Deforestation (REDD). Of these options, only two were deemed to work at the time of the planning process that took place in 2009; these included continued fundraising externally and starting to work on possible tourism in the region (next chapter). Looking at where the whole landscape funding mechanisms do stand, the overall financial planning process reveals the obvious: resources to achieve ambitions laid on landscapes as conservation paradigm are notoriously absent; the work to support this laudable concept would be limited to the amount needed to support protected areas that are encircled by massive portions of landscapes that would continue to be managed as previously had been the case. The only exception in this general conclusion is the case of the southern Lake Tumba where bonobos have had such a driving pull on people's emotion and several academics and other agencies are keen to fundraise for their own research and for providing support to local communities residing in the area with higher bonobo densities. Indeed, this success is also due to the fact that there is a local structure that has been working to promote the area even before bonobos in that area were scientifically known by the wider conservation community across the world (Chap. 6). The recognition of the limitations on funding is also a recognition of how limited the capacities to make landscapes become functional are. This recalls the general and well-known motto that mission equals means, and in this case, the most important limit of implementing the concept landscape will very likely be if the Central African countries are not to put the money on the table.

## 20.20 Exit Scenarios

Exit is often forgotten when organizations plan for their work in conservation and development globally. However, exit should be part of the planning process; particularly, exit should be taken to be a measure of success. Indeed, the work of international aid agencies as well as that of conservation and sustainable development non-governmental agencies should be to empower national institutions to take over their roles in implementing plans for development in general and conservation particularly. Projects are, definition, limited in their temporal scopes while programs too have specific objectives that are delineated in time and space. With this understanding in mind, an essential step toward sustainability is for international organizations to prepare to leave when their mission is accomplished. This is why over the planning process of the Lake Tumba Landscape, there was a great need to plan for a possible exit within a foreseeable future.

The exit plans for the Lake Tumba Project was part of the thinking process and was included in the planning process itself. These plans were all based on the fact that the role played by international non-governmental organizations (i.e. WWF and its other international partners) in the landscape was to help the Congolese government achieve its conservation objectives. Therefore, all achievements and programs would have been planned to be given to the government or its representative in the landscape once their missions were formally ended. A step-wise approach has been adopted for different activities and clear responsibility transfers have been also identified. These represented a sort of blueprint to be followed in a time span of 10–15 years from 2010.

## References

Bajracharya SB, Dahal N (2008) Shifting paradigms in protected area management. National Trust for Nature Conservation, Kathmandu

Bennett EL (2002) Is there a link between wild meat and food security? Conserv Biol 16:590–592

Blom A (2004) An estimate of the costs of an effective system of protected areas in the Niger Delta – Congo Basin Forest region. Biodivers Conserv 13:2661–2678

Bond I, Davis A, Nott C, Nott K, Stuart-Hill G (2006) Community-based natural resource management manual. WWF WorldWide Fund for Nature Southern African Regional Office, Cape Town, South Africa

Bruner AG (undated) How much will effective protected area systems cost? Center for Applied Biodiversity Science, Conservation International, Washington DC

Bruner AG, Gullison RE, Rice RE, da Fonsesca AB (2001) Effectiveness of parks in protecting tropical biodiversity. Science 291:125–128

Chambers R (1983) Rural development: putting the last first. Longman

Colom A, Bakanza A, Mundeka J, Hamza T, Ntumbandzondo B (2006) The socio-economic dimensions of the management of biological resources, in the Lac Télé – Lac Tumba landscape, DRC segment: a segment-wide baseline socio-economic study's report. Submitted to the World Wide Fund for Nature

Eagles PJF, McCool SF, Haynes CD (2002) Sustainable tourism in protected area: guidelines for planning and management. The International Union for Conservation of Nature

Feeny D, Berkes F, McCay BJ, Acheson JM (1990) The Tragedy of the Commons: Twenty-Two Years Later. Hum Ecol 18(1):1–19

Ghimire KB, Pimbert PM (2000) Social changes and conservation: an overview of issues and concepts. In: Ghimire KB, Pimbert PM (eds) Social Changes and conservation. Earth Scan, London, pp 1–45

Hall JS, White LJT, Williamson EA, Inogwabin BI, Omari I (2003) Distribution, abundance and biomass estimates for primates within the Kahuzi-Biega lowland and adjacent forest in eastern DRC. Afr Primates 6(1–2):35–42

Hardin G (1968) The tragedy of the commons. Science 162:1243–1248

Harms R (1989) Fishing and systems of production: The precolonial Nunu of the middle Zaïre. Cahier des Sci Hum 25:147–158

Inogwabini BI (2007) Can biodiversity conservation be reconciled with development? Oryx 41:2–3

Inogwabini BI (2010) Conserving great apes living outside protected areas: the distribution of bonobos in the Lake Tumba landscape, Democratic Republic of Congo. PhD thesis, University of Kent at Canterbury, United Kingdom

Inogwabini BI, Dianda M, Lingopa Z (2009) The use breeding sites of **Tilapia congica** (Thys & van Audenaerde 1960) to delineate conservation sites in the Lake Tumba, Democratic Republic of Congo: toward the conservation of the lake ecosystem. Afr J Ecol 48(3):800–8006

Inogwabini BI (2014) Conserving biodiversity in the Democratic Republic of Congo: a brief history, current trends and insights for the future. PARKS 20(2):101–110

Inogwabini BI, Leader-Williams N (2012) Using the keywords to explain the bonobo distribution as an effect of human perception of the species. Am J Hum Ecol 1(4):102–110

Inogwabini BI, Hall JS, Vedder A, Curran B, Yamagiwa J, Basabose K (2000) Conservation status of large mammals in the mountain sector of Kahuzi-Biega National Park, Democratic Republic of Congo in 1996. Afr J Ecol 38:269–276

Inogwabini BI, Omari I, Mbayma AG (2005a) Protected areas of the Democratic Republic of Congo. Conserv Biol 19(1):15–22

Inogwabini BI, Omari I, Mbayma AG, Zasy NG (2005b) Protected areas of the Democratic Republic of Congo: a habitat gap analysis to guide the extension of the network. Endanger Species Update 22(2):71–82

Inogwabini BI, Matungila B, Mbende L, Abokome M, Tshimanga WT (2007) Great apes in the Lake Tumba landscape, Democratic Republic of Congo: newly described populations. Oryx 41(4):532–538

Inogwabini BI, Abokome M, Kamenge T, Mbende L, Mboka L (2012) Preliminary bonobo and chimpanzee nesting by habitat type in the northern Lac Tumba landscape, Democratic Republic of Congo. Afr J Ecol. https://doi.org/10.1111/j.1365-2028.2012.01323.x

Inogwabini BI, Mbende L, Bakanza A, Bokika JC (2014) Crop damage done by elephants in Malebo Region, Democratic Republic of Congo. Pachyderm 54:59–65

Karanth KU (2001) Debating conservation as if reality matters. Conserv Soc 1(1):64–66

M'bokolo E (1992) Afrique Noire: histoire et civilisations. Tome II: XIX$^{è}$ et XX$^{e}$ Siècles. Hatier-Aupelf

M'bokolo E (1995) Afrique Noire: histoire et civilisations. Tome I: du XV$^{e}$ aux XXVIIIe siècles. Hatier-Aupelf

Madhusudan MD, Raman TRS (2003) Conservation as if biological diversity matters: preservation versus sustainable use in India. Conserv Soc 1:49–59

Institut National de la Statistique (INS) (2015) Annuaire Statistique 2014. Ministère du Plan et Révolution de la Modernité, République Démocratique du Congo

Ndaywell NI (1998) Histoire générale du Congo: de l'héritage ancien à la République Démocratique. Duculot

Oates JF (1999) Myth and reality in the rain forest: how conservation strategies are failing in West Africa. University of California Press, Berkeley

Programme des Nations Unies pour le Développement (PNUD) and United Nations Office of Project Services (UNOPS) (1998) Programme de national de relance du secteur agricole rural (PNSAR) 1997–2001: Monographie de la Province de l'Equateur. Ministères de l'Agriculture et de l'Elevage, du Plan, de l'Education Nationale et de l'Environnement, Conservation de la Nature, Forêts et Pêche. République Démocratique du Congo

Sayer J, Campbell B, Pethera L, Aldric M, Pere MR, Endamana D, Dongmo ZLN, Defo L, Mariki S, Doggart N, Burgess N (2006) Assessing environment and development outcomes in conservation landscapes. Biodivers Conserv. https://doi.org/10.1007/s10531-006-9079-9

Steurer R (2010) The role of governments in corporate social responsibility: characterising public policies on CSR in Europe. Policy Sci 43(1):49–72

Vansina J (1973) The Tio kingdom of the Middle Congo 1880–1892. Oxford University Press, New York

Wilkie DS, Sidle JG, Boundzanga GC (1992) Mechanised logging, market hunting, and a bank loan in Congo. Conserv Biol 6(4):570–580

Wilkie DS, Carpenter JF, Zhang Q (2001) The under-financing of protected areas in the Congo Basin: so many parks and so little willingness-to-pay. Biodivers Conserv 10(5):691–709

# Chapter 21
# Setting Habitats Aside for Biodiversity Conservation

**Abstract** Creating new protected areas is a contentious issue because the impact of protected areas on humans is thought to be negative. However, with the alarming declines in biodiversity and the diminishing soil fertility, everyone recognizes that protected areas are the backbones of biodiversity conservation and they are the most viable option to preserve habitats and ecosystems. The chapter argues that the philosophical incommensurability between needs to preserve biodiversity and needs to care about the livelihoods of local communities are rather simply felt and can be reconciled if methods to create new protected areas are cleared of the legacy of past failed experiences. Without negating mutually negative effects of protected areas on humans and of increased human populations on biodiversity, the past experience of protected areas in Democratic Republic of Congo made of imposition without community participation in the decision-making process and, ipso facto, without community free and informed consent lays at the heart of the opposition biodiversity versus humans. The creation of new protected areas within the Lake Tumba Landscape used a process that lessens the opposition between biodiversity conservation and fulfillment of human needs. This process is based on broad participation involving communities immediately adjacent to protected areas and a variety of stakeholders with true stakes in areas to be preserved. The end product was an agreement on rights to access some key resources that would classically be infringed by protected areas without people consenting to that effect, attached to sensible trade-offs such as people participating in deterring massive commercially driven poaching.

**Keywords** Protected areas · Community participation · Creation of protected area · IUCN category VI · Human needs · Human livelihoods

## 21.1 Introduction

Protected areas have been vilified by many people across the continent because they are thought to run against human needs (Rainforest Foundation UK (RFUK) 2014). However, even with contentious statements about the impact of protected areas on humans, everyone recognizes that in order for biodiversity to thrive there is need to preserve habitats and ecosystems that can support the lives of different species and



B.-I. Inogwabini, *Reconciling Human Needs and Conserving Biodiversity: Large Landscapes as a New Conservation Paradigm*, Environmental History 12,
https://doi.org/10.1007/978-3-030-38728-0_21

functions that help to keep life continued on earth. With this in mind, the opposition between biodiversity conservation and fulfillment of human needs, particularly the needs of local communities that are located within or adjacent to protected areas, would be reduced to the simple questions of how to proceed when trying to set aside lands and forests to preserve biodiversity. A second question of importance is that of what legally authorized to be done within these protected areas. Finally, it is sensible to state that the initial felt incommensurability between needs to preserve biodiversity and needs to care about the livelihoods of local communities can be reconciled if methods are cleared of the legacy of past failed experiences. Stating this does not equate to negate potential and effective effects of creating protected areas on human communities or negating effects of increased human populations on biological diversity. These mutual negative effects, as has been argued by Curran et al. (2009), are inevitably there and without doubts and undoubtedly negatively impacts on some individuals. The experience, in the case of the Democratic Republic of Congo, has been that most protected areas, in the past, were imposed on communities without their participation in the decision-making process and, ipso facto, without their free and informed consent. This means that even without outright physical displacements of communities (Curran et al. 2009), rights to access some key resources were infringed without people consenting to take the hit of creating protected areas. The process itself was part of the problem and could not lead to sensible trade-offs. Indeed even if some critical benefits were proposed to local communities, there is high likelihood that they would have been rejected if processes were flawed and less participative.

This chapter is about definitions, processes, and alternative ways of setting habitats aside to conservation biodiversity in the Lake Tumba Landscape. Although it is not the aim of the chapter to focus on the protected recently created in the Lake Tumba Landscape, the general principles that will be described in this chapter will be illustrated by what has been done during the processes that led to setting aside of given particular areas. This will be done in order to show how these principles and definitions work in reality.

## 21.2 Definitions, Principles, Processes, and Alternative Ways of Saving Habitats

Defining biodiversity conservation areas is of the utmost importance in the process of delineating the habitats to be saved for biodiversity they shelter. With that in mind, conserving habitat boils down to selecting a type of habitat and strive to protect it if its conditions are still relatively good enough to sustain the biological diversity it originally held or, if conditions have deteriorated, restore it for the same purposes.

At this point, it should be clearly said that habitats for biodiversity are more than just areas made of composites of wild plants and animals. Indeed, the tendency to narrow down biodiversity to numbers of species has not been doing biodiversity the justice it deserves but had contributed significantly to obscuring the debate about what should be conserved, why and how. Indeed, as discussed in Chap. 4, the history of the concept biodiversity is that the definition of the concept has changed over time and the ethical implications and responsibilities toward biodiversity change depending on the understanding of the word (Bosworth et al. 2011). Narrowly defined, biodiversity is equated with the number of species or what is called the 'species richness' found in a given location (Morgan 2009). However, during the past decades, this definition has moved from this narrow understanding to include living organisms and the complex interactions between as well as interactions between living organisms and their abiotic environments. In this sense, accounting for the last evolution of the debate about what biodiversity is, biodiversity is the totality of living organisms and functions that ensure that species and life are maintained on earth. This definition decomposes biodiversity in three main components: composition, structure, and function (Bosworth et al. 2011) and implies that biodiversity should not be viewed only as of the total numbers of species; it has to be expanded to include functions that inter-relate different organisms and sustain life on earth. This repetition of what was said earlier (Chap. 4) is worth in several respects. The first one is that habitats become part of the biodiversity itself; they are not to be protected only as a shelter for other species they are meant to help protect. Habitats are a major part of biodiversity, structurally speaking, and plays a non-negligible role functionally speaking. Hence, it is perfectly logical to set a given type of habitat aside for the preservation of its own and the sake of its own value regardless of what it does shelter. I will come back to this point in the discussion that will follow toward the end of this chapter.

Apart from the above, definitions of protected habitats have been debated over years; today a consensus has emerged to identify protected habitats with standard categories defined by the International Union for Conservation of Nature, which currently recognizes 7 categories in total. Rather than going back to redefine all the seven categories, suffice it here to say that their categories are defined themselves on the basis of the type of management requirements, accessibility by humans, and degrees at which human livelihood activities are permissible. Based on this assertion, it can be said that usages of habitats by humans is provided the guiding principles that define levels of protection accorded to habitats; these criteria run from total prohibition or rather stringent demands or controls on how to access and use some of the resources (Strict Nature Reserve or Category Ia) down to loosely controlled usages but still managed and traditionally exploitable areas (Protected area with sustainable use of natural resources or Category VI).

In essence, these categories provide alternative ways of trying to preserve the habitats across the globe and offer means to contextualize the needs for biodiversity conservation. The same categorical definitions, viewed as means to contextualize the nature and the management scheme for habitats and wildlife species and species of wild plants are provided with principled approaches to follow in order for setting habitats aside for biodiversity conservation purposes. The essential principles can be summarized in three main areas: the value of the area, the land need assessment, and the community consent.

The value of an area can be biological, cultural, and scenic. However, the biological value of any given space can be tricky to define because different people would attribute values differently. An ornithologist would attribute a much higher value to an area because an endemic and endangered species reside in one completely trashed piece of land. This would be equally the case if a group of distinguished herpetologists identifies a completely new species of toad in a series of ponds in a given farm; so would also be the same if an entomologist identifies an area with the world's highest density of insect species. Also as indicated above, habitats have value that goes beyond species they shelter; they play functions that are not all known but that are felt to be essential for keeping life on earth and deserve being maintained. In this sense, even a bare piece of land can harbor essential life-maintaining functions that should, if known by biologists, be kept safe. Unfortunately, there will be no way for humans to discover all these functions and lock all areas presenting such life-maintaining functions from human accessibility. What is being said about functions that biodiversity plays to maintain life on earth is also true for species; no one has the total knowledge of what species are where. Hence, in order to assess the biological value of any given area, it is conventional to use surrogate approach in combination with the precautionary approaches. In general, however, the value of a given habitat would be an assemblage of considerations and metrics, which as defined in Chap. 1, will lead to provide an approximation of what would be sensibly called the eigenvalue of the area (as stated in Chap. 1 of this volume).

The notion of eigenvalue is a mathematical concept that needs to be conceptualized and properly defined in the context of biodiversity conservation for it to bear sensible meanings. The first and the simplest way of understanding the notion of eigenvalue in the context of biodiversity conservation is to look at what differentiates an area with its nearest neighbor or its homologue area located in similar climatic and ecological conditions. Mathematically, a simple calculation of simple dissimilarity measures would be the basic idea to support the notion of the eigenvalue for any given area. As Iceland et al. (2002) say the index of dissimilarity measures the evenness with which two groups are distributed within given environments; and conceptually ranges from complete similarity (0) to complete dissemblance (1). As such and contextualized in biodiversity conservation, it measures the percentages of what neighboring habitats or habitats located in similar ecological or environments conditions would share and what they do not share. In short, the index of similarity conveyed the inverse message of the commonly used Jaccard Similarity index. Technically, with a slight modification on interpretation of variables used by the U.S Census Bureau (2002), the index of dissimilarity can be calculated using the following formula:

$$ID = \frac{\sum_{A=1}^{n} [[t_A | P_A - P|]]}{[[2TP(1 - P)]]}$$

In the above formula, $t_A$ equates to the total species found in area $A$; $P_A$ is the ratio $\frac{X_A}{t_A}$ or the proportion of the species of $A$ that is endemic. $T$ is the total species in the neighboring areas $A$ and $B$; hence, if $t_B$ designates the total species found in area $B$ adjacent to area $A$, $T = t_A + t_B$. $P$, which is the ratio $\frac{X}{T}$ represents the species of $A$ and $B$ that are specific to each of these areas separately and is mathematically rendered by $\frac{X}{T} = [\frac{X_A}{t_A} + \frac{X_B}{t_B}][t_A + t_B)^{-1}$. A wise conservation biology practitioner would calculate the Jaccard similarity index and take the inverse values to represent the dissimilarities. Of course not every piece of land has endemic species so this notion could be easily extended to factor the species that are rare or endangered using the IUCN criteria or species of specific conservation concerns that reside in those areas.

A much simpler method would consist of drawing maps of the given areas and identifying different habitats a landscape is composed of, identifying different species these habitats shelter and draw species potential critical ranging areas and potential ranging areas, draw the maps of human activities, map cultural and social values, etc. Once these layers are completed, one would have to overlay different layers and define the value of each habitat. In this sense, the eigenvalue of a given habitat is provided by the number of biodiversity layers that go to a given spot minus the area dedicated to different human activities. In mathematical terms, the operations involve calculating weighted averages of different categories (wildlife species diversity, ornithological diversity, habitat diversity, fish diversity, etc.) while excluding the weight of human-induced alterations of habitats and species diversity. Of course, such a process would miss the point of conserving biodiversity if it would equate human-induced alterations of habitats with the paucity of biodiversity as, indeed, there is a positive correlation between species richness and endemism and human population density at coarse spatial scales (Fjeldså and Burgess 2008), at least in the case of Central Africa. Indeed, areas such as the Western Albertine Rift that have the highest human densities at the regional scales are also areas where there are higher rates of endemism. This is a somewhat barefoot conservation practitioner method and would not need so much of resources; one can simply use the translucent papers and come up with an area that harbors traits that have more conservation values, which give to an area its own intrinsic value that is its eigenvalue. This is what was actually done in the meeting held in Libreville (Chap. 1) where experts met to put the knowledge they possessed of each part of the Central African forests together. Overlaying knowledge of different taxonomic groups on maps (Fig. 1.3) led to determining the values of each area. In this sense, the eigenvalue of each area was defined taking into account not only endemism but also numbers of species and the diverse habitats. A second way of defining the eigenvalue of each area was also its uniqueness even though it is rather difficult to find an area that is truly unique in nature across Central Africa.

The above method is also used in much more sophisticated manner using modern tools such as computer software packages. Indeed, given the fact that conservation planning happens for landscapes is an exercise for areas that larger than conventional protected areas and on which conservation practitioners possess only very minimal knowledge, it is sometimes efficient and acceptable to use computer software to help throughout the process of defining specific values of each given area located within the landscape. The exercise of computer-guided planning is now becoming a generic practice in biodiversity conservation; indeed this has begun a few decades back. For example, Henson et al. (2009) used a computer package for identifying suitable areas for a suite of species conservation targets basing their definition of suitability of habitats for wildlife species on large, relatively intact forest blocks and used bonobo a species whose habitat requirements would help cover the needs of other wildlife species in the region. Henson et al. (2009) also modeled areas most suited for human activities and future human expansion in functions of the influence of existing settled or cultivated areas, mobility, and access to markets. Despite the fact that this is not clearly said in their paper, the procedure consisting in defining suitable habitats for wildlife species and overlaying the intact habitat blocs with the modeled human present and future activities is equal to the operations to calculate weighted averages of different categories (as indicated above) that have to be juxtaposed with the weight of human-induced alterations on habitats and define the value of the area with the only difference that in the case of the use of computer software allows the speed and the operations are not done manually by the biodiversity conservation practitioners. As a model of overlaying different values, the eigenvalue of an area, in this case, is given by the mathematical function of union (∪) while the areas found to intersect (∩) between human habitat suitability and wildlife suitability models redirect the planners to areas of potential conflict between human needs and biodiversity conservation needs.

Overlaying maps of different categories, using mathematical functions such as union or intersection, etc. are already part of the land need assessment. Indeed, the land assessment has two parts in one process: the habitat suitability for biodiversity conservation and the assessment of habitats needed for human activities. Habitat suitability for biodiversity conservation is carried out in many different ways, which are all reducible to two general approaches that are broadly categorized as field inventories (surveys or census) and computer-based modeling. The inventories are the traditional methodological approaches to assess habitats and the status of biodiversity these habitats shelter (Sutherland 2000; White and Edwards 2000; Sutherland 1999; Hayek and Martin 1997). There is a range of methods to do this; but in the case of forest habitats, which is the case of Central African forests where the Lake Tumba Landscape is located; both direct and indirect counts methods have been used. Indeed, direct count methods, which are essential of the family of visual (Nautiyal et al. 2015; Heyer et al. 1994) or generally sensual (perceptual) surveys, strive to inventory the habitat types or species of interest directly (Bibby et al. 1992); data collected are made of records of instances of sensual perception of the type of habitat or species of interest. In practical, visual contacts or sounds are recorded, following the data collection protocols and records are brought back for analysis and report's writing. The indirect methods family is made of the records of indirect yet undeniable signs of the species of interests (Sutherland 1999, 2000; White and Edwards 2000; McClanahan 1986). Another important divide between the methods is on whether attempt is being made to count everything (individuals or signs) of interest, which is the meaning of census, in strict sense (Anderson and Ohmart 1986; Weber and Vedder 1983; Harcourt and Groom 1972) or whether the approach relies on sampling a representative population of the given habitat or species of interest. Given numerous difficulties to conduct direct total counts in forested environments such as the one in the Lake Tumba Landscape, most studies in Central Africa rely on indirect sampling methods, which are clustered in different approaches, including distance sampling methods, plot sampling methods as well as point sampling methods. Finally, a specific additional method for assessing the suitability of habitats for biodiversity conservation relies on the usage of aerial photography and satellite images. Both aerial photographs and satellite images, particularly when they are of higher resolutions, are useful in that they provide ready-to-use basic information on habitat types and clearly and easily discriminate vegetation cover and denuded land (used space) (Dechka et al. 2002). Both aerial photographs and satellite images are particularly an efficient tool in studies whose objectives are to provide coarse general conditions and help disentangle and visualize different habitat types (Tiner 1990); they can thus help produce first general physical maps of habitats. Indeed, many studies in Central Africa have used these images in their sampling designs to help stratify habitat types and allocate survey efforts (Inogwabini et al. 2012; Hall et al. 1998a, b). However, they are insufficient in themselves to produce maps that would be ecologically sensible. Despite some laudable efforts yet inclusive temptations such as the one that Kortlandt (1995) made in trying to correlate habitat descriptions from images with wild populations of bonobos and despite significant technical improvements which can lead to identifying individual trees in the multitude of other habitat's features (Persson et al. 2002), to draw many ecological conclusions solely based on imagery techniques remains a dubious exercise because differentiating some habitat types may be subject of significant errors throughout the mapping process. That is why ground truthing, which is a fact-checking process through true field surveys are always necessary to qualify the realities of coarse data generated by aerial photography and satellite images.

Using one of the above approaches dependents on the objectives of the study but the final results are essential to help define borders of the habitat that is suitable for biodiversity conservation in the process of the land assessment. The other essential part of the land assessment is to quantify the needs in habitats for human activities.

## References

Anderson BW, Ohmart RD (1986) Vegetation. In: Copperider AY, Boyd RJ, Stuart HR (eds) Inventory and monitoring of the wildlife habitat. The United States Department of Interior, Bureau of Land Management, pp 639–660

Bibby CJ, Burgess ND, Hill DA (1992) Birds census techniques. British Trust for Ornithology and the Royal Society for the Protection of Birds. Academic Press

Bosworth A, Chaipraditkul N, Cheng MM, Gupta A, Junmookda K, Kadam P, Macer D, Millet C, Sangaroonthong J, Waller A (2011) Ethics and biodiversity. Asia and Pacific Regional Bureau for Education UNESCO Bangkok

Curran B, Sunderland T, Maisels F, Oates J, Asaha S, Balinga M, Defo L, Dunn A, Telfer P, Usongo L, Von Loebenstein K, Roth P (2009) Are central Africa's protected areas displacing hundreds of thousands of rural poor? Conserv Soc 7(1):30–45

Dechka JA, Franklin SE, Watmough MD, Bennett RP, Ingstrup DW (2002) Classification of wetland habitat and vegetation communities using multi-temporal Ikonos imagery in southern Saskatchewan. Can J Remote Sens 28(5):679–685

Fjeldså J, Burgess ND (2008) The coincidence of biodiversity patterns and human settlement in Africa. Afr J Ecol 46(Supplement 1):33–42

Hall JS, Saltonstall K, Inogwabini BI, Omari I (1998a) Distribution, abundance and conservation status of Grauer's gorilla. Oryx 32:122–130

Hall JS, White LJT, Inogwabin BI, Omar I, Morland HS, Williamson EA, Saltonstall K, Walsh P, Sikubwabo C, Bonny D, Kiswele KP, Vedder A, Freeman K (1998b) A survey of gorillas (**Gorilla gorilla graueri**) and chimpanzees (**Pan troglodytes schweinfurthi**) in the Kahuzi-Biega National Park lowland sector and adjacent forest in eastern Democratic Republic of Congo. Int J Primatol 19:207–235

Harcourt AH, Groom AFG (1972) Gorilla census. Oryx 28:59–70

Hayek LC, Martin AB (1997) Surveying natural populations. Columbia University Press

Henson A, Williams D, Dupain J, Gichoh H, Muruthi P (2009) The heartland conservation process: enhancing biodiversity conservation and livelihoods through landscape-scale conservation planning in Africa. Oryx 43(4):508–519

Heyer WR, Donnelly MA, McDiarmid RW, Hayek LC, Foster MS (1994) Measuring and monitoring biological diversity: standard methods for amphibians. Smithsonian Institution Press

Iceland J, Weinberg DH, Steinmetz E (2002) Racial and ethnic residential segregation in the United States: 1980–2000. In: Paper presented at the annual meetings of the population association of America, Atlanta, on behalf of the U.S. Census Bureau, Georgia, 9–11 May 2002

Inogwabini BI, Abokome M, Kamenge T, Mbende L, Mboka L (2012) Preliminary bonobo and chimpanzee nesting by habitat type in the northern Lac Tumba landscape, Democratic Republic of Congo. Afr J Ecol. https://doi.org/10.1111/j.1365-2028.2012.01323.x

Kortlandt A (1995) A survey of the geographical range, habitats and conservation of the Pygmy chimpanzee (**Pan paniscus**): Ecological perspective. Primate Conserv 16:21–36

McClanahan TR (1986) Quick population survey method using faecal droppings and the steady state assumption. Afr J Ecol 24:37–39

Morgan GJ (2009) The many dimensions of biodiversity. Stud Hist Philos Biol Biomed Sci 40:235–238

Nautiyal S, Bhaskar K, Khan YDI (2015) Biodiversity of semiarid landscape: baseline study for understanding the impact of human development of ecosystems. Springer

Persson Å, Holmgren J, Söderman U (2002) Detecting and measuring individual trees using an airborne laser scanner. Photogramm Eng Remote Sens 68:1–8

Rainforest Foundation UK (RFUK) (2014) Protected areas in the Congo Basin: failing both people and biodiversity? Briefing Paper, Rainforest Foundation

Sutherland WJ (1999) Ecological census techniques: a handbook. Cambridge University Press

Sutherland WJ (2000) The conservation handbook: research, management and policy. Blackwell Science

Tiner RW (1990) Use of high-altitude aerial photography for inventorying forested wetlands in the United States. For Ecol Manag 33(and34):593–604

U.S. Census Bureau (2002) Racial and ethnic residential segregation in the United States: 1980–2000: appendix B. Measures of residential segregation. http://www.census.gov/hhes/www/housing/resseg/pdf/app_b.pdf

Weber AW, Vedder A (1983) Population dynamics of the Virunga gorillas 1959–78. Biol Cons 26:341–366

White LJT, Edwards A (2000) Conservation research in the Central African rain forests: a handbook. Wildlife Conservation Society

# Chapter 22
# Protected Areas: Defining the Optimum Law Enforcement Resources

**Abstract** The chapter's aim is to identify different strategies that are implemented in Africa and elsewhere to ensure better protection of protected areas newly created within the Lake Tumba Landscape and to find the most realistic approaches to ensure effective law enforcement, taking into account its social, economic, cultural, and historical contexts. Two models (guard density model, strategic deployment model) were used and theoretically tested. Guard density and strategic deployment scenarios were developed to project the number of staff needed by the Congolese Conservation Agency. Using the output of these scenarios, we developed a budget for optimal action in preserving the network of the Congolese protected areas. Both guard density and strategic deployment models indicated that PA in the DRC was globally understaffed, with current personnel representing respectively only 26% projected human force. The study suggested higher costs to render DRC's PA functional, which meant that this should be borne in minds when creating new protected areas in Lake Tumba Landscape. Without minimizing the efforts of different international conservation initiatives in DRC and in Lake Tumba Landscape, the chapter conveys the message that effort consented in conservation in both areas are below minimum investment needed to make conservation work.

**Keywords** Protected areas · Law enforcement · Guard density model · Strategic deployment model · Personnel · Budget

## 22.1 Introduction

Protected areas, in their diversity of categories, form the core of the conservation effort around the world (Emerton et al. 2006). That is why the landscape, as primarily a biodiversity conservation paradigm, has not excluded protected areas as part of the biodiversity conservation equation. Rather, the landscape approach's focus is on how to make protected areas accepted, acceptable, and part of the daily experiences of the management of natural resources. Apart from the acceptance of the protected areas by the communities, which should be gained throughout the long process of participatively conducting land assessment, land rights and agreeing on the objectives



B.-I. Inogwabini, *Reconciling Human Needs and Conserving Biodiversity: Large Landscapes as a New Conservation Paradigm*, Environmental History 12,
https://doi.org/10.1007/978-3-030-38728-0_22

for any given protected area, success in conserving biota, species, and other conservation values found in these areas is a function of multiple variables, among which the most important are sound knowledge of the conservation targets, sufficient financial resources, and sufficiently trained personnel (Inogwabini et al. 2005a, b). There is a fair agreement among managers of protected areas, conservation practitioners, and conservation biologists that sound scientific knowledge of the ecosystem functions as well as that of targeted species ecology are among the most important keys to a successful conservation program (Gray and Kalpers 2005; Blake and Hedges 2004; Primack 2000). There seems also to be a large consensus on the fact that without adequate funding levels, conservation of biodiversity will remain a mere wish, or at the very best numerous protected areas will remain paper parks and reserves (Blom 2004; Hanks and Attwell 2003; Wilkie et al. 2001; Bruner, undated). Although there is a great number of suggestions on what levels of personnel are adequate to protect specific conservation targets, there are little standardized methods to come up with these numbers. Until now, all the theoretical effort in this sense has been limited to the density of guards in given protected areas (Inogwabini et al. 2005a, b; IUCN 1992), which has been demonstrated to be the most important factor influencing the effectiveness of parks in the particular case of tropical forests (Bruner et al. 2001). Even though this is a handy logical approach to project quantified numbers of law enforcement staff, there seem to be issues that are, at least, equally important in the staffing process that is not accounted for in this approach. These may include levels of threats, geographic deployment of law enforcement effort, probability of detecting illegal incursions into a given protected area, and how much resources are needed for each strategy. The complete mission too is on how to engage communities to become the wardens of their own resources, of which protected areas should be an integral part. In order to include these important metrics into the planning processes whose aim was to properly design, create, and implement the existence of new protected areas within the Lake Tumba Landscape, a theoretical exercise was started to provide a sound theoretical approach that would guide the management decision for the future staffing of the Institut Congolais pour la Conservation de la Nature (ICCN, the national organism in charge of protected areas) in the newly created protected areas in the Lake Tumba Landscape. The first objective of this exercise was to identify different strategies or scenarios that are suggested or implemented in Africa and/ or elsewhere. The second objective of the same exercise was to apply these scenarios in the case of the newly created protected areas within the Lake Tumba Landscape with the sole reason for applying these scenarios in the case of the Lake Tumba Landscape being to find the most realistic approaches that could be adopted for the landscape's protected areas taking into account its social, economic, cultural, and historical contexts.

## 22.2 Scenarios, Materials, and How to Use the Approaches to Allocating Patrol Efforts

With three major habitat types (swampy forests, terra firma forest, and the southern savannahs (see Chap. 5), the general context of the Lake Tumba Landscape is one that runs in parallel with the context of the Democratic Republic of Congo, a country whose extensive geography has imposed a diversity of biodiversity, habitats (Sayer 1992), and cultures (Wufela 1992; Obenga 1987). This single factor (diversity) imposes a hierarchical approach. However, for apparent reasons of coherence in the system, two approaches were developed firstly: the guard density model (GDM) and the strategic deployment model (SDM). A third model based on the probability to detect illegal intrusion (PDI) was also discussed but was not fully developed as data to support such an exercise were not available from any park and were much less from new protected areas that were to be created within the Lake Tumba Landscape. Adjustments of these models were then tried taking into account local parameters, such as realistic chances of a recommendation being implemented, which are ultimately linked to finances, political potential to issue a binding engagement for proposed reforms, etc.

## 22.3 Staffing for Effectiveness

### *22.3.1 Guard Density Model*

This is the most common approach, developed in Africa and in many other tropical zones (Gray and Kalpers 2005; Inogwabini et al. 2005a; Bruner et al. 2001; IUCN 1992), and the most suggested by conservation practitioners in the Democratic Republic of Congo. The approach is based on the simple inferred assumption that there is an optimal number km$^2$ ($\Omega$) to be allocated to each guard or ranger in the total protected zone of $\Psi$ (km$^2$) (Gray and Kalpers 2005; Inogwabini et al. 2005a; Bruner et al. 2001; IUCN 1992). With that, the adequate number of guards $\sum$ per PA is straightforwardly obtained by the simple division of the total area by the threshold number $\Psi$ or simply $\sum = \frac{\Psi}{\Omega}$.

The problem with this approach, however, lays on difficulties to set a $\Omega$ that is agreeable by different conservation practitioners and, particularly, for each type of habitat. $\Psi$ varies widely, depending on sources (Gray and Kalpers 2005; Inogwabini et al. 2005a; Bruner et al. 2001; IUCN 1992), ranging from 1guard/4 km$^2$ to 1 guard/270 km$^2$; most often these numbers being simply what is found to be already in place (IUCN 1992). For the purpose of this exercise, the $\Psi$ values were those taken from legal decrees that created given protected areas in the Lake Tumba Landscape despite the fact that actual boundaries of the two protected areas in the Lake Tumba Landscape are inaccurate because, as it the case of all protected areas of the Democratic Republic of Congo (Inogwabini et al. 2005b), of lack of means to delineate these limits on ground. Setting $\Omega = 33$ km$^2$ for one guard (Bruner et al. 2001), as this seemed to be a mean value and has been demonstrated to be a good threshold in the 15 most effective parks in the tropical zones (Bruner et al. 2001), the scenario with the number of potential optimal number of guards was calculated for the two protected areas of the Lake Tumba Landscape.

### *22.3.2 Strategic Deployment Model*

The strategic deployment model (Tucker 2003) is not well developed as an approach in the current protected area network's planning and strategies yet. However, it can be easily adapted and would be a potentially valuable approach. Theoretically, the approach presupposes that protected areas are defined as any given operational territory and are perfect geometrical figures. These areas are supposed to have the same values for conservation; because of that, efforts should be allocated evenly across any protected area. Each conservation zone within the larger protected area has a headquarters (HQ) situated at the center of the grid (or a circle) and is surrounded by four other stations geographically deployed at equidistant locations from the HQ, and supported by a mobile unit . Each station is then staffed with two patrol units comprised of six guards each. The reason to have two sections per conservation zone is to allow rotation and keeping the camp safe but the total number of 12 guards by outpost makes an army section. With known conservation zones, the optimum number of guards $\sum$ per PA is straightforwardly obtained by the simple multiple of the total zones $\Gamma$ by $\Omega = 12$ guards or $\sum = \Gamma \times \Omega$ (=12 guards).

## 22.4 Budgeting for Effectiveness

The search of the existing literature on different funding levels and models for different systems of protected areas in Africa (e.g. Emerton et al. 2006; Blom 2004; Eagle et al. 2002; Wilkie et al. 2001; Wilkie and Carpenter 1999; Bruner undated) was conducted to see what would be applicable to the DRC context, and particularly to the case of the protected areas being created in the Lake Tumba Landscape. Throughout the reading of all relevant, available, and accessible literature, the most prominent approach was that of investment km$^{-2}$. Optimal numbers suggested by those idealistic case studies were US\$200 km$^{-2}$ per year (minimum) and US\$230 km$^{-2}$ per year (Eagles et al. 2002), all higher than any realistic expectation for even the coming years in the Democratic Republic of Congo. Nevertheless, using these figures, it has been possible to develop two budget scenarios based on these norms. This has been done so just to indicate what it would take to implement proper conservation programs on ground. To adjust these idealistic numbers to the realities of the country, it was therefore thought to be logical to base the planning reflections on existing cases of the recent experiences of funding provided to the five world heritage sites, which were financed by the United Nations Foundation (UNF) through the United Nations Educational, Scientific and Cultural Organization (UNESCO). The mean investment km$^{-2}$ of the UNF Program in these five sites was estimated to US$44.00 (Inogwabini et al. 2005a). This figure had been doubled to US$88.00 km$^{-2}$ in the proposed optimal scenario because the UNF–UNESCO financial support neglected some important traditional budget line items, which then reduced the efficiency of the law enforcement personnel, and subsequently impeded a proper implementation of conservation activities. Stating so does not reduce how very critical importance the support from the UNF–UNESCO was for protecting the five world heritage sites of the Democratic Republic of Congo during the social turmoil from which the country was emerging when the planning for the Lake Tumba Landscape was occurring. With a fixed amount of money to be allocated to each km$^2$, the expected maximum budget for each protected area was easy to obtain. Total budget per protected area was then divided into different budget line items, including personnel salaries and bonuses following a hierarchical professional categorization wherein the ICCN law enforcement personnel was aligned to the national nomenclature. Then the salary scale that was agreed upon by the Mbudi agreement (Gouvernement of the DRC 2004) was used in order to align all other charges with the official nomenclature of benefits and fringes that are demanded by the Congolese labor law . The provisions of the labor law in the Democratic Republic of Congo set the pension at 8% of the base salary per category, health care at 4.5% of the base salary for each family member officially under the tutorial support of the employee. To avoid the elasticity in the term family members, which can be extended to include even members of the lineages, a typical family in the landscape was considered to comprise eight people (one couple plus six children under the age of 18). Apart from these compulsory charges, the operation budgets per site were also included. These operational expenses were mostly consisting of patrol rations, vehicles, lubricants, maintenance, drivers, and mechanics for each site. Apart from these obvious operational costs, it was also important to include other important budget lines such as training (para-military, law enforcement, biomonitoring, building patrol posts, etc.). Finally, because the proposed budget should include all the costs of running conservation activities in each site, a number of projected resources was allocated to the component of community-based conservation, particularly for micro-project programs in each site, taking into account specificities of each of them. The addition of the community-based conservation was felt to be an important investment as any of the above investments because biodiversity conservation was agreed by different stakeholders to become part of the total economy of the landscape and, as such, protected areas were expected to contribute to the well-being of communities as much as they can. Given the limitations of possibilities for protected areas in the Lake Tumba Landscape to generate financial incomes from other sources, it was felt genuinely fair for conservation organizations and the state structures in charge of biodiversity conservation to contribute to the budget that would provide alternative sources of incomes to communities.

## 22.5 Effective Law Enforcement in the Newly Created Protected Areas

There are two newly created protected areas within the Lake Tumba Landscape; there are the Tumba-Lediima (7500 km$^2$) and the Ngiri Triangle Reserve (1000 km$^2$). With these figures on the surface areas of the Tumba-Lediima and Ngiri, the guard density model predicted a total of 255 guards for both of the protected areas of the Lake Tumba Landscape. The Tumba-Lediima Natural Reserve, with its 7500 km$^2$, would need 225 eco-guards in this model while the Ngiri Triangle Natural Reserve (1000 km$^2$) would need 30 eco-guards. When the strategic deployment model was considered, it predicted a total of 4 HQs and 24 patrol posts for the Tumba-Lediima Natural Reserve, which generated a total of 288 guards for the sole Tumba-Lediima. applied on the Ngiri Triangle; the strategic deployment model came up with 2 HQs and 12 patrol posts, which numbers converted into 144 eco-guards.

Using different budgeting scenarios, the amount needed to render fully operational these two protected areas dependent on the managerial decisions. But for them to be optimally managed, factoring the figures used in other African countries come out with a budget ranging between 1,500,000 US$ and 1,725,000 US$ per year to make the Tumba-Lediima Natural Reserve functional. The same financial model, projected a budget falling between 200,000 US$ and 230,000 US$ for the Ngiri Triangle Natural Reserve. However, when figures from the UNESCO experience in the five world heritage sites in the Democratic Republic of Congo were used to model projected budgets, these were significantly reduced and were within the ranges of 330,000 US$–660,000 US$, and 44,000–88,000 US$ annually, respectively, for Tumba-Lediima and Ngiri.

Accounting for the realities of the Democratic Republic of Congo, which are characterized by limited sources of funding and the low ranking of conservation activities among national priorities (for this topic), the discussions among stakeholders proposed an adaptive approach for critical numbers of law enforcement personnel, i.e. for the two protected areas of the Lake Tumba Landscape. The first scheme was to adopt optimum numbers generated by the guard density model for the Tumba-Lediima Natural Reserve (i.e. 225 eco-guards). For the Ngiri Triangle Natural Reserve, it was suggested that the best option would be to take the mean values between the guard density model and the strategic deployment model because this option would give both sufficient staff members without being extremely costly at the same time. These options were raised to cope with the reality that the mission would need the resources pondered by the realities of the country; the chosen options were deemed appropriate because costs are linked to the numbers of people employed and this model gave the lower numbers for each of the protected areas of the Lake Tumba Landscape. As for the hierarchy of operational model, it was proposed that in both cases the base of the pyramid of functions is larger than the upraising hierarchy; in particular, it was suggested that the base of the pyramid should include more than 85% of the surveillance personnel.

## 22.6 The Protected Areas of the Lake Tumba and the Broader Scenes of Protected Areas

Many parks in the Central African countries have been created without prior strategic assessment and have been managed either by goodwill, informed guess, or simple trial and error methods. However, managing protected areas is now acknowledged as a technical question, which needs to be addressed by sound planning tools (Thomas and Middleton 2003). The necessity of planning exercise does not imply only defining a set of clear objectives but also defining a set of means to achieve those objectives (Thomas and Middleton 2003). This chapter was about resource planning for the protected areas network located in the Lake Tumba Landscape. It addressed issues on how to strategically plan the deployment of human and financial resources to achieve different minimal conservation goals in the Lake Tumba Landscape, one of the 12 most important landscapes for the conservation of the biodiversity in the Democratic Republic of Congo and Central Africa as a region. At a time when the Democratic Republic of Congo, aided by international conservation organizations, has taken steps to honor its international commitment to increase its network of protected areas from ca. 10–15% of the national territory by creating new protected areas such as the Tumba-Lediima and Ngiri, exercises such as these were the most needed to complement biological definitions of highly important areas. Indeed, according to Bruner et al. (2001), the availability of different kinds of resources and the management structures correlate with the effectiveness of protected areas as a means to save biodiversity. Because of that correlation, strategic thinking on what types of resources are needed is necessary to help achieve sensible and tangible biodiversity conservation results.

However, models such as these need to be carefully assessed against the realities of the field. For example, personnel-wise, it is can be argued that the guard density model yielded a low number of the eco-guards for the Ngiri Triangle Natural Reserve. This is particularly true when one would have to consider the habitats of the Ngiri, which are mostly swamped and would demand more patrol efforts from the personnel than those operating on a terra firma environment. This reality should lead to a reduction in the number of days to be spent under tents while teams are patrolling. The reduction of the time to be spent out for patrolling means an increase in the number of the staff to cover for each other throughout an appropriate rotation system. Indeed, the suggestion that for the Ngiri Natural Reserve the optimum number of eco-guards is the one derived from the strategic deployment model was justified by including this factor in the analysis. Indeed, the guard density approach favors large entities while it has a negative bias for smaller protected areas such as the Ngiri Natural Reserve. To correct for this bias, strategic deployment models are the most indicated, in particular cases of smaller protected areas.

The second example that is needed to be treated in this circumstance is that protected areas need more than eco-guards; they need administrative personnel to fill in support functions such as accountants, logisticians, radio operators, and health staff. When factored in the global picture in combination with the actual eco-guards, except the higher-rank command personnel who are comprised within the remaining 15% of the 100% law enforcement staff, the resources modeling above is giving just a minimum figure. To incorporate these other categories of needed human resources needs another way of factoring support needs. Indeed, throughout the assessment, it came out from field experienced conservation practitioners that one patrol team would need to mobilize two drivers and one logistician in addition to one field-based permanent radio operator to coordinate safety and security of the team movement. These additions mean that for the Tumba-Lediima, there would be 24 staff members to add on the projected number to accommodate the 24 patrol posts. The same logic applied to the Ngiri leads to 12 additional members to be effective.

Protected areas of the Democratic Republic of Congo have been described by Inogwabini et al. (2005a, b) as being all understaffed in their majority. Inogwabini et al. (2005a, b) have attributed this phenomenon of having fewer guards than needed for many protected areas to the lack of strategic assessment for type and needs of employment before the creation and during the implementation of protected areas. This situation has been largely worsened by the general collapse of the role of the state in many aspects of the national life of the Congolese State since the early 1990s. The collapse in the economy decreased state-generated revenues, which would help pay salaries to workers employed by state agencies such as the Congolese Institute for the Conservation of Nature (ICCN). As a strategy to cope with that incapacitating situation, the government has congealed hiring new personnel, and to compensate those who were hired before the generalized crisis by paper promotions have become the only bonus. In many protected areas, there are many officers than patrol units to command. The Congolese Institute for the Conservation of Nature has more directors than it really needs. For the biodiversity conservation to work within the Lake Tumba Landscape, stakeholders agreed to use the strategic thinking through the participative meetings to make certain that the natural reserves of Tumba-Lediima and Ngiri would become a reality.

Comparing budgetary projections at the global scales showed that the network of protected areas of the Democratic Republic of Congo would be between 5,927,944 and 41,568,360 US$ annually to make all protected areas functional but without the Tumba-Lediima and Ngiri being included. The maximum amount of 41,568,360 US$ per year for the entire country represents 48% of the amount budgeted for the conservation of different ecosystems of the Niger Delta and Congo Basin Forest (Blom 2004); it is about 4% of the total amount it would take to properly manage the network of protected areas across the world, which Buner et al. (2003), Vreugdenhil (2003) estimated to be 1,100,000,000 S$. The same maximum amount of 41,568,360 US$ (annually) represented about 69% of the total budget provided by the Central Africa Regional Program for Environment (CARPE) for 12 priority conservation landscapes across the Congo Basin for 5 years (ca 60,000,000), ca. 52% of the announced Congo Basin Forest Funds (80,000,000 US$) and 76% of the intervention of the African Bank of Development in the Congo Basin Forest Partnership (ca. 55,000,000 US$). These figures reveal things that are different, viewed from different angles: first, they pose the question of whether it was wise financially speaking to increase the number of protected areas in the Democratic Republic of Congo and particularly adding protected areas in the Lake Tumba Landscape. This question is appropriate given the fact that funding to support the Tumba-Lediima and Ngiri Natural Reserve would come as additional needs while the already existing protected areas do not even receive a quarter of what they need. The other issue revealed by the figures on the global needs for the network of protected areas in the Democratic Republic of Congo is that, in comparison, the Tumba-Lediima and Ngiri Natural Reserve would not add a greater financial burden, has the country been financially solvent. Indeed, the respective ranges of annual funding needs for Tumba-Lediima and Ngiri Natural Reserves 1,500,000–1,725,000 US$ and 200,000–230,000 US$ are rather meager compared to the global figures and needs for other protected areas of the country and the region.

But these budgets should not be considered as the maximum amount of funding needed for making these natural reserves functional. They are solely an exercise to ensure proper law enforcement. There is no data to quantify the socioeconomic costs as a means to ensure acceptable trade-offs with communities. Indeed, Leader-Williams and Albon (1988) were right in stating that the socioeconomic background of protected areas is relevant when looking at conservation schemes. And confronting theoretical budgeting considerations as it is done in this chapter would be a violent blow to the very principles advanced by the landscape approach. In fact, the approach is that not only the biodiversity should be taken care of but also the needs of the communities and other stakeholders. The models of budget projections ideally picture only the law enforcement side and should be augmented by some significant magnitudes of socioeconomic budgets to contribute to the local sustainable livelihood. But this contribution needs to come from a completely different assessment method, which is the rural development appraisal.

Bruner et al. (2001) suggested that shortages in budgets were the single most important explanation for the degradation of protected areas around the world and for weak motivation of personnel and corruption of managers have been identified by Inogwabini et al. (2005a, b) as contributing to disempowering protected areas in the Democratic Republic Congo and argued that the semblance of protection that has been going in many protected areas of the Democratic Republic of Congo could, therefore, be attributable only to the heroism of committed conservation workers (guards and their families) that have been doing their work over years. But what missed in these perspectives was the fact that protected areas would only work if they are part of the reality of the economic and social landscapes where they are located. Taking this perspective on board, it is sensible to argue that protected areas in the Democratic Republic of Congo have also survived because, in many ways, they have become by the force of history part of global social entities that people take for granted.

Models used in this for allocating efforts were entirely based on law enforcement approaches, even though our approach in budgeting included some funding for community conservation. Conclusions drawn on budget and necessary personnel are therefore minimal. Because the DRC has a lot more pressing demands for economic development such as human health care, education, food security, and national security, conservation will only emerge at the bottom of the national priorities. Results from this exercise call for a square look at the reality. To affront this reality, several options are to be sought. The first one is for the DRC government to scale down its ambitions and promote new mechanisms of conservation such as downgrading strictly protected areas to loose categories such as IUCN Category VI and combine these with community-conservation practices and private conservation areas. Such approaches will relieve the burden of law enforcement from the shoulders of the government and place local communities and/or private owners face to their responsibilities. Of course, there should remain a core conservation network of PAs that will fall under the entire control of the state. The second possible option, and maybe the most viable option, is to define with international conservation organizations mechanisms to identify sources of continuous and significant financial resources. Such mechanisms would include environment service payments and international trust funds. One reality that the international conservation community has to face is that DRC will not be able to correctly bear its responsibilities in a few year's time and hopes about tourism providing that many resources are, at least for the time being, dreams that will take long before becoming reality. Therefore, while rightly pushing the country to expand its network of protected areas in order to avoid major losses in biodiversity caused by the anarchy dwelling in the country for the last 18 years, the international conservation community should be ready to pay for significant amounts of money to make for both old and newly created PAs functional and true refuges of the biological diversity.

## References

Blake S, Hedges S (2004) Sinking flagships: the case of forest elephants in Asia and Africa. Conserv Biol 18(5):1191–1202

Blom A (2004) An estimate of the costs of an effective system of protected areas in the Niger Delta – Congo Basin Forest region. Biodivers Conserv 13:2661–2678

Bruner AG (undated) How much will effective protected area systems cost? Center for Applied Biodiversity Science, Conservation International, Washington DC

Bruner AG, Gullison RE, Rice RE, da Fonsesca AB (2001) Effectiveness of parks in protecting tropical biodiversity. Science 291:125–128

Eagles PJF, McCool SF, Haynes CD (2002) Sustainable tourism in protected area: guidelines for planning and management. The International Union for Conservation of Nature

Emerton L, Bishop J, Thomas L (2006) Sustainable financing of protected areas: a global overview of challenges and options. The International Union for the Conservation of Nature

Gray M, Kalpers J (2005) Ranger based monitoring in the Virunga-Bwindi region of East-Central Africa: a simple data collection tool for park management. Biodivers Conserv 14(11):2723–2741

Hanks J, Attwell CAM (2003) Financing Africa's protected areas. In: Vth world parks congress: sustainable finance stream. Overview session: introduction to the sustainable finance stream and the policy context for protected area financing. Panel A - estimating the costs

Inogwabini BI, Omari I, Mbayma AG (2005a) Protected areas of the Democratic Republic of Congo. Conserv Biol 19 (1):15–22

Inogwabini BI, Omari I, Mbayma AG, Zasy NG (2005b) Protected areas of the Democratic Republic of Congo: a habitat gap analysis to guide the extension of the network. Endanger Spec Updat 22(2):71–82

International Union for the Conservation of Nature (IUCN) (1992) Protected areas of the world: a review of national systems, vol 3. Afrotropical. Gland

Gouvernement de la RDC (2004) L'accord de Mbudi ou le contrat social de l'innovation, est un accord signé entre le Gouvernement de transition de laRépublique démocratique du Congo et les syndicats de l'Administration publique et interprofessionnels. https://www.droitcongolais.info/files/810_accordmbudi.pdf

Leader-Williams N, Albon SD (1988) Allocation of resources for conservation. Nature 336(6199):533–535

Obenga T (1987) Le Zaïre: Civilisations traditionnelles et culture moderne. Présence Africaine, Paris

Primack RB (2000) A primer of conservation biology, 2nd edn. Sinauer Associates, Inc. Publishers, Sunderland

Sayer JA (1992) Zaire. In Sayer JA, Harcourt CS, Collins NM (eds) The conservation atlas of the tropical forests: Africa. MacMillan, UK, pp 272–282

Thomas L, Middleton J (2003) Guidelines for management planning of protected areas. International Union for Conservation of Nature

Tucker SC (ed) (2003) United States Army: The encyclopedia of U.S. military history. ABC-Clio Publisher

Vreugdenhil D (2003) Modeling the financial needs of protected area systems: an application of the "Minimum Conservation System" design tool. Paperpresented at the Fifth World Parks Congress; 8–17 September 2003, Durban, South Africa

Wilkie DS, Carpenter JF (1999) Can nature conservation help finance protected areas in the Congo Basin? Oryx 33(4):332–338

Wilkie DS, Carpenter JF, Zhang Q (2001) The under-financing of protected areas in the Congo Basin: so many parks and so little willingness-to-pay. Biodivers Conserv 10(5):691–709

Wufela YA (1992) A la recherche d'une identité: Littérature, langues et recherche scientifique face au processus du développement du Zaïre. ILCAA African literature series, vol 3

# Chapter 23
# Planning the Mobilization of Resources via Sustainable Tourism

**Abstract** Questions around the viability of protected areas and conservation landscapes evolve around their financial sustainability. Hence, the chapter addresses this question for the Lake Tumba Landscape via conducting a thorough assessment of its ecotourism potential, which was identified as the most possible and viable independent source of financial revenue. The study identified several natural assets to support the ecotourism (viewing wild bonobos, birds, landscape sceneries, bush trekking) and cultural assets (traditional dancing fair and net hunting). Additionally, the study identified navigating Kasai-Congo River and kayaking along the rapids found in the landscape around Malebo as additional assets. The chapter discusses the n*ecessary* investments in infrastructure, logistics, and human resources and proposes the joint-venture (DRC Government, conservation, private sector, and communities) as the route to ensure successful ecotourism in the Lake Tumba Landscape.

**Keywords** Ecotourism · Income sharing · Wild bonobos visioning · Birding · Landscape sceneries · Bush trekking · Traditional dancing fair · Net hunting

## 23.1 Introduction

The main objective of the ecotourism initiative in the southern Lake Tumba landscape is to find an independent source of financial revenue that can support conservation activities, including the support to the field teams, maintenance of the field research site, and community conservation activities in the region of Malebo and adjacent territories.

Since 2006, the Lake Tumba Landscape, led by the World Wide Fund for Nature, had confirmed the information provided by its partner local ONG named Mbou-Mon-Tour (MMT) that bonobos were present in the Malebo area. The field work conducted by the teams of the World Wide Fund for Nature found higher densities of the bonobos in several locations (see above) and other species of conservation concern such as forest elephants and lions. The teams have also documented the presence of five monkey species, forest buffaloes, and, preliminarily, some hundred species of birds. The power of the bonobo as flagship species, the presence of other species of conservation concern, and the fact that bonobos can be seen in their natural habitat in



B.-I. Inogwabini, *Reconciling Human Needs and Conserving Biodiversity: Large Landscapes as a New Conservation Paradigm*, Environmental History 12,
https://doi.org/10.1007/978-3-030-38728-0_23

such a short distance from Kinshasa had dictated the decision for the implementation of the bonobo habituation program, which apart from its scientific reasons, aims at the community conservation program. The community conservation program is based on the understanding that it is only when people gain some alternative sources of either revenue or protein that they can become fully immersed into the conservation of species of conservation value. Even though in the case of the Malebo region, the conservation of the bonobos by local communities has preceded the community conservation activities, World Wide Fund for Nature believed that global community needs such as finding alternative sources of protein via fish culture, domestication of animals for meat, improved health care, and transportation need to be supported by incomes generated from the ecotourism that focuses on the bonobos and other wild adventures that can be found only in this region.

## 23.2 Natural Assets: Bonobos, Birds, Landscape Sceneries, and Bush Trekking

Natural assets on which the ecotourism program in the Malebo region will be based are (1) habituated bonobos in their natural habitat, (2) bird-watching session, (3) bush trekking along rivers and small monkey viewing, (4) landscape beauty vision, and (5) cultural tourism with local communities. This last aspect is divided into two distinct categories: (a) ancestral dances and fairs and (b) net hunting by young people. The sixth tourism asset is a long yet pleasant river tour by boat on the way down from Makayabo to Kinshasa.

The bonobo habituation program has started in March 2006 concentrating on four groups of bonobos (2 groups: Nkala; 1 group: Mpelu and Mbee, respectively). Since then, one group has become 60–80% accessible to visitors. The distance of vision between bonobos of the Nkala and visitors ranged between 20 and 35 m. Even though bonobos are tracked on a daily basis by the field staff, viewing them by tourists would be at its best when fruits are ripe on trees, which is in early March, June, and September–December. Despite this, however, the most physically fit tourists would easily view bonobos almost every day. Though there was still some work to be done on visibility distance, the second group (Mpelu) is also becoming more accessible while the work on the third group at Lenga has planned to increase their visibility for several months from the end of 2009. It was felt back in 2009 that when all three groups would become accessible, the program would be able to receive 15 people in a day for 4 days per week.

The region of Malebo, with its forest-savanna mosaic offers a great assemblage of birds, including species that were never thought to occur in this tropical region. Therefore, birding is an asset and would be an activity that would serve mostly in early mornings and late in the afternoons. This activity can almost start at the moment as birds present at the region offer a nice scene every morning and every afternoon. The activity will, however, become more interesting for people interested in birds at the north–south migration of birds as the region is in bird fly-away zone. More birds have been observed in the months of June and July, which corresponds to the dry season in the region.

The gallery forests of the region offer a magnificent view of the quiet forest itself and several other species that can be viewed only by quiet forest admirers walking through the forest in the silence of early mornings and late in the afternoons. Bush trekking has become a tourism product in the recent past. People simply leave the crowds of major towns and go exercise for fitness in a natural environment with pure air. The ecotourism program in the Malebo region will offer some of the incomparable assets in this type of activity. Currently, the project has opened forest circuitries in forests of villages where we daily track bonobos, which could be used by tour operators for visitors. The best of our circuitry is the 10 km-long trail near along the Bambo River where we have our field base and hope to install one of our main tourism camps. This trail offers beautiful views of scenes of small falls of the Bambo River while providing a spectacular view of the stands of forest. People will see small monkey groups, including species such as the red tail monkey *Cercopithecus ascanius*, Allen's swamp monkey *Allenopithecus nigroviridis*.

The southern portion of the landscape is located in the forest-savannah mosaic, which offers beautiful natural scenes. Visitors can be driven on an open vehicle and let to wander around to take pictures of large expanses of beautiful savannahs intersecting with forest edges. This tour would also allow taking pictures of large birds that can be viewed from long distance.

## 23.3 Cultural Assets: Dancing Fair and Net Hunting

Beyond unusual events such as the coronation of new traditional chiefs and/or birth and deaths, the Bateke have dancing events several times per year. This is part of their culture. They have dancing for 'thanksgiving' events at the end of the dry season (August–September) when the cultivating seasons end. They also have another fair event when comes the moment to celebrating their production. These events are accompanied by dances involving all social classes; people share food and drinks. These events can be important in easily attracting a lot of interest from tourists. The tourism program for the Malebo region will be on this asset and would request the tour operators to organize such events more frequently (for example, four times a year compared to the actual frequency of two times a year) to make tourists live that beautiful cultural experience while paying money directly to local communities.

Net hunting is also another cultural trait of the Bateke of the Malebo region. Net hunting is a collective effort to provide meat for the village community as a whole. Traditionally, the meat of the prey killed during the net hunting exercise was not sold but divided to feed families of all participants and old people and local traditional authorities. Net hunting is also part of traditional initiation for young men and is still practiced nowadays. This cultural exercise would be used for tourists willing to feel the spirit of the initiation and those that would love to practice with traditional hunting tools. Net hunting has been also used in other areas of Central Africa and would be a great tourism asset for the Malebo. In addition to just putting nets and communicating with dogs, the net hunting exercise is always followed by singing and dancing, which provides a completely different world for people coming from other cultures. The money generated from such an exercise would be used for an educational fund to support the most brilliant students from the region, while the meat would be given for distribution to villagers.

## 23.4 Navigating Kasai-Congo River and Kayaking Along the Rapids

Tourists would be taken by boat to and/ or from Makayabo located near the mouth of the Kwa to reach the Malebo region on the vehicle. The trip between Makayabu and the WWF field base where most of the infrastructure is being built is 91 km even though the conditions of roads may make it a four-hour drive. With some repairs and maintenance of the road, this trip could easily be shortened. However, it should be noted that traveling along the Congo River on a VIP type of boat is, in itself, an evasion to nature, a wonderful experience that most nature-driven tourists would simply love. The trip could be combined with sport fishing along the Congo, which would be an intriguing exercise for most adventurers.

The WWF Field Base in the Malebo region is situated at 200 m from the Bambou River. This 10–15 m wide river is brown and cool and offers a diversity of scenes, including rapids and flooded zones. The tourism would benefit to people that would like to make dugout canoes and kayaks to navigate a 5 km distance from the bridge downstream up to the point where the water goes under a huge rock. Tourists would be net and/or hook fishing on their way down and terminate their trip on this huge rock that cuts the river in two different halves. This huge rock would serve as a picknick point where people could cook fishes captured on their way down or simply eat what they have brought with them.

## 23.5 Necessary Investments: Infrastructure

To provide a viable tourism routing in a world made of competition, the above-mentioned natural assets need to be combined with high quality infrastructure. This means building high profile bungalows, high profile restaurants, and other necessary accommodations at different locations on the route to Malebo. The Lake Tumba Landscape aimed at building its operational base near the Bambou River, 5 km from the landing strip of Malebo. This spot is where the most important logistical support and the first set of 15–20 bungalows are planned to be built at a distance of about

200 m from the WWF operational base. An open-sky restaurant and a swimming pool are also in the investment plans to accommodate tourists.

The World Wide Fund for Nature operational Field base is composed of an administrative building with a large room composed of offices for field staff, a conference room and two VIP visitor rooms, and a modern kitchen. Other buildings included two dormitories of five rooms each which serve to accommodate the World Wide Fund for Nature field researchers and a storehouse for fuel and other pieces of important equipment. Also, the location of the site has been chosen in a way that a large compound has been cleared under the beautiful *Uapaca guineensis* trees to allow to set more than 20 tents for camping. Bungalows were planned to be built in an area adjacent to the savannah bordering the Uapaca guineensis forest while the swimming pool will be placed about 100 m from bungalows in the middle of the savannah. The restaurant was planned to be located near the Bambou River in order to allow visitors to view the savannah on the other side of the Bambu River after some cutting some of the small trees and shrubs that border the Bambu.

Other than those principal infrastructures, there was a need to build three outposted facilities to accommodate the tourists while visiting the region. The first one is to be built at Makayabu to accommodate people who will be traveling along the Congo at their arrival. This facility should be constructed at the top of the plateau, about 1000 m from the Kwa River. At this place, there is a beautiful view of the river from the large green meadow that allows cool wind to circulate in the late afternoon and early mornings. Below the location of the chosen camping site, there is a water cascade that will also be used as a natural asset for tourism. This transit outpost will need a nice restaurant, five VIP rooms, and big camping tents to accommodate tourists in transit. In order to function properly, there is a need to plant some fast-growing trees in order to provide some shading for the camping site.

The third location where there was a need for infrastructure was the village of Nkala, precisely at the savannah near Mbou-Mon-Tour headquarters. This area needs a light infrastructure, essentially composed of a café, and nice toilets and communication equipment. The establishment of those assets is necessary because the village of Nkala is where we have the first habituated bonobo group and it is located at an equidistant position from all other sites where bonobos are being habituated. The café should provide enough space, enough variety of beverages and enough food to accommodate tourists while they return from their visit to see bonobos.

Fourthly, at the top of Kenge, the highest point in the region of Malebo, there was a need to build a bar for evenings when people want to relax. This should be a bar with the spirit of the region, meaning built-in local materials and offering local food as well as imported beverages. This would be an ideal place to organize the traditional dance fair, with the support of local traditional authorities and local NGOs, particularly our field partner Mbou-Mon-Tour.

Fifthly, there was a need to accommodate people in Kinshasa when they come from abroad and before departing to the field. There are three main options on how to handle this issue. The first and most obvious is to have people stay at different hotels in the town and take them in charge only while they are ready to leave. The second option is to have an agreement with one single resort that can accommodate all people arriving before they leave and when they return from the field. A third option is to acquire a property and build a resort that can accommodate people while in transit through Kinshasa. This latter option may seem expensive but in the long run, it might reveal to be the best as this would secure a safe place for visitors and would be included in the package to be sold.

## 23.6 Joint-Venture: Government, Conservation, Private Sector, and Communities

The tourism in the Malebo region was envisaged as a multi-stakeholder joint-venture, with World Wide Fund for Nature playing a leading role. The role devolved to World Wide Fund for Nature was to facilitate the broad stakeholder participation, provide technical expertise, and share its knowledge in governance, management, and ecotourism in other regions of the world. World Wide Fund for Nature was to provide the administrative and political backstopping; it was also to ensure that revenues are equitably shared between different components of the alliance.

The alliance will need a private sector investor to serve as the principal manger of tourism. The role that devolved this private sector investor will include to provide complementary financial resources to those already identified and/or invested by World Wide Fund for Nature, and expand the project beyond the Bambu camp (i.e. participate in the construction of other infrastructures as indicated above). The private sector investor will manage the tourism program in all senses of the term, including hiring trained personnel, marketing the tourist assets of the Malebo zone, and caring and accommodating tourists. The most important expectation that the alliance has from this private sector investor is to generate revenue from the tourism activity. The private sector investor was to join the alliance after presenting a business plan to the World Wide Fund for Nature and its other partners, principally the government of the Democratic Republic of Congo.

The alliance will also include partners that will focus on participatory natural resources management, with particular attention paid to local communities' participation in the overall tourism program. The role devolved to this partner will be to initiate local communities in collegial ways of making decisions on how to spend their share of tourism revenues. A further important role will be to encourage people grow greens and other staples and produce local art products that can be sold to the tourist resort. A fourth partner in the alliance was to be Mbou-Mon-Tour. Its role in the alliance will be to sensitize local communities and lead the process of revenue sharing through the identification of activities and/or investments that are of common interest.

## 23.7 Revenue Sharing

The most important element in this issue is that only the benefits were to be shared. World Wide Fund for Nature proposed, at that time, that the repartition key is as follows, based on the experience in the region. Fifty percents were to go to the private investor, 20% for World Wide Fund for Nature to support its field activities in the region, 25% reinvested for the expansion of the program, and 5% be allocated to local communities' general interest activities. The 5% allocated to local communities were to be invested in such activities as covering school roofs with teens, provisioning local dispensaries with drugs, repairing and maintaining roads, and improved agriculture.

## 23.8 Investment Plans: Invested Funds and Available Assets and Further Needs

The Damseaux's family graciously donated a piece of land where the World Wide Fund for Nature built the field base and five tourist bungalows. The value piece of land donated by the Damseaux is estimated at 90,000 US$. The actual infrastructure at the field base received initial funding from Nancy Abraham, the manager of Alexander Abraham Foundation and internal funding from the World Wide Fund for Nature. Both financial sources provided 335,000 US$ for initial investments, including the field operation office and five finished bungalows. Back in 2006, it was projected that the World Wide Fund for Nature should invest an approximated amount of 720,000 US$ to sustain the bonobo habituation program and for operations in the Malebo region. This amount was only to cover activities related to the research and habituation of bonobos.

The annexed investment for the whole joint-venture to become fully operational was rather very high. For example, tourism and its appending bonobo habituation program needed a global investment of 1,926,700 US$ deployed between 2008 and 2012. Of that amount, 222,000 US$ were needed essentially to build tourism infrastructure comprising bungalows, restaurants, and cafeterias while a greater chunk of the money (258,700 US$) was needed for running initial operations. Besides the above invested resources, there was a need to gather an amount of 1,255,760 US$ from 2009 to 2012 to get every piece of the tourism program in place. While the conservation and, particularly the bonobo habituation program, was covered by internal funds that World Wide Fund for Nature provided, the tourism component needed an immediate capital investment of 692,760 US$ over the period of three years from the private sector to make the tourism a reality. This capital investment, as denoted on the detailed budget, included increasing the number of bungalows, building cafeterias, one swimming pool, two bars, and tourism operations. That investment did also include hiring and training qualified tourism personnel.

## 23.9 Timeframe of the Investment

It was envisaged that every piece be in place within the next five years following 2006 when the first group of bonobos started to be habituated to human presence. Therefore, the plan has targeted 2012 as the time mark by which time every piece should have been in place and tourism fully operational. This global timeframe does not, however, mean that tourism cannot start in the near future. The launching of tourism is planned to be gradual, depending on available assets and the capacities to accommodate people while visiting the site. The table below provides a clear picture of the deployment of activities across a five-year timeframe.

## 23.10 Tour Operators and Immediate Tourism Activities

Using the minimum infrastructure already present on the site, it was planned to launch small scale tourism before the end of 2009. In order to make this happen, the project had planned to proceed by an open process of selecting tour operators that would be ecologically sensitive and economically viable partners. The process of selecting tour operators should involve different stakeholders within the World Wide Fund for Nature that was interested in the questions of bonobos and tourism in Lake Tumba Landscape. The process was also to involve other people with critical knowledge of tour operators and businesses in Central Africa. For the selection to be productive, there was a need for willing tour operators to plan trips to the field to have firsthand knowledge of the site and its natural assets and provide their guidance on the investment plan and suggest needed improvements for existing infrastructures.

At the time when the plans were being crafted, there were five bungalows at Malebo, a descent food restaurant, and it is possible to visit bonobos, organize bush trekking exercise, birding, and natural scenes viewing. Based on the existing facilities, it is possible to receive five people (or 10 if people would be sharing bungalows) per week for three nights. This would be equivalent to hosting 20 people (or 40 people maximum) monthly. The flight costs to Malebo for Malebo are estimated at 7,400 US$ (round trip) and cost per person for food and lodging are estimated at 125$ per day (US$100 for lodging and US$25 for food: breakfast and two meals included). Tracking bonobos in the forest is estimated at 100 US$ per person, including vehicle transportation while going to the bonobo sites. Other assets such as cultural events would cost US$25 per person per day but birding and forest trekking will be initially for free.

# Chapter 24
# Decent Knowledge for Future Directions in the Landscape Management

**Abstract** Need in knowledge was and is still immense for the Lake Tumba Landscape. Collecting data to document the biodiversity of and the role of humans in shaping the Lake Tumba Landscape was essential before embarking on any conservation activity. Beyond these primary questions and gross-resolution preliminary studies, however, ecological functions and conditions of the habitats in the landscape were to be studied in a multi-resolution assessment of amphibians, birds, fishes, insects mammals (large and small), and plants. Despite the fact that acquiring fine-tuned data may appear idealistic, the long-term acquisition of the knowledge on these biological and social components is important in setting conservation goals and monitoring targets. The chapter proposes some methods on how to go about these studies and comprehensively addresses gaps in knowledge and provides epistemological reasons why such data should be collected, where and when.

**Keywords** Biodiversity assessment · Systematic herpetological surveys · Fish species diversity and biomass survey · Insect systematic survey · Long-term ecological studies

## 24.1 Introduction

Central African landscapes are large areas composed of different microhabitats and human microcultures that should have been understood prior to any temptation to create a large-scale management plan. However, given the need for a conservation organization to alleviate the serious problem of alarming rates of biodiversity losses, action in implementing large-scale landscape programs has begun without sufficient knowledge. The need for sound knowledge appears to be the most important step for the days to come if sound management is what is needed. This chapter provides a coherent program for acquiring scientifically reliable information. Drawing on lessons learnt from the ongoing efforts to fill in knowledge gaps, it will concentrate on ensuring an integrated research program and how that has to be implemented throughout the landscape. In so doing, the major challenge is to reach an optimum point where the needs of field data do not obstruct the field conservation action and the urgency of field conservation action does not precipitate efforts to activities

 
B.-I. Inogwabini, *Reconciling Human Needs and Conserving Biodiversity: Large Landscapes as a New Conservation Paradigm*, Environmental History 12,
https://doi.org/10.1007/978-3-030-38728-0_24

unfounded on scientific evidence. A major challenge in conserving biodiversity in large areas such as those called landscape in the Central African conservation jargon is the paucity of field biological data to support the allocation of the conservation efforts. This chapter is written in the spirit that only long-term efforts will provide sufficient knowledge to tackling difficult issues related to biodiversity conservation in the Lac Tumba Landscape. Our efforts here are dedicated to showing what data are needed and, possibly, how to collect them. Of course, this book is not about detailed methods on any subject; it simply indicates broad views and ideas on how to proceed. However, this chapter is a prospective proposal for a large-scale and long-term research program that has five main objectives described below. The chapter provides one thinking process that could be of general application in many other circumstances, even though it is based on the concrete need for knowledge from the particular case of Lake Tumba.

The thinking process that leads to the identification of needs in necessary knowledge for a sound management of a large area such as Lake Tumba is based, as it is the case for any scientific inquiry, on questions being asked. In principle, these questions are similar in many ways in conservation biology, with the special question being only a minor part of the process. All data were to concur with this primary objective and to contribute to the logical framework that will map out the different steps for the implementation of the management plan for each functional zone.

## 24.2 Needs in Knowledge for the Lake Tumba Landscape

Needs in knowledge that we felt were of importance for Lac Tumba were necessary to respond to the principal questions above. The first one being why is the Lac Tumba an important biodiversity conservation landscape? To answer this question, the first objective for data collection is to document the current biodiversity of the Lac Tumba Landscape. The other important objectives included in this question are the assessment of ecological functions and conditions of the habitats in the landscape. The second important question that the research of knowledge tries to answer is what is the human role in shaping this landscape? The answer to this question leads us to the third objective of the proposed research scheme, which is to describe human interactions with biodiversity in the region. To initiate pilot long-term biodiversity and environmental monitoring program for aspects of the biodiversity and their habitats that need to be monitored over a long time needs efforts to build national capacities. As multi-resolution assessment, the biodiversity component of the proposal focuses on describing major biodiversity segments: (1) amphibians, (2) birds, (3) fishes, (4) insects, (5) mammals (large and small), and (6) plants. All this may look too idealistic yet landscapes are primarily biodiversity conservation units in a geographic sense. Therefore, the long-term acquisition of the knowledge on these biological components is important in setting conservation goals and monitoring targets. The list of what is needed from the biological diversity assessment also means that landscape conservation programs need to be viewed as long-term conservation endeavors but not short-lived shows.

Also important is to describe environmental conditions prevailing within the landscape. For the particular case of Lac Tumba, describing environmental conditions should have a particular focus on freshwater habitats, water quality, and delineating the terra firma habitats. The proposed long-term scheme will also aim at screening human dynamics (past and present) in the region in order to determine how human society has shaped trends currently observed in the biodiversity of the Lac Tumba region. Further data sets should include actual land-use patterns in combination with expected population expansions in order to project future human needs and allocate areas that will be used to respond to these needs. Finally, based on the results of the proposed scheme, beyond gaining the knowledge on different dynamics operating within Lake Tumba, the aim is to initiate a long-term monitoring to gauze the effects of landscape conservation programs.

## 24.3 Biological Diversity Knowledge

To start collecting those data at the initial phase of any given program, as was the case of Lake Tumba, it is fine but demands a lot of realism; otherwise, most time needed for action would be spent collecting data without concrete action on ground. In order to find this equilibrium point, it is not necessary to conduct a wall-to-wall kind of research. This is expensive, needs a lot of time, and will swallow most of the conservation resources. Several traditional sampling methods can be useful in solving this problem; habitat stratification being the most used. In the case of Lake Tumba, we propose a shift in the use of conventional methods. This is founded on the fact that preliminary studies have given us a decent appraisal of both the importance of each habitat and a differentiated picture of where to find new information. That is why it is proposed to take those areas of the landscape that present some unique features and concentrate effort in collecting more scientific data from them. These areas are where most of the biodiversity, not necessarily in higher densities, is likely to found. The zone consists of five areas have been chosen to collect six categories of data as identified below: (1) Ngiri Triangle, (2) Bolombo-Losombo, (3) Ngombe, (4) Congo Island, (4) Mabali, (5) and Lake Tumba and adjacent forest .

### *24.3.1 Systematic Herpetological Surveys in the Lake Tumba Landscape*

There has never been any thorough assessment of herptiles, small mammals, large mammals, birds, and fishes in Lake Tumba. Therefore, the proposed multi-resolution study would aim to launch the first large-scale program to collect specimens in those five zones. The following paragraphs summarize the backgrounds and methods to be used for each of these systematic surveys. Except where otherwise stated, these surveys will aim to build species lists and provide indicators for future rapid ecological assessments. They aim at collecting the fundamental knowledge of the species richness and abundances in the Lake Tumba landscape. For all collected specimens, appropriate authorizations will be obtained from the Government of the Democratic Republic of Congo and all international regulations will be followed. For species whose specimens can be easily obtained, the aim is to leave one specimen locally for each species. Amphibians seem to be declining worldwide for reasons that are difficult to comprehend (Wyman 1990) and recent research has documented two cases of mass extinctions in the rainforests (Berger et al. 1988). Despite the fact that amphibians are known to be good indicators of the effects of climate change, the knowledge of the amphibians in the Democratic Republic of Congo is very scanty (Groombridge and Jenkins 2002). Some 53 species of endemic amphibians and some 33 species of endemic reptiles have been described in the Democratic Republic of Congo (Groombridge and Jenkins 2002). However, no systematic surveys of amphibians have been undertaken in the Lake Tumba Landscape. Amphibians are thought to be good indicators for deciphering effects of the global climate change, and therefore good biological indicators of the ecosystem's health. This is due to the fact that most of them are, at least in the tropical conditions, very seasonal, tending to disappear during hot seasons, and becoming more visible and active in the rainy seasons (Howell 2002). Despite their acknowledged importance, however, very little is known about species of amphibians in the region of Lake Tumba Landscape. It is ironic that Hughes and Hughes (1992) suggested that aquatic frogs are abundant in this region but could only identify tadpoles taking refuge in vegetation near the banks of the lake among the 12 known species of frogs that rely on the lake. They have also identified two endemics (*Cryptothylax minutus* and *Phlyctimantis leonardi*), both members of the Hyperoliidae family. Several large aquatic reptiles and mammals live in the ecoregion. Equally less documented is the diversity of reptiles of the region. Only large species such as crocodiles have been documented, albeit in sporadic and opportunistic ways. Even more ironic was the fact that evens a large-bodied vertebrate such as the dwarf crocodile (*Osteolaemus tetraspis*) widespread in the region does not appear on all the documentation as being present in this region.

Adult amphibians and their larvae and reptiles should be sampled using various standard methods, including drift fence and pitfall traps, canopy walkway trap, and snake trapping. In the area adjacent to Lake Tumba, a 5 km long transect will be laid out from the shores of the lake to the adjacent swamps. Sampling units composed of forest litter plots of 4 $m^2$ (2 m × 2 m) will be located randomly (Scott 1982) and species of amphibians and reptiles collected. These forest litter plots should alternate with other methods along the same transects, i.e. drift fence and pitfall nests and canopy walkway traps. Both are trapping methods but differ only in materials used and how they are laid out. The basic principle of the *drift fence and pitfall traps* method is that animals walking on the forest floor are drifted away from their travel direction when they encounter a barrier. Rather than crossing the barrier, either by burrowing under it or breaking through it, animals choose the least resistant path following the fence, leading them into a pitfall trap (Howell 2002). The canopy walkway involves constructing a runway of mosquito-sized mesh into which is sewed a funnel-shaped bag. The entire construction is placed into the canopy and left for reptiles using the walkway to fall into the funnel traps (Vogt 1987). All specimens caught will be collected and brought to the museum to be identified. For readily recognizable species, only some tissues should be collected and fixed separately for further laboratory works, including the DNA and for the analysis of pesticides, metals and other pollutants (Lambert 1993).

### *24.3.2 Systematic Bird Survey in the Lake Tumba Landscape*

The Ngiri Triangle, now a protected natural reserve, is located within the Lake Tumba and is one of the world's important areas for birds. As such, it has been declared as the world's largest Ramsar site (website). Ngiri is a remote area of swamp forest situated between the Ubangi River in the west and the Congo River in the east. Large numbers of water birds breed including Long-tailed Cormorant *Phalacrocorax africanus*, African Darter *Anhinga rufa,* and Purple Heron *Ardea purpurea.* It is the only known site in the country for Congo Sunbird *Cinnyris congensis.*

Again, few systematic data exist to document its importance. Preliminary assessments have indicated significant differences between bird species in the northern portion of the landscape and those in the southern edges. These differences can be attributable to differences in habitat types. Even though birds are the best-known group of vertebrates (Bennun and Howell 2002), the recent documentation of the Nominate Lesser Black-backed Gulls by Kylin et al. (2010) indicates that more bird species still remain to be identified in this region, therefore calling for systematic surveys that will establish a solid baseline data for future bird inventories. According to the African Bird Club (2008), there are vast areas of this zone that have not been surveyed ornithologically and many important sites probably remain to be discovered and documented. Birds are also considered good useful indicator groups either for monitoring environmental changes (Furness et al. 1993) or assessing biodiversity importance (Stattersfield et al. 1998).

A bird survey is always based on both visual and vocal identification (Bennun and Howell 2002). However, this is daunting in forest environments. In each of the five locations identified as subsets of the study region, line transects should be laid randomly to conduct the bird systematic and relative abundance surveys. Transects seem appropriate to account for the need to diversify microhabitats within the sampled region, as forest birds are correlated with some specific niches. Both timed species counts and timed transect methods should be used because one compensates the weaknesses of the other and vice versa. They provide simple method sets that are readily usable by people in the field. In some circumstances and if the needs prevail, specimens should be collected and taken to an appropriate museum for identification. Along transects, microhabitats will be recorded so will be bird calls and sightings. Bird identification should be done using conventional methods such as the use of illustrated field guides even though there is no guide book specific to this region or even forested Central African region.

### 24.3.3 Fish Species Diversity and Biomass Survey in the Lake Tumba Landscape

Descriptions of fish biodiversity in the Lake Tumba Landscape are limited because so little data has been acquired since the study carried out by Poll (year). Figures of species residing in the landscape range from several hundreds down to some dozens. There are problems of synonymy at some levels, but a critical problem is that of designing an optimum sampling strategy for the region. The waters of Lake Tumba Landscape harbor many different types of habitats and microhabitats, which would suggest that there are many specialized species of fish. Lake Tumba Landscape contains large and small lakes, stagnant water ponds, large and differentiated swamps. The aim of this part of the study should be to describe the fish diversity in Lac Tumba. It should also include how species found therein are connected to the overall diversity of fish species in the Congo River using biogeographical linkages and how fish species identified in that zone could be used as a biological indicator for the environmental health. While linking the species diversity in the region and that of the main tributaries of the mighty Congo River, one part of the fish diversity assessment should be able to help identify economic values of the species and how they can become part of numerous economic scenarios for improving human livelihoods in the region. The study should also seek to understand the thresholds of offtakes in fish communities, linking therefore with conservation strategies being developed to help the management of the landscape providing precautionary approaches to local communities.

This study should be designed to determine the abundance, diversity, and biomass of fishes within different reaches of the system during the wet and dry seasons. The survey should cover a range of habitats, from small creeks to large turbid rivers, swamps, islands, and lakes. Specific collecting/survey methods should be selected depending on the habitat type. These habitats should be sampled once during the wet season and once during the dry. Sampling points should be selected in a way that they would capture any existing habitat gradient and overall sampled twice in each season. To gain the most comprehensive understanding of species presence or absence, a diversity of sampling techniques should be used, and including electro-fishing device, seines, nets, flood traps, large river bottom trawling, and rotenone in areas with flowing waters (Umali 1950; Von Brandt 1964; Shrestha 1995). Systematic quality habitat description should be recorded using the conventional Central Africa habitat types developed by the program of the Monitoring of Illegal Killing of Elephants (MIKE) that has been endorsed by most research organizations in Central Africa. Those qualitative habitat descriptions will have to be tied with GPS coordinates, weather condition and a rough site map was drawn. To determine if water quality has an impact on species distribution, water chemistry data will also be collected. These will include such elements as water pH and turbidity. Fish specimens should be preserved in 75% ethanol following fixation in 10% formalin.

### *24.3.4 Insect Systematic Survey in the Lake Tumba Landscape*

Insects play key ecological role in tropical ecosystems. They aerate the soil, pollinate flowers that give birth to fruits, and participate in the process of decomposition of dead materials thereby reintroducing nutrients into the soil. They can also serve to control pests. Burrowing insects create channels through the soil, which enable the circulation of water in the underground and can be beneficial to plants. Insects also fertilize the soil with the nutrients from their droppings. Many insects are primary consumers and are eaten by a variety of secondary consumers, including humans. Despite that critical role, insects are the least known group of biodiversity and have not been thoroughly documented across the Democratic Republic of Congo.

Insect diversity should be examined at different ecological niches, based on identified major forest strata. The study should capture all species encountered in the region of Lake Tumba Landscape. Because captures depend on individuals' behavior, activity level, and size, the study should use different mass-collecting methods. These should include pitfall traps, Malaise traps (tent-like structures that funnel flying insects into a collecting jar), light traps, sweep nets, canopy fogging with insecticides, and electrified rings. In some cases, insect specimens should be stored unmounted insects in an insect envelope with the collection data printed on it. They should need to be rehydrated frequently to maintain their colors and forms. Other specimens should be stored in jars of 70% ethyl alcohol. The alcohol jars should be topped off over time as the alcohol will evaporate evenly with the lid on. They will be transported to where identification will be made. Collaboration should be sought with people with entomological expertise.

## 24.4 Mammalian Studies in the Lake Tumba Landscape

Mammals are among the most studied and documented animal group in the Lake Tumba Landscape. Interest in detailed ecological studies has begun to accumulate in species such as bonobos in the southern Lake Tumba Landscape, diurnal primates around the Mabali Scientific Reserve and chimpanzees in the Ngiri Triangle Natural Reserve. However, huge knowledge gaps still remain, including the nature of the ecological processes that link species distribution, abundance and their habitats. Therefore this study will launch three major endeavors: (1) collection of small mammals (bats, rodents, and insectivores), (2) systematic survey of medium-sized and large mammals present in Lac Tumba, and (3) an investigation of factors that determine the distribution and abundance of species of conservation concern, in order to establish long-term monitoring programs.

### 24.4.1 Mammalian Diversity Assessment, Abundance and Distribution

This study should be designed to determine the diversity of mammalian species in Lake Tumba Landscape. For large mammals, it will lead also to the determination of relative abundance for most species and densities for those species that are of conservation concerns. The study should also include the association between large mammal species and different habitat types in the landscape. As such, a variety of methods should be used. For collecting small mammals, live-trapping, mist nets, harp traps, and removal trapping techniques will be used. Collected specimens will be handled appropriately. They should be immersed and stored in 70% ethyl alcohol. Before being wrapped in muslin and being put in storage jars, stomach contents should be evaluated and recorded. Live tissues should also be collected to allow DNA analysis in the laboratory. In some cases, only skins will be needed. In those instances, skins will be sun-dried and stored to be taken to the museum for identification.

### 24.4.2 Comparative Ecology of Bonobos and Chimpanzees

This part of the study should consist of (1) comparing the ecology of bonobos and chimpanzees in the northern part of Lake Tumba Landscape and (2) determining the key foods of forest elephants and factors that determine their distribution.The association of wildlife species and habitat types is one of the most fundamental subjects of study in ecological research. Therefore, habitat parameters are important because physical elements of the habitat such as canopy cover, understorey composition, tree distribution and density, and foliage density play an important role in habitat selection by wildlife. However, a few studies exist in this area for the great apes of the Lake Tumba Landscape because these great ape populations have only been relatively recently described. Important for great apes are the distribution and quality of food, particularly for bonobos and common chimpanzees. The two species of great apes eat mostly ripe fruit, apart from insects, leaves and meat of other mammals. The availability of these food items is felt to dictate spatial associations of communities, which are variable across different locations. The proposed research program should focus on three major themes: (1) assessment and quantification of the food resources available to the bonobo and chimpanzee populations in two study sites; (2) assessment of the contribution of each food resource to the bonobo and chimpanzee diet by season; and (3) assessment of the relationship between the food resources availability and the distribution of bonobo and chimpanzee populations across different forest strata.

All bonobo and chimpanzee food leftovers (plant species, parts eaten, species of insects and animal eaten) should be recorded along transects or along feeding trails, after the stratification of the two study sites into major vegetation zones. Fecal samples should be searched and collected, washed in 1 mm mesh sieve and dried, and their fiber contents and other recognizable parts quantified. Interviews of local trackers about what they know as food of the bonobos and chimpanzees is necessary because they will allow us to broaden our knowledge to include the resources that cannot be identified during the course of the study. Counts of fruits along three 1 km-transects in each stratum should be conducted following previous studies, with transects walked on a fortnightly basis, at different seasons, and numbers of ripe and unripe fallen fruits of all species encountered on a 1 m strip will be counted. Keystone food resources for bonobos or chimpanzee should be identified and their distribution, productivity, and their temporal availability studied using standard techniques. The use of a GIS program, satellite and radar images, and maps and their combination with forest descriptions found in the literature will help stratify the two study sites in its major vegetation strata. Because most of these maps have never been ground-truthed, a more precise stratification of food resources available across the two study sites can be drawn after the survey through a post hoc strategy. Maps obtained from this strategy will help the overall definition of habitat structures and food availability across the two study sites. Bonobo and chimpanzee populations' survey will need a pre-study stratification. Strata will be defined based on a combination of their habitat structure (as identified on the maps) and their respective distances from access routes and villages. Stratum point densities will be compared to examine differences between different strata.

### *24.4.3 Forest Elephants and Key Food Species*

The general goal of this study is to describe the ecology of the forest elephant in both the Ngiri Triangle and the Bolombo-Losombo area . Both locates in the northern part of the Lake Tumba Landscape. The research program that seek to demonstrate that elephants, as the largest mammal in Central African forest ecosystems, play a key role in the ecosystem. This part of the long-term study sets out particular objectives: (1) document the current abundance and distribution of elephants, (2) establish the frequency of the key fruit resources available to elephants in relationship with elephant distribution. The study will also thrive to understand ecological roles elephants play in the ecosystem in terms of dispersal the key fruit species, and ways these key species sapling and trees are distributed in relationship to the elephant trail system.

The survey design should be generated by the use of standard distance sampling methods after a pilot study has determined elephant dung encounter rates. Line transects should be used to provide an unbiased estimate of mammalian abundance, allow simple spatial modeling of elephants in relation to ecological and human influence covariates while removing spatial and habitat bias thus allowing for valid abundance estimates by habitat type. The key data to be recorded are dung piles along the transect line with accurate measurement of perpendicular distance from the transect center-line. Habitat, fruit fall, and human signs, relief and other ecological variables will also be recorded. Data should be collected in randomly stratified habitat types, with strata being defined based on human influence rather than ecological or geographical boundaries. The analysis of the data collected from line transects will use distance.

A key parameter that this study should seek to quantify is the abundance of key fruit species that are available as food to elephants, to determine which ones are really dispersed by elephants. Furthermore, the seasonal availability of these key species of fruits in the forests should be quantified to see if they have some influence on elephant movements and their interactions with human activities within and around the two study sites. Counts of key fruit species along transects should be conducted during the survey along the same strata as counts of elephant dung piles. Perpendicular distances will be measured between clusters of key fruit species and transects. Key fruit trees should be identified in the study area and should be monitored monthly to record fruiting events.

## 24.5 Botanical Systematics in the Lake Tumba Landscape

Botanical surveys should be conducted to identify plant species present in the selected zones. They should also help determine environmental effects on all botanical resources, including species distribution, species abundance and species special status plants (rare, threatened, and endangered plants), and plant (vegetation) communities. For each selected zone in the Lake Tumba Landscape, a floristic survey should be conducted along transects placed perpendicular to any felt habitat gradient to allow sampling to occur in diverse habitat types. It will require that every plant observed be identified to the level of species, subspecies, or variety as applicable. To properly characterize the site, a complete list of plants observed on the site should be produced for every section of forest visited by other surveys conducted through this multidimensional study to aid with the knowledge of the vegetation communities that define each habitat type. Two specimens for each species will be collected: one will remain in a national collection and the other shipped overseas for proper identification. Permanent plots wherein forest biomasses will be measured will be randomly placed and visited in the long run on a yearly basis in order to study the forest dynamics of the region. Data from these permanent plots will help deduce the quantities of $CO_2$ sequestrated by different types of forest.

## 24.6 Assessment of the Biophysical Environment in the Lake Tumba Landscape

### *24.6.1 Assessing Freshwater Habitats and Conditions in the Lake Tumba Landscape*

The waters of the Congo Basin are one of the world's major freshwater ecosystems, yet also one of the least understood. In Lake Tumba Landscape, these waters are very important for local livelihoods. Changing land use and climate will influence these waters and the ecosystem services they provide. Sustainable development strategies need to be based on knowledge of the aquatic ecosystem, and its sensitivity to different influences. Therefore, an aquatic assessment program is urgently needed to provide that knowledge. The research program should develop an assessment framework with indicators for major land use and climate influences tailored to the aquatic ecosystems of Central Africa.

The waters of the landscape are characterized by humic substances leaching from surrounding wetland soils. The heavily weathered soils (mainly kaolinite), and high annual runoff contribute to the weak buffering capacity of the water. As a result, the waters are low in nutrients, acidic, and very brown. In turn, water chemistry determines the biota from bacteria to fish that can survive under these conditions, which may change due to climatic variation or human impact. The research program should seek to establish a baseline for characterizing the chemistry of the water, and the base of the food chain, including bacteria, algae, plankton, and invertebrates.

This component of the proposed research program should provide valuable baseline information about the inland surface waters of the Lake Tumba Landscape and build national expertise to use this baseline in monitoring and assessment of the aquatic ecosystems. The information that should be gathered, however, needs to be placed in a framework for environmental assessment with appropriate indicator systems for human impact and climate change. The basic framework for such an assessment can be built around those that have emerged in response to the demands of the European Water Framework Directive for measuring progress toward the achievement of the overall goal of '*Good Ecological Status*'. Therefore, two assessments should be carried out: (1) what are the conditions of the water with respect to the production of ecosystem services and human influence (status of freshwaters) and (2) how far from a 'natural' reference state has the aquatic system moved in response to human pressures—land use, pollution, and climate change.

Key steps in the development of the assessment framework are characterizing the range of physical, chemical, and biological conditions present, identifying the appropriate indicators of human stress/climate influence, and then developing appropriate indicator metrics and classification schemes. Methods range from relatively simple algorithms or indices to combinations of multiple indices (multi-metric approaches), and to relatively complex multivariate approaches for pattern recognition and prediction. All three approaches are commonly used in biomonitoring and assessment studies to detect ecological change. The specifics of the stresses to be assessed, the appropriate indicators and reference levels need to be based on the information collected in this program. The use of this information in crafting the assessment/indicator systems requires both local knowledge of the environments in question and expertise in the statistical/ecological basis of assessing ecosystems. Appropriate collaboration between internationally renowned institutions and the national expertise on the data collected in the pilot program should be able to create an indicator system for assessing the ecological status and degree of human influence (as well as the likely causes of that influence) for the Lake Tumba Landscape. This system and the assessment will serve as the basis for management recommendations and also serve as a model that can be applied to the waters of other landscapes in Central Africa.

The research program should use bacteria, algae, plankton, and invertebrates to construct both a baseline data and a long-term monitoring system. Planktonic algae (phytoplankton) have been used since the late 1800s for water quality assessment. They react rapidly to environmental changes due to their short generation times. As fundamental components of the pelagial food-web (primary producers of organic matter, oxygen producers, food resource for grazers, compartment of the microbial loop), they initiate a chain reaction successively reflected in other organism groups as zooplankton, benthic fauna, fish, and birds. Changes in the physical and/or chemical status of the water are traced after some weeks through alterations of dominant planktonic species. Long-lasting changes in the water quality may be reflected from one season to another. The group cyanobacteria formerly known as blue-green algae are toxin producers and affect both the health of cattle and human beings and many other fractions of the food web. In many tropical regions, cyanobacteria and the toxin levels reached by those are more and more of general concern. In most monitoring work in Europe and the North American continent, planktonic algae indices are used not only for water quality assessment primarily in a trophic gradient but also to test for organic and metal pollution. Algae have also been an important indicator used for studies in acid lakes reflected in structural changes both concerning life-forms and overall diversity.

Currently, the use of benthic macro-invertebrates constitutes the backbone of many bio-assessment programs (Rosenberg and Resh 1993). Methods range from relatively simple algorithms or indices, to combinations of multiple indices (multimetric approaches), and to relatively complex, multivariate approaches for pattern recognition and prediction. All three approaches are commonly used in biomonitoring and assessment studies to detect ecological changes (Johnson et al. 1993). Also known today is the fact that macrophytes (aquatic immersed vegetation) include filamentous algae, mosses, charophytes, and vascular plants affect and are affected by biological and hydro-geochemical processes. Worldwide, macrophytes are used as indicators of water quality since they represent an integrative measure of, for example, the trophic status of lakes and rivers. According to their life-form, macrophytes are commonly divided into helophytes, elodeids, nymphaeids, lemnids, and isoetids. These life-forms respond differently to environmental gradients. The rooted short-stemmed isoetids prefer nutrient-poor lakes with high water transparency whereas the entirely floating lemnids prefer nutrient-rich environments. In addition, macrophytes are important for lake and river ecosystem functioning since they provide among others nursery habitats for many fish species.

## 24.7 Aims of This Part of the Multi-resolution Study of Lake Tumba Landscape

The main goals of this part of the proposed multi-resolution assessment of the Lake Tumba Landscape should be as follow. Describe the benthic algal composition and biomass (both diatoms and other algae communities) of the three areas (Lake Tumba, Lake Maindombe, and rivers of the landscape. It should be intended to calculate a Diatom Index based on genus level to assess water quality status. It should be tried to correlate the whole benthic algal community with water quality to develop a benthic algal method for assessing water quality. Bacterial parameters are not suggested for a long-term monitoring program but knowledge of the magnitude of bacterial production in relation to fluctuations in organic carbon loading and to potential bacterial grazers during wet and dry seasons (simple carbon budget) should aid in the planning of the monitoring program, and assessment of the results. The study should also acquire the knowledge on species composition of macrophytes and related ecosystem structure and functioning in Lake Tumba, Lake Maindombe, and streams. This last question should answer the following questions: (1) which species and life-forms are the dominant macrophytes in the studied systems; (2) are certain macrophyte species suitable as environmental indicators; (3) is macrophyte composition affected by surrounding land use; and (4) what are the potential implications of macrophyte community structure for ecosystem functioning, especially related to fish nursery habitats?

The suggested monitoring program initially covers a broad range of water chemistry and biological indicators. The chemical parameters should include nutrients (nitrogen and phosphorus compounds), pH-value, major ions (positively charged calcium, magnesium, sodium, and potassium as well as negatively charged sulfate, nitrate, and chloride), and organic matter (brown, humic substances). The biological parameters should include macrophytes, plankton, invertebrate bottom fauna, and benthic algal samples with a focus on diatoms and heterotrophic bacterioplankton biomass and production. Stratified random sampling should be used to determine the spatial variability of benthic invertebrate assemblages in Lake Tumba and streams. Samples should also be taken seasonally to determine within-year variability. Taxonomic and functional composition (e.g. traits) should be compared with water chemistry, substratum, and riparian characteristics to determine the main predictors of invertebrate assemblages. Metrics commonly used in Europe to assess nutrient enrichment (e.g. ASPT, BQI, O: C) and acidity (AWIC, MILA/MISA) will be calculated using the species by site data and compared with water chemistry indicators.

Major ions should provide a 'finger print' on the water sampled, e.g. effects of various land use and thus complement the interpretation of nutrient sources and status. Two long-term sampling programs are envisaged. Some 25 different rivers will be sampled, representing three types of land use: Ranching, logging, and low-intensity traditional use. Each month 50 water chemistry samples will be generated. The basic ambition level includes funds for reconnaissance surveys to collect data from 10 lakes and one hundred rivers and flooded forests in the region, once in the rainy season and once in the dry season. The purpose of this is to get a better sense of the spatial variability in the landscape. The monitoring program will also include specialized analyses for mercury and trace elements that may be affecting the ecosystem in combination with comprehensive biological monitoring.

Diatom indices should be calculated using usually techniques at the species level but would also use genus level indices that should be tested in this study as a first indicator of water quality as it has been the case in Europe. Also, the amount of malformed valves can be a good indicator of toxic substances including pesticides, herbicides, and metals, and the study should especially check for such malformed valves to test their use as an indicator. Samples for bacterial biomass should be fixed with formaldehyde (for example, 2 ml formaldehyde to 18 ml of sample) and should be stored in a cool environment and sent away for analysis after fixing. Bacterial cells should be counted and measured with epifluorescence microscopy while their production should be measured using the method developed by Smith and Azam (1992). Benthic invertebrate assemblages from littoral, sublittoral, and profound habitats in Lake Tumba and riffle/pool habitats of streams will be sampled in different seasons to determine spatial and temporal variability. All samples will be collected and processed according to European and international (EN/ISO) standards. Littoral samples will be collected using standardized kick-sampling (20 s $\times$ 1 m) with a hand net (mesh 0.5 mm). Wind-exposed and wind-sheltered habitats will be sampled. Soft-bottom sublittoral and profounder samples will be collected using an Ekman sampler and sieved through a 0.5 mm mesh. Stream riffle samples will be collected using a Surber sampler (0.25 mm mesh) and pools will be sampled using standardized kick-sampling. All invertebrate samples will be preserved in 70% EtOH and identified (to the lowest taxonomic unit possible) and counted using a dissecting microscope. Samples of benthic algae should also be collected simultaneously from both lakes and streams using the European standard method. Samples of benthic algae should be preserved with Lugol's solution and stored until the water chemistry analyses indicate which sites are most interesting for further study. Then, the samples should be analyzed both for algal bio-volume and for taxa identification, including diatom identification, giving the possibility to correlate both to the good water chemistry background, including the analysis of toxic substances. In that way, it should be possible to develop an indicator for long-term monitoring of the study area. Macrophytes should be sampled at 10 localities in Lake Tumba and at 10 localities in streams. Sampling should be performed by standard transect methods. At each locality, three transects should be permanently placed perpendicular to the shoreline. Along transects, macrophytes should be sampled at a binary scale to the depth of maximum colonization. Information on land use at the landscape scale at each locality should be derived from satellite images and analyzed in a geographic information system (GIS). Using uni- and multivariate techniques, macrophyte community structure should be related to physical–chemical and landscape parameters. Macrophyte samples from each species should be stored for expert determination at the species level.

## 24.8 Climate Vulnerability of Freshwater Resources in Lake Tumba Landscape

The seasonal fluctuations in water levels are an important foundation for the fisheries of the landscape. High water levels expand fish habitat into flooded forest areas and promote increases in fish stocks. Lowering of the water levels forces fish into the rivers and lakes where they are more accessible for harvesting. Climate-related changes in the water balance and seasonality of flows will, therefore, have direct ramifications for fisheries. The quality of waters in the landscape is also strongly influenced by hydrological dynamics. High concentrations of organic matter that are leached from the flooded forests are a strong influence on the aquatic habitat, as this organic matter limits light, consumes oxygen, and acidifies the water The heavily weathered soils (mainly kaolinite) and high annual runoff contribute to the weak buffering capacity of the water. As a result, the waters are low in nutrients and ionic strength, acidic, and brown.

Biotas from bacteria to fish are regulated by these physiochemical conditions. These conditions are susceptible to climatic variation that may be compounded by human impacts such as logging or plantations, as well as polluted effluents from settlements on the waterfront. This part of the study of the Lake Tumba Landscape seeks to establish a baseline of hydrological fluctuations over almost a century to aid in characterizing the hydrological system and the chemistry of the water, and the base of the food chain (bacteria, algae, plankton, and invertebrates) and the fishery status. the long-term records of water level, precipitation, air, and water temperature from the southern tip of Lake Tumba that have been collected at the Mabali research station since the 1930s are the key to this study. This long-term record should be the basis for a hydrological characterization of the landscape over time. The parameters determined here should be used to predict how changes in rainfall patterns and temperature will alter the hydrology using an ensemble of climate predictions for the region. This project should lay the groundwork for a more thorough hydrological characterization using hydrological modeling approaches supported by field data that include isotopes of water as natural hydrological tracers.

## 24.9 Hydrological Characterization

To assess the climate vulnerability of the Lake Tumba system, we should be able to continuously assess the major water balance components and storage changes of the system. Currently, there is no systematic hydrological definition of the storages and residence times of water within different compartments of the landscape, although there is a multi-decadal record of water levels and other climatological measures in Lake Tumba itself that could be a valuable benchmark to be used in the context of a regional model to recreate major features of the water cycle in response to meteorological drivers over much of the past century.

Long-term historical records of climate and water level will be retrieved from records held in Belgium and more recent decades of data held by the Mabali station on Lake Tumba. These data will be assembled; quality controlled, and then analyzed statistically to look for changes over time and correlations the variables. The latter should help to make an initial estimate of the water balance dynamics for the Lake Tumba Landscape. This should then be the basis for a follow-on project which will seek to create a model of the system with which the sensitivity of this system to climatic fluctuations can be predicted. This will begin with the water storage dynamics of Lake Tumba and the connected flooded forested wetlands. The methods used in this work should include water level measurement in the lakes and watercourses of the region at selected locations from the sites being sampled for water quality. And the stream/river sties, stage-discharge relationships should be established so that the water level records can be transformed into flows of water. This will lay the groundwork for a long-term hydrological monitoring program.

Remote sensing will be used to map flooded areas over time using data gathered by the CARPE program that one of the investigators has been involved in for the past five years. The extent of the flooding will be linked to digital elevation models (DEM) enabling estimates of the changes in water storage. A suitable modeling approach will then be implemented to estimate both future and historical flooding/water level fluctuations. In the anticipated follow-on project, the satellite-based GRACE determinations of water storage on the land surface can provide a means of independently testing the ground-based modeling. Another focus of the hydrological work that can follow on from this initial work will be evapotranspiration and the interaction between surface water and groundwater. Evapotranspiration (ET) is one of the major drivers of the water balance in this system. The ET is also strongly linked to climate change. ET should be assessed at meteorological stations at the Mabali (Lake Tumba) and its affiliates on Maindombe using standard meteorological methods.

To get a better understanding of the spatial variability and to backup the meteorological calculations, a long-term research should measure stable isotopes of water ($^{2}H$, $^{18}O$) in combination with chloride to estimate evaporation fluxes from open water and soil water between monthly sampling intervals. In addition, water isotopes should be collected in the rainfall at the Malebo lab and its affiliates. The interaction of surface water with groundwater and soil water may be an important link for nutrient transport and water quality. With the help of stable isotopes, source water (end members) can be differentiated and hence the mixing volume and proportion of water can be derived (i.e. whether water is from rainfall or from flooding, etc.). The project we are seeking here will enable the recovery of important historical records and initial analysis of these records that can be used as the basis for more sophisticated follow on projects.

## 24.10 Human Perceptions and Biodiversity in the Lake Tumba Landscape

Past human migrations have led to great cultural diversity within the region, and to great social unrest, including armed conflicts. In turn, these have affected the size and composition of households and their social interactions. These changes may have had noticeable effects on local perspectives on biodiversity conservation. Transformation in the ethnic composition of the landscape can be traced back to migrations in search of dry land, ethnic wars of the late 1800s–early 1900s, and displacement of villages during the colonial period, and the recent war. The search for subsistence and economic alternatives is currently contributing to some seasonal fishing camps evolving into permanent villages. Governmental representatives and traditional authorities coexist in all villages. In most cases, the traditional authority plays de facto a the role that should in principle be enforced by modern state institutions. Access to local natural resources is mostly controlled by traditional regulations relating to agriculture, hunting, fishing, and collection of non-timber forest products (NTFP) that are juxtaposed with the new forestry code. The arrival of logging concessions, which are authorized by the national government, is a perplexing situation and is perceived as a mixed blessing: both an opportunity for development and a threat to the survival of local traditions and major sources of income. Consumption of and commerce in fish are both widespread, as well as awareness of decreasing fish stocks associated with an intensification of the activity in terms of time, utensils, and methods. These facts weaken the traditional vision of conservation policies, bushmeat trade, and overpresent authority of the state. They suggest that there must be some other possible routes to achieving conservation goals. They call for new insights in implementing democratic means to manage natural resources.

Demographic and socioeconomic data should be collected from both household and focus groups across the Lake Tumba Landscape to see how the social and economic patterns of the landscape influence local attitudes toward and uses of the natural resources in their areas in the landscape. A semi-structured questionnaire should be developed to assemble answers from households and markets. That questionnaire should be backed by focus group sessions that will have as a *raison d'être* to triangulated answers from the questionnaires and gather issues that would not be covered by the questionnaires. Answers should be analyzed in a way that would allow making links between biodiversity in different local areas and the human perceptions of the value of their forests and what these forest have as resident species.

# References

Bennun LA, Howell K (2002) Bird sampling techniques. In: Davies EG (ed) African forest. biodiversity: a field survey manual for vertebrates. Earthwatch Institute 131–161

Berger L, Speare R, Daszak P, Green DE, Cunningham AA, Goggin CL, Slocombe R, Ragan MA, Hyatt AD, McDonald KR, Hines HB, Lips KR, Marantelli G, Parkes H (1998) Chytridiomycosis causes amphibian mortality associated with population declines in the rainforests of Australia and Central America. Proc Nat Acad Sci 95:9031–9036

Brandt (von) A. (1964) Fish catching methods of the world. Fishing. News (Books) Ltd., London

Furness RW, Greenwood JJD, Jarvis PJ (1993) Can birds be used to monitor the environment ? In: Furness RW, Greenwood JJD (eds) Birds asmonitors of environmental change. Chapman & Hall, pp1–41

Groombridge B, Jenkins MD (2002) World atlas of biodiversity: earth's living resources in the 21st Century. University of California Press, California

Howell K (2002) Amphibians and reptiles: the herptiles. In: Davis G (ed) African forest biodiversity. A field survey manual for vertebrates, Earthwatch Institute, pp 17–44

Hughes RH, Hughes JS (1992) A directory of African wetlands. International Union for Conservation of Nature

Johnson HB, Polley HW, Mayeux HS (1993) Increasing CO2 and plant-plant interactions: effects on natural vegetation. Vegetatio 104(105):157–170

Kylin H, Louette M, Herroelen P, Bouwman H (2010) Nominate lesser black-backed gulls (Larus fuscus fuscus) winter in the Congo basin. Ornis Fenn 87:106–113

Lambert MRK (1993) Effects of DDT ground-spraying against tsetse flies on lizards in NW Zimbabwe. Environ Poll 82:231–237

Rosenberg DM, Resh VH (1993) Introduction to freshwater biomonitoring and benthic macroinvertebrates. In: Rosenberg DM, Resh VH (eds) Freshwater biomonitoring and benthic macroinvertebrates. Chapmann & Hall, pp 1–9

Scott Jr. NJ (1982) The herpetofauna of forest litter plots from Cameroon, Africa. In: Scott Jr. NJ (ed) Herpetological communities: a symposium of the society for the study of amphibians and reptiles and the herpetologists' league, Washington DC, pp 145–150

Shrestha J (1995) Enumeration of the fishes of Nepal. Department of National Parks and Wildlife Conservation, Ministry of Forests and Soil Conservation. Biodiversity Profiles Project Publication No. 10

Smith DC, Azam F (1992) A simple, economical method for measuring bacterial protein synthesis rates in seawater using 3H-leucine1. Mar Microbial Food Web 6(2):107–114

Stattersfield AJ, Crosby MJ, Long AJ, Wege DC (1998) Endemic bird areas of the world. Priorities for biodiversity conservation, BirdLife International

Umali AF (1950) Guide to the classification of fishing gears in the Philippines. US Fish and Wildlife Services

Vogt RC (1987) You can set drift fences in the canopy! Herpetol Rev 18:13–14

Wyman RL (1990) What's happening to the amphibians? Conserv Biol 4(4):350–352

# Part V
# Ethics of Biodiversity Conservation and the Needs of Local Communities

This part of the book has a single entry and that entry is about the conflict between preserving biodiversity and fulfilling basic human needs of local communities residing within and near protected areas. I argue that local human communities should be allowed to exploit resources within and near protected areas even if that exploitation by human local communities is harmful to biodiversity. I will claim that all types of biodiversity are not of equal value and do not matter at the same level; there are differences in functions; some species or ecosystems matter more than others. In order to defend this view, I will appeal to biological concepts such as trophic levels, Keystone species, hotspot areas, and the species redundancy, which all hierarchize biodiversity in different degrees of importance. I will then address the question of why conserving biodiversity on earth matters. I will claim that the most convincing argument in defense of biodiversity conservation is to appeal to instrumental values. Biodiversity matters more to humans for the services it provides to human well-being than for any other reason. Furthermore, I will claim that even if biodiversity has **intrinsic value** (the view that biodiversity should be preserved for its own sake), intelligent life such as human life is necessary to acknowledge it, if that intrinsic value is to be recognized and that even if intrinsic values were admissible as such, all species do not have equal intrinsic value. I also will discuss ethical reasons why harmful human activities by local communities should be allowed. I will use several strands of arguments, including the biological concept of species redundancy, the ethical notion of the sacredness of human life (right to life), the priority view of the distributive justice, *Jus Soli* and the rights of the first occupants and environmental justice to support my main claim that regardless of circumstances, human local communities should be allowed to use natural resources. Finally, I will discuss possible reconciliation between human basic needs and biodiversity. My argument will be that partial reconciliations are possible in most cases by hierarchizing harms, accounting for scales, and identifying trade-offs by agreeing with local communities, in a participative process, on what can be extracted by local communities.

# Chapter 25
# Are There Ethical Reasons to Preserve Biodiversity Against Local Communities?

**Abstract** The chapter is about the conflict between preserving biodiversity and fulfilling the basic human needs of local communities residing within and near protected areas. The chapter argues that local human communities should be allowed to exploit resources within and near protected areas even if that exploitation by human local communities is harmful to biodiversity. A claim is made that all types of biodiversity are not of equal value and do not matter at the same level; there are differences in functions; some species or ecosystems matter more than others. As a defense for this view appeal is made to biological concepts such as trophic levels, keystone species, hotspot areas, and the species redundancy, which all hierarchize biodiversity in different degrees of importance. The chapter also claims that the most convincing argument in defense of biodiversity conservation is to appeal to instrumental values. Biodiversity matters more to humans for the services it provides to human well-being than for any other reasons. In this sense, the argument runs that there might be ethical reasons why harmful human activities by local communities should be allowed. Several strands of arguments, including the biological species redundancy, the ethical notion of the sacredness of human life (right to life), the distributive justice (*jus soli,* rights of the first occupants and environmental justice) are used in the support the main claim of the chapter that regardless of circumstances, human local communities should be allowed to use natural resources. The chapter finally discusses the possible reconciliation between human basic needs and biodiversity, arguing that partial (at least) reconciliations are possible in most cases through possible hierarchizing harms, accounting for scales, and identifying trade-offs by agreeing with local communities, in a participative process, on what can be extracted by local communities.

**Keywords** Biodiversity conservation · Human basic needs · Local communities · Distributive justice · Reconciling incommensurable demands

## 25.1 Introduction

Ecology has emerged as one of the most important concerns of the end of the twentieth century and will continue to be an area where the global world community will have serious disquietedness over a large part of the twenty-first century. Preserving the



B.-I. Inogwabini, *Reconciling Human Needs and Conserving Biodiversity: Large Landscapes as a New Conservation Paradigm*, Environmental History 12,
https://doi.org/10.1007/978-3-030-38728-0_25

world's biodiversity is part of that ecological disquietedness. Worries to preserve the world's biodiversity have become one of the most stringent narratives of the modern global society. However, too often, answering to the questions '*what is biodiversity*' and '*what kind of biodiversity should matter*' is not as easy as it would appear at the first glance and constitutes one of the major reasons why so often discussions on biodiversity conservation lead to fierce opposition between pro-conservation and their opponents. Furthermore, fundamental ethical questions such as '*why conserving biodiversity on earth matters*', '*at what optimal levels*', and '*for what values*' are often left unsettled. Also unsettled are questions that are related to human uses of biodiversity, particularly in cases where and when concerns to preserve biodiversity conflict with basic human needs for communities residing in areas where biodiversity needs to be preserved. This essay is an effort to respond to this later question. It looks at good ethical reasons that might help promote the preservation of biodiversity at the expense of the needs of local communities who could benefit from exploiting those natural resources. The main claim is that in cases of irreconcilable conflicts between the values of human life and the value of preserving biodiversity, basic human needs should be given precedence for local communities. The precedence to basic human needs for human local communities is justified because of the human right to life; humans have an intrinsic value that cannot be traded with any other value. I also recognize the possibilities of partial reconciliation between human needs and the needs to preserve; they possibilities arise from the principle of natural biodiversity resilience, which means that in some instances and within a threshold harmed biodiversity will bounce back and continue to persist.

Schematically, I first define biodiversity as being more than sheer numbers of species in any given area. For the sake of clarity, I define some key concepts used in the area of biodiversity conservation such as trophic levels, keystone species, hotspot areas, and species redundancy. Based on these key biodiversity concepts, which hierarchize components of biodiversity in different degrees of importance, I claim that all types of biodiversity are not of equal value; there are differences in functions; some guilds or species matter more than others. Secondly, focusing on the Meso level, which constituted of entities that people can perceive, using and enjoying as part of the daily preoccupations such as individuals, species, habitats, ecosystems, and landscapes, I present different values (**instrumental** and **intrinsic**) that biodiversity is said to have. I argue that all species do not have equal intrinsic value because some species are more inviolable than others. I claim that the most convincing argument in the defense of biodiversity conservation is to appeal to instrumental values because biodiversity is more valuable for the services it provides to human well-being, inclusive of maintaining life itself than for any other reasons. Thirdly, there are human activities that truly threaten biodiversity and its very persistence over time in some parts of the world. However, I defend the view that even in these cases where human subsistence activities are detrimental to biodiversity, human local communities should be allowed to conditionally use resources within protected areas because human life is the most sacred form of life and humans have right to life, which has precedence over other species. Additional reasons for this claim are fulfilling basic human needs and the need for justice and/or fairness. Fourthly, I discuss how to reconcile (at least partially) cases where fulfilling local human communities' needs necessarily implies infringement on biodiversity and the interests of both parties. The research presents what are possible means to be used to reconcile that conflict (at least partially) and justify why it is morally sound to give precedence to human needs in cases where reconciliation is impossible.

## 25.2 What Is Biodiversity and What Kind of Biodiversity Should Matter?

**Biodiversity**: The definition of biodiversity has changed over time and the ethical implications and responsibilities toward biodiversity change depending on the understanding of the word (Bosworth et al. 2011: 2). Narrowly defined, biodiversity is equated with the number of species or the 'species richness' found in a given location (Morgan 2009: 235). However, this definition has moved from this narrow understanding to include living organisms and the complex interactions between these living organisms and their abiotic environments. In this essay, biodiversity is the totality of living organisms and functions that ensure that species and life are maintained on earth. Hence, biodiversity consists of three main components: composition, structure, and function (Neem et al. 2008, cited by Bosworth et al. 2011: 2), which implies that biodiversity should not be viewed only as the total numbers of species; it has to include functions that interrelate different organisms and sustain life on earth. In this view, for example, if we theoretically have two worlds wherein the first one has five species and the second one has 2 million, the first one would be said to be less diverse based on species numbers (species richness) but if life could be maintained in both, these two planets would be equivalent in their respective functions to maintain life. Therefore, what matters the most is the number of species within the totality that would suffice for life to continue thriving on each of these worlds. This leads me to argue that all species are not of equal value because they play different roles in maintaining life on earth; some have a more critical role to play than others. Consequently, there are species that matter more than others so far as diversity in functions to maintain life on earth is concerned.

**Ecological Trophic Levels**: Ecological science admits the existence of different trophic levels, which are positions occupied by each species or guild within the food chain. Trophic levels comprise (1) primary producers (plants and algae that make their own food), (2) primary consumers (herbivores that eat plants), (3) secondary consumers (carnivores that eat herbivores), and (4) tertiary consumers (carnivores that eat other carnivores). The fifth trophic is the apex predators that sit at the top of the food chain and are not preyed by other species.

**Keystone Species**: It refers to a species whose extinction within a given ecological community will cause the loss of many others (Mills et al. 1993: 219). Keystone species are so important in determining the ecological functioning of a community that they warrant special conservation efforts to (Mills et al. 1993: 219). Mills et al.

(1993: 220) link keystone species with the trophic levels and define five categories of keystones species, which are associated with the respective effects of their removal on their ecosystems. Of these five categories, three are of the food web while the remaining two are functional categories. There are keystone predators whose removal will increase one or several predators (increased consumers and competition over limited resources), which in turn will extirpate prey species. There are also keystone prey species whose removal will lead to the extinction of prey species that are more sensitive to predation, which in turn will result in the crash of predators. There are plant keystone species whose removal from ecosystems will result in the extinction of dependent wildlife species (i.e. pollinators and seed dispersers). There are key functional links and key functional modifiers; the removal of the first would result in failure of reproduction and recruitment in certain plant species with cascades of other losses; the removal of the key functional modifiers would lead to the loss of structures (or materials) that affect habitat types and energy flow and the disappearance of successional habitat-dependent species.

**Species Redundancy**: Principles of species redundancy are that despite being phylogenetically singular, species are rarely singular in terms of functions each species plays in the ecosystems (Naeem 1998: 42). Species are redundant in that species of the same trophic level can play functions interchangeably with other species, albeit with different efficiencies. This means that species in the same group can be equivalent functionally. That is why species' extinctions occurred throughout the natural history of the earth and ecosystems always managed to remain resilient. If ecosystems depend on functions played by different species, it logically follows that extinct species are functionally replaced by others to continue surviving after extinctions. This stems from the compensatory abilities of species within functional groups; local extinction within functional groups is often followed by compensatory growth of others, which effectively leads to a replacement of the contributions of lost species to the overall group functioning (Naeem 1998: 42). Species, regardless of their importance, are not necessarily functionally irreplaceable.

**Biological Hotspot Areas**: These are areas featuring exceptional concentrations of endemic species and experiencing exceptional loss of habitat (Myers et al. 2000: 853). This definition implies that these areas have more species than other areas of the world; losses of hotspots would be more detrimental to the overall biodiversity than losses of some other areas. Strangely enough, biological hotspots which represent 12% of the terrestrial surface are also home to 20% of the world population (Cincotta et al. 2000: 990). This explains why habitats in hotspots are particularly vulnerable.

Cells, tissues, organs, organ systems, organisms, populations, communities, ecosystems, landscapes, biomes, and ecospheres are levels of organization (Odum and Barrett 2004: 1 and 5). This layering is complex and can confuse laypeople whose concerns are far from being those of environmentalists and ecologists. This raises another set of questions that should be clarified to help advance the debate opposing different views on whether humanity should preserve the biodiversity or not. This type of news questions include the following: what kind of biodiversity matters? Are all the species and organisms that constitute biodiversity of equal value? etc. This strand of questions is discussed below.

## 25.3 What Kind of Biodiversity Matters? Are All the Species and Organisms of Equal Value?

Biodiversity is the totality of living organisms and functions that ensure that species and life are maintained on earth, and it is a complex. More palatably, this biodiversity should be broken into several elements and levels of organization (Odum and Barrett 2004: 1 and 5) to reflect the diversity of life on earth because it encompasses many levels (genes, cells, tissues, organs, organ systems, organisms, complex species, life-forms, populations, communities, functional roles, and spatial patterns for biological communities such as ecosystems, regions, landscapes, biomes, and ecospheres (Odum and Barrett 2004: 1 and 5; Colwell 2009: 257). These layers are clustered down to three major layers: Micro, Meso, and Macro levels. The micro level will consist of cells, tissues, organs, organ systems, and organisms. They all concur to sustaining life on earth but cells are important at a level that is different from where tissues stand; organs are important at a layer where organ systems are functionally more important; organ systems have levels where they are more critical as ecological entities than organisms. The Meso level comprises individuals, species, habitats, ecosystems, and landscapes; species and populations occupy specific niches and different trophic levels. The battleground of proponents and opponents of biodiversity preservation is at this level, principally on species and, in some cases, on habitats and ecosystems. Ecosystem implies a physical description of a community in its habitat […], with an emphasis on physical processes connecting different components of a given areas (Sarkar 2014). Both (proponents and opponents) agree that because these entities are perceived, used and enjoyed by humans, they are the focus of conservation. The Macro level includes biomes and ecosphere, entities that are remote from the laymen's daily preoccupations. Ecosphere is a planetary system that contains all elements that can allow life to flourish and comprising the assemblage of the atmosphere, the geosphere, the hydrosphere, and the biosphere. Ecosphere is a synonym of the biosphere and also refers to zones of the universe where life should be sustainable (Huggett 1999: 425). Biomes are intertwined with ecosystems but differ from ecosystems in that they typify scales that are larger than ecosystems and encompass large communities fashioned by similar environmental conditions and patterns, including climate and geology.

The classification of life in three layers is an indication of differential actions and the importance of biodiversity at different organizational levels. The Meso Level being the battleground for different views on biodiversity conservation, my discussion in what follows will mainly focus on this level. I will discuss concepts of trophic levels, keystone species, and species redundancy in given ecosystems and biological hotspots with the aim to see what values can be inferred from these concepts.

Different roles played by different species in given ecosystems are often classified in the pyramid of different trophic levels, defined above in terms of food webs and energy pathways. Primary producers (feeding on vegetation and transforming the energy directly from greens) occupy the base of the pyramid of and top predators (species that are dependent on other species to survive) occupy the top of the pyramid. Top predators are hardly replaceable while species at the bottom play equivalent roles. In a case of the complete extinction of one species, another species would play the role of the extinct species. So, role-wise, there are species that can be replaced in their respective ecological roles and others cannot. Keystone species is a simple concept and definable through its two hallmarks (Mills et al. 1993: 219): (1) keystone species are crucial in maintaining the organization and diversity of their ecological communities and (2) implicitly keystone species are exceptionally important in relation to the rest of the community. Clearly, as keystone species play greatly important roles that render them irreplaceable, their disappearance from ecosystems where they occur would lead to catastrophes such as chains of extinction of other species. Hence, species have relative values; they all do not play similar functions and are not of the same value, at least biologically. Habitats also vary in the wealth of biodiversity they shelter. Hotspot often relates to a location where something out of the ordinary occurs (Nelson and Boots 2008: 556). There is a diversity gradient; biodiversity decreases with increased distance from the equatorial habitats, with a few exceptions. Ecologists define areas of exceptional biodiversity and higher rates of endemism (Médail and Quézel 1999: 1510; Cincotta et al. 2000: 990), which should be preserved because major biodiversity losses and subsequent consequences will ensue after their destruction. The general gradient indicates that all habitats are not of equal value in terms of biodiversity and '*something out of ordinary*' is in terms of rarity of occurrence or of uniqueness in composition (Bosworth et al. 2011: 5). Biological hotspots show that ecosystems are differentiated because of their uniqueness, rarity, and functional roles. Hence, ecosystems have relative values; there are those that are more valuable than others.

The above raises the questions: (1) are all species of equal value? (2) are all species equally worth conserving? Trophic levels, keystone species, and biodiversity hotspots incline to say no because species play different roles and are of relative values, biologically. Those that play essential roles are given more weight than others. This view is not that of groups promoting deep ecology, for example. One thing that needs to be clearly stated here is that different walks of ecologists accept concepts of trophic levels, keystone species, species' irreplaceability, and biological hotspots (even if they use them differently). The assertion that keystone species and hotspots weight more than nonessential species, habitats and ecosystems, raises the question of what it is the fate of the nonessential components of biodiversity? Are 'nonessential' species, habitats, and ecosystems dispensable? These are difficult questions and answering these questions depends on several factors. Further clarifications are needed on terms being used almost interchangeably up to now before responding. These terms are important in the thinking of biodiversity conservation debate. The terms 'preservation' and 'conservation' need to be differentiated (Minteer and Miller 2011: 945). Biodiversity preservation is a trend of thinking that holds that biodiversity should be preserved in its entirety; hence the proclaimed aims to protect biodiversity from direct human actions. Conservation is the thinking that biodiversity can be used wisely to accommodate human needs. This distinction dates back to the debate between John Muir and Gifford Pinchot (Callicott 1990: 16). Muir advocated for the total preservation of biodiversity using the argument of a transcendental philosophy of nature. Pinchot advocated for prudent use of nature's bounty. A prudent use opposes the unrestricted extraction of natural resources. The above questions come down to asking the question: are species that are nonessential in ensuring the sustainability of functioning ecosystems dispensable? The answer will vary according to whether people follow Muir or they promote Pinchot's views. Preservationists would argue that species should be preserved intact regardless of the roles of living systems and regardless of their value to and for humans. They would use the intrinsic value argument and the precautionary principle (based on the limitation of the human knowledge on biodiversity). Conservationists agree with the idea that even 'nonessential' species should be protected, mainly using the precautionary principle. I will argue later that even the nonessential species are instrumentally valuable, hence should be protected.

## 25.4 Why Does Conserving Biodiversity Matter?

This section is about values (instrumental and intrinsic) that biodiversity may hold. It will stress the reasons why biodiversity matters and should be conserved. I argue that the most convincing argument in the defense of biodiversity conservation is its instrumental values because biodiversity is more valuable for the services it provides to human well-being than for any other reasons. Against the claims of intrinsic value that say that biodiversity should be preserved for its own sake, I argue that for that intrinsic value to be recognized as such it needs intelligent beings to acknowledge that value. I propose that even claims for intrinsic value are essentially anthropocentric; they reflect what pleases humans and what is useful for human well-being (material, moral, and spiritual); they are, in relative degrees, instrumental and utilitarian claims. Because all species do not have equal value (see above), some species are more inviolable than others. This is but a confirmation of the biological hierarchy.

## 25.5 Instrumental and Intrinsic Values of Biodiversity

Something has instrumental value when it is valuable as a mean rather than an end (Bosworth et al. 2011: 25) while something has a utilitarian value if its use increases human pleasure and it maximizes the good or it brings about 'the greatest amount of good for the greatest number'. Biodiversity is used as a means for humans to achieve their well-being in many instances. This is simply related to the fact that people eat, heat with, build with, and otherwise consume many living beings, so some degree of diversity is required if these varied human needs are to be met (Callicott 1997: 30). People use biological goods (such as meat, fruits, and greens), biological services (such as the provision of chemical molecules by plants, and water recycling by plants), and information (such as genetic information) which are found in nature and particularly in living organisms. These goods, services, and information constitute the natural capital in opposition to human-created capitals (financial capital and social capital). Hence, conserving biodiversity matters because biodiversity contributes economically to human welfare. Biodiversity offers natural assets that are irreplaceable (at current levels of human knowledge). The second type of instrumental value of biodiversity consists of the value of experiencing nature, which is indispensable for humans because biodiversity's esthetical values that offer leisure and recreation to humans. For example, humans enjoy watching birds, flowers, or other natural sceneries for their recreation time, which contributes to increasing human pleasure. In this sense, birds, flowers, or other natural sceneries are used as a means for humans to achieve their own good, which is recreation and leisure. Watching and observing nature can, is also, often used to increase human knowledge on biodiversity; hence biodiversity is used as a means for humans to augment their own knowledge. Interacting with biodiversity via recreation or scientific observation has been often thought of as intrinsic values because of the deep feelings it leaves on people; these feelings are in some cases closer to religious feelings. Experiencing nature is often thought of as reading a message from something inherent to biodiversity. Esthetic value is given great consideration in ethical theory and it is frequently used to justify biodiversity conservation as something of common interest (Bosworth et al. 2011: 25). The use of esthetic value to justify biodiversity conservation is so because of the presupposed potential universal or intrinsic qualities of biodiversity. Different people have different experiences with nature. Some communities attribute great value to biodiversity because of their culture; many deities are incarnated in some species in some cultures. But when people interact with deities in seeking grace or other benefits, they do this to increase their own pleasure, forces, their own chance, etc. Despite these deep feelings and the delights that one might reach when experiencing nature, the results are anthropocentric. Humans value these interactions with nature because of the benefits they get from experiencing nature. Hence esthetical values attributed to biodiversity are part of instrumental values because they contribute as a means for humans to achieve their economic well-being and increase the pleasure for humans. One way to illustrate my thoughts here is to take the example of diamonds on Mars: suppose astronauts happen to discover diamonds on Mars, would it be sensible to say that these diamonds had value before being discovered by humans? The obvious answer is yes they were valuable. However, that value is linked to the human history of knowing what value diamonds possess. Without that past knowledge about what humans can achieve with diamonds, saying that diamonds in Mars held their own inherent value seems to be off-reality. The point here is that diamond might have had its own value but this did not matter until people found its usefulness for the improvement of their own conditions; a dormant value becomes value only when there is some usefulness attributed to it by humans.

Something is intrinsically valuable [...] if its value is independent of what people happen to enjoy or want [...] for them (Dworkin 1993: 71). Hence, intrinsic values are inherently possessed and are not externally conferred; possessors of intrinsic value are valuable for their own sake. If biodiversity is intrinsically valuable, then its value is independent of the usefulness of the nonhuman world for human purposes (Naess 1995: 68). Saying that biodiversity has an intrinsic value means that preserving biodiversity does not need to refer to other valuable goals external to biodiversity. In this sense, conserving species, habitats, and ecosystems should be carried out for their own sake because of their intrinsic value. This view is supported by biocentrism, broadly encompassing approaches and ethical theories reclaiming moral considerations of all life (Bosworth et al. 2011: 12). Assessing biocentrism depends on (1) the sort of things that may be said to possess intrinsic values and (2) whether an intrinsic value exists objectively or is conferred subjectively (Callicott 1997: 32).

Appeals to instrumental and utilitarian values are easily understandable because biodiversity provides humans with the means to achieve their own goals. This cannot be said of the intrinsic value of biodiversity. The question is often raised about whether all sorts of biodiversity possess intrinsic values? While many would attribute some intrinsic value to species such as great apes and elephants, most people find it mind-boggling to attribute intrinsic values to all species of animals, plants, habitats, and ecosystems. To solve the dilemma posed by the question of what species should be recognized as intrinsically valuable, attempts have been made to extend intrinsic values to sentient animals (Donner 2002: 99). Demands of what criteria have led to delimiting the sphere of intrinsic values to sentient beings, excluding other living beings such as plants and non-sentient animals, led to obvious objections. For some thinkers, all contemporary forms of life are [...] relatives [...] of one extended family (Donner 2002: 100); thence they all are equal members of one society, the biotic community. Because of this, values that are intrinsic to some members of this kinship should be extended to all the members. The kinship is justified by both ecology and Darwinian evolution. For this group of thinkers, biodiversity is intrinsically valuable because human is naturally *biophilia* (Takacs 1996). Biophilia is the idea that humans are genetically predisposed to love nature (Wilson 1990). Despite the strength of the biophilia hypothesis, protecting species such as mosquitoes that actually take life from humans remains hardly convincing. This is because of the hierarchy in the biological importance of species, which needs to be translated into values attributed to different species. To deal with this situation, Dworkin (1993: 70) distinguishes two categories of intrinsically valuable things: those that are intrinsically valuable because of their instrumental value and those that are intrinsically inviolable (sacred). This distinction, combined with the idea of degrees of sacredness (Dworkin 1993: 80), is a better way to grade species according to their importance and provides a solid ground for selecting which species should be totally inviolable and which one can be partially violable. Indeed, there are degrees of sacredness [...] and our convictions on inviolability are selective (Dworkin 1993: 80). So, even when claiming that biodiversity is intrinsically valuable, only some species are inviolable. That is why a few people would care about the extinction of mosquitoes, for example, while they will be in great sorrows to hear that great apes have gone extinct.

The claim that intrinsic values are objective equates to saying that these values are independent of the consciousness of their existence. The question of how values residing in the valued species and other biological entities are valued is a pendant to this claim (Donner 2002: 103). At best, there is no clear-cut answer to the question. What remains certain is that intrinsic values are conferred by humans. In a sense, humans project their values onto things. It is the human consciousness that perceives and describes living entities and gives meanings and values to species, habitats, and ecosystems. It is plausible to counterargue that humans discover dormant values in living organs; and dormant values in living organs independently preexisted in human's discovery. In a sense, these values can be thought of as being intrinsic because they preexisted the human interventions. Yet, it remains true that the dormant values become values to humans only once they are unearthed by humans as argued above. This is of the same nature as the question of essence and existence of values: '*does the essence of values attributed to biodiversity precede the existence of these values when attributed by humans?*' The answer to this question depends on different thinking traditions. Someone from the empirical tradition would argue that a value is attributed to something only as it is seen, described, and its usefulness asserted by the intelligent being whereas another tradition would see things as holding values even when these things are not known and described. Many living organs are not known and described yet; logically their usefulness cannot be asserted. In that sense, the intrinsic values are more subjective and are truly human-dependent in their metrics and would be conditional to human abilities to depict them. The process to envisage something even when that something has no material reality is cognitive (depends on intelligence); so in many ways, it can be asserted that intrinsic values of biodiversity are values only as they respond to the criteria defined by humans. They do not exist on their own as they depend on the definition of values by humans.

Intergenerational fairness and unidentified biodiversity's values are among other considerations of biodiversity's values. The intergenerational fairness argument runs that current humans have the duty to leave a thriving planet to future generations of humans. A thriving plan will be instrumentally valuable to future generations because they will need services offered by biodiversity to survive and because future generations will need a planet capable of holding life on it. And biodiversity is what maintains life on earth; it should, therefore, be preserved. Intergenerational fairness raises questions and counterclaims. Principally, these counterclaims are (1) argument from ignorance, (2) disappearing beneficiaries, and (3) temporal location. The argument from ignorance stresses because current human beings know little about people of the future; current human societies do not have moral duties to unknown people (Desjardins 2006: 75). The disappearing beneficiaries' argument holds that current humans can have no obligation (of maximum happiness or of duties to respect the rights) to bring future generations into existence because future generations are no particular people; current humans have no responsibility to direct to these future generations (Desjardins 2006: 75). The temporal location argument is that humans cannot have responsibilities today toward people who might not exist for many years [to come] (Desjardins 2006: 77). Desjardins (2006: 70–93) lengthily and convincingly discusses these counterclaims and affirms that humans do have ethical responsibilities to future generations. The ignorance's claim was rebutted by parallelism between civil law and reasons why current humans have responsibilities to future generations (Desjardins 2006: 77). Compellingly, as civil laws are designed not to deal with the immediate breaches of commonly agreed social codes but to prevent potential future lawbreakers to account for acts that will happen in the future, present generations have responsibilities for future harms to people and the extent of harms that are presently unknown (Desjardins 2006: 77). The disappearing beneficiaries' argument was rejected because although any potential beneficiary may disappear under alternative decisions, the relative amount of suffering or happiness does not. Hence, the obligations of current humans are to recognize a certain minimal requirement of moral responsibility. The example of a trip on a crowded long-distance train where a toxic package is discovered convincingly illustrates how humans react and, using that known human behavior, how the future generations might react against the actions of current humans. If humans react the way they do against unknown people that threaten the very basis of their lives, future generations (who will be humans like us—albeit maybe with different social and economic contexts) would have the same reaction to the mess being created for them today. All that said, it is sensible that the intergenerational duty becomes a strong argument for conserving biodiversity.

It is often rightly argued that unidentified biodiversity values and species interconnectedness should lead humans to be precautious in how they use biodiversity. Certainly, the knowledge humans possess on biodiversity is limited. The uncertainty argument (precautionary principle) says that because we are uncertain of threats and harms human actions cause to the environment, humans should be prudent in dealing with the environment. The species interconnectedness was defined above (Donner 2002: 100) and consists of seeing species as part of extended kinship. Because of these, ecologists claim that humans need to value whatever is there because it might hold values that are unknown but possibly could be revealed in the future. Uncertainty and species interconnectedness justify the precautionary principles. Humans have more propensities to conserve biodiversity. The persistence of currently living entities will sustain established natural equilibriums.

## 25.6 Is It Ethically Admissible to Exploit Natural Resources in Detriment of Biodiversity?

In this part, I will leave aside human activities that pose no threats to biodiversity. Indeed, in some cases, disturbances are good and increase the species richness. Also, species redundancy species take over the roles of their disappeared counterparts. Hence, I will focus only on activities that truly threaten biodiversity by destroying biodiversity resilience. I will argue that even in cases where human subsistence activities are detrimental to biodiversity, human local communities should sustainably use resources within protected areas. For human life is the most sacred form of life, it has precedence over the biodiversity valuation. Human life, being the most sacred, basic human needs of local communities residing near and outside important areas should be fulfilled even if doing so hurts biodiversity for reasons of justice, fairness, and right to life. As it is implicit in the notion of sustainability, local communities should know and agree upon the limits of offtakes. They should use biodiversity when alternative resources lack; they should avoid unnecessary over-consumption. In all cases, if the resources are of vital value for humans to thrive, they should be used by communities.

I need to frame the situation being discussed to articulate my argument. Protected areas are often created as goods of public interest. In many cases, they are dubbed to serve the global human community, including nations, provinces, and other geographic subdivisions where they occur and the whole world. Hence, protected areas are given moral precedence over the interests of local communities. However, protected areas are created in areas where most people live under conditions of abject poverty and rely on offtakes from nature (see above and Callicott 1997: 30) to live on and have no other alternative sources of goods. Hence protected areas clash with the needs of local communities. Anti-anthropocentrism and interests of the global community are the two major reasons explaining why human communities are prevented from using the resources within and near protected areas. Anti-anthropocentrism is around the idea that biodiversity's value is similar to human value. The importance of service rendered by biodiversity worldwide is higher as compared to the local interests. In the following sections, I offer to defend my claims by rejecting each of these arguments.

Against the anti-anthropocentric argument, even if human subsistence activities are detrimental to biodiversity, human local communities should be allowed to conditionally use resources within and near protected areas. *Firstly*, species redundancy, which is essentially a functional attribute, says even when species are singular in their phylogenies, they are functionally replaceable (Naeem 1998: 42). Within the same trophic level, species play roles of their disappeared counterparts. When human activities harm a particular species and lead it to extinction, another species will carry out the ecological functions of the extinct species. This will ensure the continuity and persistence of the system. That is why biological systems are sometimes autoregulated systems. Once more, however, positive feedback loops generated by species redundancy will function only up to a certain threshold, beyond which the system will break. This argument, purely biological, may seem weaker in terms of the ethics of biodiversity conservation but it is a natural reason why local communities should be left to exploit natural resources. At the scales of local communities, the main issue of biodiversity destruction may not be as challenging as sometimes presented; cases of real harmful human effects are in reality much rarer than it sometimes seems.

*Secondly*, human subsistence activities should be allowed because human life is more sacred than that of other species (Dworkin 1993: 80) for two main reasons. The first reason is the religious one and says that humans have been created '*in the image of God*'. Humans are, therefore, more than just another creature. Obviously, this is too anthropocentric as an argument to those who are not religious or those who think that humans are part of the ecosystems wherein they occur. Far from being just too anthropocentric, this view just expressly says that humans are a different type of species. They have genuine claims to rights, they confer meanings to things around them, they can think about the fate of the planet and other species, they can think of the future, etc. Even if all things are inspired by God, [*and should*] be honored as such (Kohen 2006: 236), humans who are the image of God should be given more space and more weight while balancing its survival against other species. The second reason is based on the claim by Dworkin (1993) that whether religious or not, people have instinctive convictions about […] why […] any human life is intrinsically valuable (Kohen 2006: 236) and why human life weights more than any life's form. This is depicted in cases where other living organisms cause pain to humans; the immediate reaction from humans is to preserve human life. Humans do not protect other species in the name of biodiversity if these species threaten human lives. This is why those who would see my argument as too anthropocentric would combat mosquitoes, amoebas, and other disease-bearing species, for example, to protect human life. They do, indeed, value human life when it faces destruction caused by other species. There is a somewhat Darwinian quest for the survival of the species that makes humans fight to preserve their own species than they would fight for other species. Dworkin (1993: 80) is right to introduce degrees of sacredness, which *ipso facto* introduces a hierarchy of the importance of species and justifies us in being selective on which species should be totally inviolable and which one cannot be seen as such. To any anti-anthropocentrism which would ask '*why give precedence to humans who are part of biodiversity?*' the rebuttal to this is so because humans value human life more than any other life's form. Humans show the moral propensity to protect humans when they fight other species that cause pain to humans; humans do not want to see other humans suffering and when humans suffer because of other species, the normal reaction from any sensible human is to rescue humans but not to help other species butcher human life. That is why, even when conceding that humans are part of ecosystems where they happen to live, it is a reality that the value conferred to species, habitats, and ecosystems is measured against humans. We believe that earth without humans would be a different place, and biodiversity would lose its own value as well because it will lose one of the major species that currently shape the current distribution of species. To paraphrase Dworkin (1993: 80), it would be regrettable if a beautiful species of exotic bird is destroyed, but it would be even worse if we stamped out humans from earth.

The sacredness of human life is linked to the human right to life, as being enshrined by article 3 of the Universal Declaration of Human Rights: '*everyone has the right to life* […]'. Human welfare is, in this sense, the first right of the above provision. Local communities have rights over the resources located in their areas. Not accessing these resources equates to intentional extermination of the lives of individuals and whole communities. Lives of individuals would be intentionally exterminated if they are not allowed to access essential goods (such as food, clean drinking water, and shelter) that are of the utmost importance for their subsistence. Most local communities draw essential goods from the ecosystems. Preventing people from accessing essential goods is like killing individuals; killing a person infringes her (his) right to life. Once more, human life is the most sacred form of life (Dworkin 1993: 80); since these local communities are made of individual human beings holding the right to life, the lives of communities ought to weigh more than other biodiversity components.

Because it is not acceptable to terminate human life intentionally, it follows that local communities should exploit natural resources located within and near protected areas.

*Thirdly*, there is a moral obligation for communities to *fulfill basic human needs*. This justifies local communities to exploit resources even if human subsistence activities infringe biodiversity. Basic human needs are about goods, processes, and strategies for humans to achieve the characteristics of a good life and include needs for subsistence, reproduction, security, affection, understanding, participation, leisure, spirituality, creativity, identity, and freedom (Costanza et al. 2007: 268–269). Basic needs that humans satisfy by directly using biodiversity include subsistence needs (food, shelter, vital ecological services such as clean air and water, [...]), health care [...], leisure needs (recreation, relaxation, tranquility, access to nature, [...], spiritual needs (engaging in transcendent experiences, access to nature, and participation in a community of faith), identity needs (status, recognition, sense of belonging; differentiation, and sense of place) (Costanza et al. 2007: 270)). Additionally, there are needs that are indirectly satisfied by use of biodiversities such as security needs (stewardship of subsistence into the future, safe distance from crossing critical ecological thresholds). The needs for accessing knowledge (information, intuition and rationality) and needs for participation (to act meaningfully in the world, contribute to and have some control over political, community, and social life, being heard, meaningful employment, citizenship) are also among needs that can only be indirectly satisfied.

Without fulfilling their basic needs, local communities will starve to death. As argued above, human life is sacred and it cannot be left to die because of lack of food while the meat, greens, tubers, and water are locked within protected areas. Local communities are at the lowest end of poverty and have fewer alternative sources of food; they cannot import their food from other areas. They are *the worst off* the global human community. They lack means and capabilities other than those provided by the use of biodiversity located in their areas. Local communities are justified to lay a morally genuine claim on biodiversity based on priority view of the distributive justice because benefits to the worst off should matter more. One can be tempted to argue that the overall community can pay for the subsistence needs of local communities. This has been tried in Central Africa, in the Republic of Congo where local communities living adjacently to the Nouabalé-Ndoki National Park were provided with food and other subsistence goods. But providing imported food is against the human needs for dignity, recognition, and freedom. Experiences such as the one in Nouabalé-Ndoki National Park can be bearable financially in the long-term perspectives only in areas with very low human densities such as it is the case in Congo. Unfortunately, most biodiversity hotspots occur in areas with high human densities where such solutions cannot hold for a long time. Also, importing food deprives people of the sense of their own dignity. Indeed, dignity is a highly valuable good and strongly demands for the bearer of dignity and people around the dignity bearer (Stoecker 2011:8) to be respected. Provisioning local communities with imported food deprives them of their dignity in that it prohibits them from exercising their duties to work in order to sustain their own lives. Work creates a sense of pride that drives many of human motivations. Local communities are fully entitled to claim the fulfillment of their basic needs, which would be a way to assert their capabilities to exercise their duties and rights to live.

Apart from the criticisms of anthropocentrism, proponents of protected areas would argue that because of the global importance of protected areas, they should serve the human community worldwide. Given the moral precedence of the global interest over the interests of local communities, access by local communities to natural resources within and near protected areas should be prevented. A *first* refutation of this view is to *appeal to considerations of justice and or fairness.* The priority view given above leads to Rawls' second principle of justice, the difference principle: 'social and economic inequalities are to be arranged so that they are [...] to the greatest benefit of the least advantaged'. Most local communities are the poorest within the global community; they are in many ways the least advantaged: they lack alternative sources of food; they cannot afford to pay for construction of modern buildings to shelter their families and are not able to pay for modern medicines to treat their sicknesses. Hence, social inequalities should benefit them. Though Rawls' view of justice fits perfectly in this discussion on its own right, appealing to 'environmental justice' (Des Jardins 2001: 240) also provides strong support of my argument. The need to conserve biodiversity has become more pressing because of the global action of humans on the earth's living system. In the global human actions, however, actions to develop the developed world have had a more important impact on biodiversity. Communities in the developed world are now so rich that they no longer need to rely on biodiversity to live. The process that led to the development of richer countries destroyed ecosystems and biodiversity worldwide. Therefore, arguing about the global public interest to justify preventing poor communities to extract natural resources in favor of the global community's benefit is but unjust. The poorest created less damage to the environment and biodiversity in areas where they live; they should not be punished because they have behaved parsimoniously. They should not bear the entire burden of the deeds of others. Des Jardins (2001: 240) is right in saying that distributing environmental benefits and burdens [equally] is prima facie unjust. Those who contributed to destroying living systems in their own areas have no moral right to prevent other poor local communities to live on the biodiversity. The appeal to Environmental Justice should distribute the burdens proportionally to the destruction caused and leave local communities to use biodiversity.

A *second* argument against the '*global importance of biodiversity*' in support of allowing human local communities access the resources within and near protected areas is that these communities hold long-standing interests and subsequent rights on these areas. The interests of the local communities are to sustain their welfare by use of resources. Their rights are on lands, forests, wildlife, fishes, and other important ecological services such as water and air. These rights are theirs because local communities have always been in these areas. This is justified because of the right of the first occupants. The welfare of these individuals should be more important than the welfare of the global community. The welfare represents the highest interest for local communities whereas the right of the first occupants justifies local communities to use the resources within and outside protected areas. Not doing so would jeopardize the interest of the communities. The consequences of lack of access to these resources (at least those that are vitally important) would be very dire for large numbers of people. Indeed, when people lack vital resources such as food and water to live on, their right to life can be seriously compromised. Examples of famine, starving whole communities around and within protected areas are numerous. These examples are more of instances of sacrificing humans as means to preserve biodiversity. But conferring precedence to biodiversity against human welfare would look like using humans as mere means to achieve biodiversity conservation goals. This cannot be accepted since humans are ends in themselves and cannot be used as mere means to achieve other ends, valuable as these ends might be.

I asserted that local communities should sustainably use biodiversity and other natural resources located near and outside of protected areas because of *the claim of a human right to life, claims to fulfill basic human needs, and appeals to considerations of justice and fairness.* But the qualification 'sustainabiliy' has several implications. Sustainability requires that not only that resources are used parsimoniously but also that the levels offtakes from the natural systems remain within the naturally supportable limits. Hence, local communities should use resources when the resources being used are of vital value for humans to thrive. Natural goods are vitally important to humans if they are indispensable for human life to continue; lacking them would be of the utmost negative consequence on human lives. A good example of this situation is access to land, which should always remain safe. The second condition is that of lack of alternative resources; when there are no other resources to live on, local communities should be allowed to use whatever they can find in protected areas to sustain their own lives. A good example of this situation is that of sources of protein, particularly meat and fish. Equally important are to avoid unnecessary over-consumption) and respect for known and agreed-upon limits of resources offtakes. The over-consumption is a consumption that goes beyond what is necessary for a mentally, physically, emotionally, and spiritually healthy life (Curry 2012: 21). This means that ecological knowledge people possess of ecosystems where they live is of crucial importance and should help frame the discussions between what should be protected and what can be used by communities. These discussions are where local communities and policy-makers, biodiversity conservation bodies and other stakeholders such as the enterprises engaged in the trade of biodiversity goods would come together and bargain the trade offs.

## 25.7 Needs of Local Communities and Causing the Least Possible Harm to Biodiversity

This is the last part of this essay, which is about the possibilities of reconciling local human activities and the need to do the least possible harm to biodiversity. I look at cases where fulfilling local human communities' needs necessarily infringe on biodiversity and whether and how it could be possible to care for the interests of both parties. Harm to biodiversity is constituted by two different possible elements: (1)

overexploitation of biodiversity, which goes beyond the natural resilience capacity of a biodiverse environment, and (2) unnecessary intentional and non-intentional destruction of biodiversity components, even if that leaves the biodiversity still resilient. Not all human activities need to be harmful to biodiversity whereas some are clearly detrimental. My main claim here is that if there is a conflict between the values of human life (as described in the concept of basic human needs) and the value of preserving biodiversity, (1) there are sometimes means to reconcile that conflict, at least partially and (2) where reconciliation is not possible, basic human needs should be given precedence. The partial reconciliation claim is justified, first, on the biological ground through the principle of natural biodiversity resilience, which means that that biodiversity can be harmed, yet sometimes harmed in a way that still allows it to persist. This means that activities that are harmful to biodiversity but which are useful for humans to fulfill their needs can be allowed up to the resilience thresholds to allow both biodiversity to thrive and human needs to be fulfilled. Secondly, the partial reconciliation claim is justified on the basis of the idea of a human right to life, which is based on the intrinsic value of human life. I will first briefly present examples of those activities that are non-harmful to biodiversity. Then I will move on to examining cases where the preservation of diversity is in stark opposition to human needs. I will use concrete examples wherever necessary to clarify positions that I develop in this part of the essay. These examples are from developing countries where meeting basic needs and sustaining livelihood are still the major concerns of the majority of people.

## 25.8 Non-harmful Human Activities to Biodiversity

There are mainly two types of activities that humans can undertake in areas where biodiversity occurs: extractive activities and non-extractive activities. Extractive activities mainly consist of the removal of members of species. In the contexts of countries in the tropics, which are both poor and endowed with large areas of biodiversity, most of these extractive activities are essential human subsistence activities. Non-extractive activities are those that consist of exploiting forest (or other kinds of the natural environment) for cultural and leisure reasons. Both extractive and non-extractive activities are non-harmful only when they are ecologically sensitive, meaning when they are done in such a way that they leave the biodiversity with its potential to bounce back to its normal status. Extractive activities that are often considered non-harmful include collecting non-timber forest products such as mushrooms, or wood for energy; they also include fishing or hunting for meat without the use of tools and techniques that can lead to the massive destruction that can bring the species hunted beyond their capacities to rebuild back its own sustainable levels. These activities are meeting basic needs because they are related to providing energy and food to communities that do not have other alternative resources for doing this.

Non-extractive activities include such cultural items as traditional initiation field trip through forests, woodlands, and savannahs where young people are introduced to the art of knowing what their duties are, what can be used for what, etc. Ecological tourism is also part of the non-extractive activities, with its good intention not to destroy the biodiversity though the end result is not often that non-harmful, for diverse reasons.

## 25.9 Biologically Harmful Activities to Biodiversity but Essential to Human Basic Needs

Extractive and non-extractive activities can be non-harmful to biodiversity only up to a given threshold. Beyond that point, they become detrimental to their resilience capacity. The particular conditions in most of the tropical developing countries, where biodiversity is more abundant in comparison to developed countries, have pushed both activities to get near that threshold point and in some cases these conditions have gone beyond the limits of sustainability. In those countries, local communities remain dependent solely on removing species which are part of a biodiverse environment for both food and energy while human populations increased dramatically over the last five decades. But, resources available in the same countries remained the same. Furthermore, the needs to develop countries exert increasing pressures on governments, which have been authorizing extractive industries to extract whatever can be extracted to get financial means to direct toward development. Mass tourism has been also allowed with the same idea behind global managers' thinking. Most of these activities have been done to the detriment of local communities' right to a sustainable livelihood.

The first example of a conflict between biodiversity and extractive human activities is that of fishing. Most technical reports have described decreases in fish stocks and rarefaction of species in the African continental waters. Reasons for this situation are multiple but can all be condensed down to the over-fishing due to increased human populations, use of new tools and techniques of fishing, as well as the trade in fish. As a reaction to this situation, most conservation institutions have been pleading for the prohibition of fishing in many areas, including areas where fishing activities are predominantly conducted by local communities for subsistence and accrued and improved livelihoods. These conservation institutions ground their claims on the harm being caused to fish biodiversity and the potential for fishing to drive species to extinction. The proponents of the local communities' needs for sustained livelihood argue that prohibiting local communities is not the solution; humans should be allowed to continue fishing for subsistence. This is a clear case where the needs to preserve biodiversity is starkly opposed to local human communities' needs. There is a triple facets conflict: (1) conservation of fish species and stocks against needs for local communities, (2) preservation of fish species and stocks against fish industry, and (3) fish industry against local communities. The point of view of conservationists is justified because continuing to exploit the fish stocks at the current levels is clearly unsustainable and does a lot of harm to biodiversity; depletion of fish stocks leads to the collapse of fish communities and possibly mass extinctions in the near future. Depleted fish stocks and possible extinctions that this depletion of fishes will bring about would disrupt the ecosystem functions, which in turn, will affect the overall life over large regions. Because humans have a duty to ensure that life is maintained on earth, the appeal for prohibiting fishing to avoid mass extinctions of species can be justified. This is so because, for example, the air that humans breathe is a product of a natural process whereby plants and other organisms play a key role in its provision. Furthermore, the food and other goods that biodiversity provides for humans to sustain their life are very dear if converted in terms of prices that caring for biodiversity is but a just and fair return for humans to pay back what it takes. The proponents of local human communities' livelihood argue that prohibiting people to fish equates to preventing them from accessing a resource that provides them with food (basic human need) as well as other means for livelihood such as generating income for other basic needs such as buying soap, paying for schools for one's children, and paying for medical treatments. The appeal for the imperative to have basic human needs be fulfilled provides a strong argument in justifying the position defended by the proponents of continued fishing activities. Basic human needs are those needs that will make it possible for humans to continue thriving locally. Local communities in those areas that were recently depleted of their fish stocks and where symptoms of rarefaction are emerging in fish species diversity have the right to food; they need to continue fishing to survive. Also, humans are part of life on earth, if there is a need to keep life diversity on earth humans too, as part of biodiversity. As such, humans are to be part of this global picture of biodiversity conservation.

Is this conflict impossible to reconcile? My claim is that this is a typical case where partial reconciliation is plainly possible. First, this is possible through layering the hierarchy of harm. Some activities have more harmful effects on biodiversity than others. In this case, fishing activities are to be delineated according to their effects on fish stocks and fish species. Scale-wise, it is known that industrial and semi-industrial fishing is more harmful to biodiversity than local subsistence fishing by a small local community. Indeed, industrial and semi-industrial fishing cause bycatches, which unnecessarily deplete fish stocks and collect species that are not necessarily needed for consumption. My point here is that in cases where preserving biodiversity is of the utmost importance, the first human activities to be banned should be those that are more harmful to biodiversity while those that are essential to sustain local communities' needs either in terms of basic needs or those of sustained livelihoods should be continued. A counterargument to this line of thought would be that preventing industrial fishing would affect the basic needs of humans who would lose their jobs, for example. A rebuttal to this argument is that while the situation of those who would lose their jobs would be worst compared to their current conditions, the conditions of the local communities are worse than theirs; hence using Rawls' difference principle ('*social and economic inequalities are to be arranged so that they are […] to the greatest benefit of the least advantaged*'), loss of jobs is comparably less problematic than prohibiting fishing by local communities. Furthermore, communities of industrial fishermen have more capabilities than the local ones; they can identify alternative opportunities for job losers. Secondly, partial reconciliation is plainly possible through conditional fishing modalities, which are defined in terms of fishing seasons, specified fishing gears, and fishing rights. Indeed, fishing is a very seasonal activity; there are seasons when fish captures per unit effort are higher than in other periods. These seasons are proxies of the times when fish reproduction seasons and maturation happen. What it would take for humans to avoid destroying fish stocks would be to practice seasonal fishing, i.e. allowing fish to reproduce and to rejuvenate the fish stocks. This would allow fish stocks to be maintained at sustainable levels while allowing human local communities to continue extracting some significant bulks of fishes for their livelihoods in limited time frames. The other conditionality would be on types of gears and techniques being used to capture fishes, which are often beyond the necessity. Often, nets' mesh sizes are so small that nets capture very young fishes, which are often unnecessary for consumption. An agreed-upon conditional fishing that would care for both biodiversity preservation and basic human needs for local communities would be the one wherein local communities are allowed to fish under the conditions that they can only use gears that are appropriate, meaning the ones that are capable of capturing sufficient fish to fulfill basic needs of local communities while allowing smaller fish to be preserved. Smaller fishes would be left to grow and become the progenitors of the next generations of fishes. Seasonality in fishing and specified gears fishing conditionality could also partially help deflect the conflict between biodiversity conservation and fishing industry. But in cases where seasonal fishing and gear-specified fishing are not the solution to keep the three parties happy, fishing rights would provide a solid ground for local fishermen and fisherwomen to continue fishing in their ancestral water basins while excluding people that have no right claims on these water basins. Local communities have rights over these water bodies per the *jus soli*, which confers full rights to people to enjoy goods and participate in creating goods in countries where they were born.

The second example of a conflict between biodiversity and extractive human activities is that of protected areas. Confronted with alarming declines in biodiversity across the world, many developing countries in the tropics created protected areas of all categories. In most cases, these protected areas were created to protect one flagship species and were officially created without proper consultation with local communities to get their consent over conservation activities (Jepson and Whittekar 2002). Local communities need to be consulted for several reasons, including the fact that the presence of the protected areas would take off the rights over lands they possessed per jus soli and the right of first occupants. This consultation is also important as a democratic principle because it would allow all stakeholders, particularly people most affected by political decisions, to have their views taken care of by decisions being taken on their lives. In most cases, protected areas come with prohibitions for people to access classified areas and to conduct any activity whether extractive or explorative within the protected areas. In some cases, these protected areas are located within high human densities and where good cultivatable lands, unspoiled forests, and watersheds are often the only areas left for biodiversity conservation. In some countries (e.g. Democratic Republic of Congo, South Africa, and Indonesia), creating protected areas has been even accompanied by displacing large numbers of communities outside of areas where their communities have lived for centuries without their consent and without compensations. In these circumstances, protected areas are in stark opposition with local communities' basic needs and the basic needs for local community's conflict with the *global importance of biodiversity*. The creation of protected areas in such regions is done through selecting the most suitable habitats because they are often biologically the last '*pristine*' (least disturbed) habitats for large herbivores such as elephants or other wide-ranging species; sometimes selected habitats are even the only forests remaining in an area that could ensure the perpetuation of the area's microclimate, which sustains life for all, including humans. A good local area microclimate is part of basic needs for local communities but other equally important basic human needs for local communities such as access to land to grow food, access to forests for subsistence hunting, and access to drinkable potable water, collection of construction materials such as woods, thatches, as well as other spiritual and cultural assets are infringed by the presence of these protected areas as they would often come with a cohort of prohibition, including *fencing* and locking resources from being accessed by local communities.

Is the presence of protected areas reconcilable with basic human needs in areas where there is only one land for both? Should the fact that beyond the sheer needs for local communities such as food, water, and thatchesbe equally recognized, there might seem to be other conflicts between different forms of basic needs of local communities as well: this is, for example, the case of a good microclimate for the local area, which would provide clean air and sufficient rainfalls for local communities. The conflict is that without conserving biodiversity, the delivery of services provided by the protected ecosystems would be equally jeopardized. Partial reconciliation, once more, is possible in this case. First, this is possible through recognizing that people's sheer basic needs have to be fulfilled as a matter of principles of justice. Rawls's Basic Rights Principle [...] requires equality in the assignment of basic rights and duties, [and] [...] holds that social and economic inequalities [...] are just only if they result in compensating benefits for everyone, and in particular for the least advantaged members of society (Rawls 1999: 13). If local communities have no food, no drinking water, and no shelters, they will simply die. This would be the case because local communities do not have other means; they are the least advantaged members of the global human society (Rawls 1999: 13); it is not acceptable for humans to leave other people to starve to death. Because it is not admissible to starve people to death, biodiversity conservationists should identify, through an equal participation process (Rawls 1999: 194), biological resources that are necessary for local communities (living within or outside of protected areas) to fulfill their human basic needs. Rights of other stakeholders, which are embedded in the notion of '*global importance of biodiversity*' for the global community, are of lighter weight in this case based on Rawls' difference principle, as local communities in these areas are 'least advantaged'. To accommodate other basic needs of local communities that seem to conflict directly with direct use of biodiversity, the principle of the human duties *regarding* the environment (Des Jardins 2001: 103) and biodiversity would call for a participative process to identify the minimum requirements for species and the habitats to persist. Once more, the participative process is justified both as a means to ensure that divergent interests for different stakeholders are thoroughly debated and addressed and as a way to ensure a democratically peaceful cohabitation.

The trade-off would be reached through agreeing on what can be extracted by local communities and under which circumstances to keep people alive and enjoying their lives while wildlife and their habitats would still continue to have the minimum sizes they need to continue their existence too. This can be justified using the '*rule of minimum wrong*' (Des Jardins 2001: 145), which seeks to make human interests compatible with the basic interests of nonhumans through allowing some damages on nature to satisfy the human interest. The *rule of minimum wrong* would also accommodate the needs of the wider human community, which is attached to the idea of the global importance of biodiversity. The interests of the wider community would be covered by the fact that the minimum wrong will allow the continuation of thriving biodiversity to produce essential services needed by the global community. The process needs to be participative and human communities need to give their free consent for reasons given above (rights owned by virtue of *jus soli* and the rights of the first occupants). The consent of communities is needed because they hold rights of lands being locked by protected areas by virtue of the *jus soli* principle and the rights of the first occupants.

A more radical example is the theoretical possibility to have just one piece of cultivatable land which would also be the only space remaining where one keystone species is to be protected. This example may be seen as removed from reality, yet it is a statistical probability in regions like the Sahel in Africa where cultivatable lands are shrinking due to effects of climate change and herds of elephants (a keystone species) that previously wandered in the region have been confined to smaller areas that are also becoming the only available agricultural lands for people. In this example, keeping people alive requires not only infringing some aspects of a given biodiverse environment but clearly increasing the likelihood of a keystone species going extinct. The question here is then what should be done? Should local communities be allowed to cultivate the remaining piece of land in order to fulfill their basic needs and let the keystone species disappear while fully acknowledging the potential consequences of that extinction on the rest of life diversity in the region?

My answer to the above question would consider the example as being in a closed system; that is, there is no possibility for importing food for human local communities and that the keystone species cannot be translocated to other regions of the world or taking individuals of the species into a zoo. First of all, it should be acknowledged that this is an extreme case even though it is one that is theoretically and factually possible. In this case, reconciling basic human needs and the need to preserve biodiversity is not possible. In such cases, fulfilling basic human needs for local communities should be given precedence over the need to preserve biodiversity. The precedence of basic human needs over the need to preserve biodiversity can be justified using the principle of the human right to life, which is based on the intrinsic value of human life. Something has rights when it has interests or good of its own to be protected by rights (Des Jardins 2001: 105). Wildlife species may be said to have the right to life too. But, in this situation the keystone species is like a drowning person who needs the rescue from someone else. Unfortunately, local communities too are in dire conditions in this example and have no means to rescue the species. As you do not throw someone with no swimming skills to rescue a drowning person to avoid losing both, there is no reason to sacrifice human lives to save the near-extinct species. Feinberg argued that only things with cognitive life, with conscious wishes, desires, hopes or urges or impulses; or unconscious drives, aims, or goals; or latent tendencies, direction of growth, and natural fulfillments can be said to have interests (Des Jardins 2001: 105). Keystone species (i.e. elephants) exhibit some of these traits (particularly urges, impulses, unconscious drives and latent tendencies, direction of growth, and natural fulfillments) that make them higher animal species and can lead some people to confer them some rights. However, although it is plausible that animals and plants have interests and that human beings talk about what is good for animals and plants, their status is not that of human beings (Rawls 1999: 442). This is so because wildlife and plants lack some of these fundamental traits, which are possessed only by humans. These fundamental right-conducing traits include, beyond those that are shared with wildlife species, cognitive life, with conscious wishes, desires, hopes, aims, or goals. These confer to humans what I could call a higher category of rights, which should be given priority in case of conflicting interests with other categories of rights. One would object that newborn humans lack these properties too; but this is not plausible because newborn humans do have these attributes in latency; they will develop as they grow up. Secondly, an appeal can be made to the principle of *self-defense* (Des Jardins 2001: 145) that favors human interests when human life is threatened because human interests are in conflict with interests of nonhumans. In this example, the life of the local human communities is being endangered by the lack of cultivatable land that would allow them to get food to live on. Because we do not want to see human health and life being jeopardized by interests of nonhumans (even if in this case we are speaking of a keystone species), we should allow the piece of land to be cultivated to provide food to local communities. The fact that we do not want to see human health and life being jeopardized by the survival of a keystone species in the extreme is an example because human life is sacred (Dworkin 1993: 80). Human life is sacred because humans are at the end of the evolutionary line (at least for the time being). Dworkin (1993) argued that some things are intrinsically valued because of time spent to bring them into existence; which means that the longer it has taken something to be crafted the more intrinsically valuable that thing is. Looking at the evolutionary process, the species that has taken too long to come into being is the human species. Hence, being at the end of the evolutionary line makes humans a special type of species, endowed with all right-conducing traits include what is described above, chief among them being living a cognitive life and formulating conscious wishes, desires, hopes, aims, or goals. Indeed, even at times when human interests conflict starkly with other species as it the case in this extreme example, humans are the only one species that try to find solutions for both species. This implies that saving humans would at least leave the windows for the alternative solution opened while if the land is to be left to elephants and if local communities would come to perish, there would unlikely be possible alternative solutions provided by elephants. This vision, even though it is disconnected with religious beliefs about human life agrees, in some extends with those who believe that human life is sacred because humans are the image of the creator.

## 25.10 Conclusions

In the first paragraph of this essay, I defined biodiversity as being more than the sheer number of species and inclusive of species richness, all living organisms, the complex interactions between living organisms, and their abiotic environments and functions that ensure that species and life are maintained on earth. With that definition, the view that there were differences in functions and that some species mattered more than others was defended and was supported by four key concepts of biodiversity (trophic levels, keystone species, hotspot areas, and the species redundancy), which all introduce hierarchies of components of biodiversity. For example, keystone species play ecologically more important roles than other species; there are areas that are more diverse (hotspots) than other areas to the point that losing hotspots will be ecologically more disastrous than when several other areas were lost. But, finally, I argued that claims of differentiation of species based on the uniqueness of each species should be toned down because many species are redundant and can be replaced by other species, which implies that losses in certain species can be compensated by other species, which reduces the potential of the uniqueness of many species. *Secondly*, the instrumental and intrinsic values of biodiversity were weighted. Ecosystems, habitats, and species have tangible instrumental value to humans because species, ecosystems, habitats and nonmeasurable (but objectively conferrable) goods such as recreational and spiritual experiences contribute to human welfare by providing goods and services that humans commonly use. Apprehending the intrinsic value of biodiversity was felt to be difficult because the notion of intrinsic value raised several questions, of which the most obvious is 'what species possesses intrinsic values'. It was argued that whether biodiversity has intrinsic values or not, the most appealing argument to conserve biodiversity is for its instrumental value. This was felt to be so because economically and culturally valued goods provided by biodiversity are of utmost conservation importance and motivate local communities to preserve them, particularly for human communities whose subsistence depends on biodiversity because their very physical existence rests on the uses of biodiversity. Where there are no supermarkets and chemical industries to produce sufficient food for entire communities, humans would value biodiversity because biodiversity provides them with sources of all kinds of goods; this would be the case for most humans in similar conditions. A question raised by this assertion was what happens to species with no instrumental values? Prizing biodiversity because of its instrumental value does not mean that humans would slaughter noninstrumentally valued species; conserving noninstrumental species is latent in the species interconnectedness. Species with no apparent instrumental value, indeed, contribute to the instrumental value of species that are used instrumentally. Therefore, these species can be protected because of that service they rounded even if its instrumental values are not quantifiable. In a sense, the interconnectedness is a way of saying there are no species without instrumental values. The third part of the essay argued in favor of allowing local communities to use natural resources located in and near protected areas even when that use threatens biodiversity. I took the view that human subsistence activities should be allowed because of several reasons. Species redundancy was that since biologically functionally distinct species are rare when human activities would bring a particular species to extinction, other species would take over its ecological functions and continue to ensure the persistence of the system, which should allow human activities to occur within protected areas. Also, even cases where human activities go beyond the resilience thresholds and prevent species to rebound, local communities should still continue to live on these resources because human life is the most sacred form of life in the hierarchy of the sacredness. The priority view of the distributive justice justifies accommodating the basic needs of communities (e.g. foods and shelter) because local communities are often the worst off and the least advantaged. Another sensible reason for that use is that of rights local communities have over the resources per *jus soli* and the rights of the first occupants. The fourth part of the essay throve to think about a possible reconciliation between local human needs and needs to conservation biodiversity. It was argued that there were truly irreconcilable cases but for most of them partial reconciliation is plainly possible by hierarchizing harms, accounting for scales and identifying trade-offs by agreeing with local communities, in a participative process, on what can be extracted by local communities. With these points, more harmful activities should be banned while those that are essential to sustain local communities' needs should be continued. Basic rights principle, the rule of minimum wrong, the jus soli and the rights of the first occupants support these claims.

## References


Bosworth A, Chaipraditkul N, Cheng MM, Gupta A, Junmookda K, Kadam P, Macer D, Millet C, Sangaroonthong J, Waller A (2011) Ethics and biodiversity. Asia and Pacific Regional Bureau for Education UNESCO Bangkok

Callicott JB (1990) Whiter conservation ethics? Conserv Biol 4(1):15–20

Callicott JB (1997) Defense of earth's insights. Worldviews 1:167–182

Cincotta RP, Wisnewski J, Engelman R (2000) Human population in the biodiversity hotspots. Nature. 404:990–992

Colwell RK (2009) Biodiversity: concepts, patterns, and measurement. In: Levin SA, Carpenter SR, Godfray HCJ, Kinzig AP, Loreau M, Losos JB, Walker B, Wilcove DS (eds) The Princeton guide to ecology. Princeton University Press, pp 257–263

Costanza R, Fisher B, Ali S, Beer C, Bond L, Boumans R, Danigelis NL, Dickinson J, Elliott C, Farley J, Gayer DE, Glenn LM, Hudspeth TR, MahoneyD F, McCahill L, McIntosh B, Reed B, Rizvi AT, Rizzo DM, Simpatico T, Snapp R (2007) Quality of life: an approach integrating opportunities, humanneeds and subjective well-being. Ecol Econ 61:267–276

Curry P (2012) Ecological Ethics—A introduction. 2nd edn. Fully revised and expanded. Polity

Desjardins JR (2006) Environmental ethics: an introduction to environmental philosophy. Belmont

Donner W (2002) Callicott on intrinsic value and moral standing in environmental ethics. In: Ouderkirk W, Hill J (eds) Land. Callicott and Environmental Philosophy. State University of New York Press, Value, Community, pp 99–106
Dworkin R (1993) Life's dominion: an argument about abortion and euthanasia. Harper Collins Publishers
Huggett RJ (1999) Ecosphere, biosphere, or Gaia? What to call the global ecosystem. Global Ecol Biogeogr 8(6):425–431
Jepson P, Whittekar RJ (2002) Histories of protected areas: internationalisation of conservationist values and their Adoption in the Netherlands Indies (Indonesia). Environ Hist 3:129–172
Kohen A (2006) The problem of secular sacredness: Ronald Dworkin, Michael Perry, and Human Rights Foundationalism. J Human Rights 5:235–256
Medail F, Quezel P (1999) Biodiversity hotspots in the mediterranean basin: setting global conservation priorities. Conserv Biol 13(6):1510–1513
Mills SL, Soulé ME, Doak DF (1993) The keystone species concept in ecology and conservation: management and policy must explicitly consider complexity of interactions in natural systems. Bioscience 43(4):219–224
Minteer BA, Miller TR (2011) The new conservation debate: ethical foundations, strategic trade-offs, and policy opportunities. Biol Conserv 144(3):945–947
Morgan GJ (2009) The many dimensions of biodiversity. Stud Hist Philos Biol Biomed Sci 40:235–238
Myers N, Mittermeier RA, Mittermeier CG, Da Fonseca GAB, Kent J (2000) Biodiversity hotspots for conservation priorities. Nature 403:853–858
Naeem SC (1998) Species redundancy and ecosystem reliability. Conserv Biol 12(1):39–45
Naess A (1995) The deep ecological movement: some philosophical aspects. In: Sessions G (ed) Deep ecology for the twenty-first century. Shambhala, pp 64–84
Neem S, Bunker D, Hector A, Loreau M, Perrings C (2008) Biodiversity, ecosystem functioning and human wellbeing. Oxford University Press, New York
Nelson TA, Boots B (2008). Detecting spatial hot spots in landscape ecology. Ecography 31:556–566 (The point cited was quoted from Ord and Getis 2001)
Odum E, Barrett GW (2004) Fundamentals of ecology. 5th edn. Cengage Learning
Rawls J (1999) A theory of justice. Oxford University Press
Sarkar S (2014) "Ecology", The Stanford Encyclopedia of Philosophy (Spring Edition), Zalta EN (Ed) http://plato.stanford.edu/archives/spr2014/entries/ecology/
Stoecker R (2011) Three crucial turns on the road to an adequate understanding of human dignity. In: Kaufmann P, Kuch H, Neuhauser C, Webster E (eds) Humiliation. Human Dignity Violated, Degradation, Dehumanization, pp 7–17
Takacs D (1996) The idea of biodiversity: Philosophies of Paradise. Johns Hopkins University Press
Wilson EO (1990) Biophilia. Cambridge University Press

## *Uncited References*

Barnes RFW, Majam JN, Asamoah-Boatang B, Agyei-Ohemeng J (2006) The survival of elephant dung piles in relation to forest canopy and slope in southern Ghana. Pachyderm 41:37–43
Bauer H, Van Der Merwe S (2004) Inventory of free-ranging lions **Panthera leo** in Africa. Oryx 38(1):26–31
Brugiere D (2005) Monkey community structure in the old growth forests of the Lope reserve, Gabon. Afr J Ecol 43:70–72
Brugière D, Fleury MC (2000) Estimating primate densities using line transect and home range methods: comparative test with the black colobus monkey **Colobus satanas**. Primates 41:373–382

Brugière D, Gautier JP, Moungazi A, Gautier-Hion A (2002) Primate diet and Biomass in relation to vegetation composition and fruiting phenology in a Rain Forest in Gabon. Int J Primatol 23(5):999–1024
Chapman CA (2000) Constraints on Group Size in Red Colobus and Red-tailed guenons: examining the generality of the ecological constraints model. Int J Primatol 21(4):565–595
Chapman CA, Peres CA (2001) Primate conservation in the new millennium: the role of scientists. Evolution Anthropol 10:16–33
Chapman CA, Lambert JE (2000) Habitat alteration and the conservation of African primates: a case study of Kibale National Park, Uganda. Am J Primatol 50:169–186
Chapman CA, Lawes MJ, Eeley HAC (2006) What hope for African primate diversity? Afr J Ecol 44:116–133
Chapman CA, Chapman LJ, Cords M, Gathua JM, Gautier-Hion A, Lambert JE, Rode K, Tutin CE, White LJ (2002) Variation in the diets of cercopithecus species: differences within forests, among forests, and across species. In Glenn ME, Cords M (eds) The Guenons: Diversity and Adaptation in African Monkeys. Kluwer Academic/Plenum Publishers, pp 325–350
Chapman CA, Balcomb SR, Gillepsie TR, Skorupa JP, Struhsaker TT (2000) Long-term effects of logging on African primate communities; a 28-year comparison from Kibale National Park. Uganda. Conservation Biology 14(1):207–217
Cords M (eds) The Guenons: Diversity and Adaptation in African Monkeys. Kluwer Academic/Plenum Publishers, pp 325–350
Ellis CM (2001) The influential role of women in the commercial bushmeat trade in Cameroon. In: The Great Apes: Challenges for the 21st Century. Conference proceedings, Brookfield Zoo, pp 227–229
Gautier-Hion A, Brugiere D (2005) Significance of Riparian Forests for the conservation of Central African Primates. Int J Primatol 26(3):515–523
Gouvernement de la République du Zaïre (1984) Droit du Travail I: Code du Travail. Kinshasa Gombe, République Démocratique du Congo, Editions du CADICEC
Gouvernement de la République Démocratique du Congo (2004) Contrat social de l'innovation entre le gouvernement et les syndicats de l'administration publique et interprofessionnels. Mbudi 12 Février 2004
Inogwabini BI (in preparation) The Ethno-botanical knowledge of Ngömbë, Democratic Republic of Congo: A preliminary assessment
Inogwabini BI (2008) Etudes des possibilités de transmission des maladies entre les animaux domestiques, les animaux sauvages et les populations humaines—protocole de collecte des données pour le programme Santé Animale. typescripted document submitted to WWF US and USAID
Inogwabini BI, Mputu (2010) Evaluation de la qualité de l'eau des rivières de basse altitude dans la zone de Malebo pendant la saison sèche: cas de Bambou. Lebomo et Bongo, Typescript report submitted to WWF and USAID CARPE Kinshasa
Kingdon J (1987) The Kingdom Field Guide to African Mammals. Academic Press
Koyo JP, Foteu R (Undated) Harmonization of forest policies and programmes in Central Africa. http://www.fao.org/docrep/009/a0970e/a0970e11.htm
Liu W, Li Y, Shaw KS, Learn GH, Plenderleith LJ, Malenke JA, Sundararaman SA, Ramirez MA, Crystal PA, Smith AG, Bibollet-Ruche F (2013) African origin of the malaria parasite Plasmodium vivax. Nat Commun 5(3346):1–10. https://doi.org/10.1038/ncomms4346
Magnuson L (2005) Conservation of African monkeys. Int J Primatol 26(3):511–513
McGraw S (1994) Census, habitat preference, and polyspecific associations of six monkeys in the Lomako Forest, Zaire. Am J Primatol 34:296–307
Mittermeier RA, Valladares-Pádua C, Rylands AB, Eudey AA, Butynski TM, Ganzhorn JU, Kormos R, Aguiar JM, Walker S (2006) Primates in Peril: The World's 25 Most Endangered Primates, 2004–2006. Primate Conserv 20:1–28
Mittermeier RA, Rylands AB, Konstant WR (2004) IUCN/SSC Primate Specialist Group Report 2001–2004. International Union for Conservation of Nature (IUCN)

N'sosso D (1997) Problématique de Gestion de L'Eléphant d'Afrique dans la Réserve de Faune de Conkouati, au Kouilou (Congo). Pachyderm 23:50–58

Oates JF, Whitesides GH, Davies AG, Waterman PG, Green SM, Dasilva GL, Mole S (1990) Determinants of variation in tropical forest primate biomass: New evidence from West Africa. Ecology 71:328–343

Plato, 360 B.C.E. Timaeus and Critias. Translated by Lee D (1972) Penguin Books Ltd.

Pope Paul VI (1967) Populorum Progressio: Encyclical Letter of His Holiness on the Human Development. Vatican

Straede S, Helles F (2000) Park-people conflict resolution in Royal Chitwan National Park, Nepal: buying time at high cost? Environmental Conservation 27:368–381

Thomas S (1991) Population densities and patterns of habitat use among anthropoid primates of the Ituri Forest, Zaire. Biotropica 23:68–83

Tutin CEG, White LJT (1999) The recent evolutionary past of primate communities: Likely environmental impacts during the past three millennia. In Fleagle JG, Janson C, Reed KE (eds) Primate communities. Cambridge University Press, pp 220–236

Uehara S, Ihobe H (1998) Distribution and abundance of diurnal mammals, especially monkeys, at Kasoje, Mahale Mountains, Tanzania. Anthropol Sci 106:349–369

Walker EP, Warnick F, Hamlet SE, Lange KI, Davis MA, Uible HE, Wright PF, Paradiso JL (1964) Mammals of the world, Volume II. The Johns Hopkins Press

White LJT (1994) Biomass of rain forest mammals in the Lope´ Reserve, Gabon. J Animal Ecol 63:499–512

White F (1983) The vegetation of Africa: a descriptive memoire to accompany the UNESCO vegetation map of Africa (United Nations Educational, Scientific and Cultural Organization)/AETFA (Association pour l'Etude Taxonomique de la Flore d'Afrique Tropicale). Paris

Whitesides GH, Oates JF, Green SM, Kluberdanz RP (1988) Estimating primate densities from transects in a West African rain forest: a comparison of techniques. J Animal Ecol 57:345–367

World Wide Fund for Nature (WWF) (Undated) Congo Basin Forest Partnership: a unifying Force at Work in the Congo Basin. Available on the website http://www.worldwildlife.org/what/wherewework/congo/WWFBinaryitem8854.pdf

The manufacturer's authorised representative in the EU is Springer Nature Customer Service Centre GmbH, Europaplatz 3, 69115 Heidelberg, Germany. If you have any concerns regarding our products, please contact ProductSafety@springernature.com

Printed and bound by CPI Group (UK) Ltd, Croydon, CR0 4YY
30/09/2026
02225449-0015